Thomas Hilger

Medium Modifications of Mesons

Thomas Hilger

Medium Modifications of Mesons

Chiral Symmetry Restoration, in-medium QCD Sum Rules for D and B mesons, and Bethe-Salpeter equations

Südwestdeutscher Verlag für Hochschulschriften

Impressum / Imprint

Bibliografische Information der Deutschen Nationalbibliothek: Die Deutsche Nationalbibliothek verzeichnet diese Publikation in der Deutschen Nationalbibliografie; detaillierte bibliografische Daten sind im Internet über http://dnb.d-nb.de abrufbar.

Alle in diesem Buch genannten Marken und Produktnamen unterliegen warenzeichen-, marken- oder patentrechtlichem Schutz bzw. sind Warenzeichen oder eingetragene Warenzeichen der jeweiligen Inhaber. Die Wiedergabe von Marken, Produktnamen, Gebrauchsnamen, Handelsnamen, Warenbezeichnungen u.s.w. in diesem Werk berechtigt auch ohne besondere Kennzeichnung nicht zu der Annahme, dass solche Namen im Sinne der Warenzeichen- und Markenschutzgesetzgebung als frei zu betrachten wären und daher von jedermann benutzt werden dürften.

Bibliographic information published by the Deutsche Nationalbibliothek: The Deutsche Nationalbibliothek lists this publication in the Deutsche Nationalbibliografie; detailed bibliographic data are available in the Internet at http://dnb.d-nb.de.

Any brand names and product names mentioned in this book are subject to trademark, brand or patent protection and are trademarks or registered trademarks of their respective holders. The use of brand names, product names, common names, trade names, product descriptions etc. even without a particular marking in this works is in no way to be construed to mean that such names may be regarded as unrestricted in respect of trademark and brand protection legislation and could thus be used by anyone.

Coverbild / Cover image: www.ingimage.com

Verlag / Publisher:
Südwestdeutscher Verlag für Hochschulschriften
ist ein Imprint der / is a trademark of
AV Akademikerverlag GmbH & Co. KG
Heinrich-Böcking-Str. 6-8, 66121 Saarbrücken, Deutschland / Germany
Email: info@svh-verlag.de

Herstellung: siehe letzte Seite /
Printed at: see last page
ISBN: 978-3-8381-3627-1

Zugl. / Approved by: Dreden, TU, Diss., 2012

Copyright © 2013 AV Akademikerverlag GmbH & Co. KG
Alle Rechte vorbehalten. / All rights reserved. Saarbrücken 2013

Contents

1 **Introduction and motivation**   1
   1.1 Dynamical chiral symmetry breaking 4
   1.2 Hadrons in the medium 6
   1.3 QCD sum rules 8
   1.4 Bethe-Salpeter–Dyson-Schwinger approach 12
   1.5 Experimental and theoretical status and perspectives 15
   1.6 Structure of the thesis 17

2 **Introduction to QCD sum rules**   19
   2.1 Subtracted dispersion relations 24
   2.2 Operator product expansion 27
   2.3 Sum rules 33

3 **QCD sum rules for heavy-light mesons**   37
   3.1 Operator product expansion 37
   3.2 Parameterizing the spectral function 39
   3.3 Evaluation for D mesons 42
      3.3.1 The pseudo-scalar case 42
      3.3.2 The scalar case 49
   3.4 Evaluation for B mesons 53
   3.5 Interim summary 53

4 **Chiral partner QCD sum rules**   55
   4.1 Differences of current-current correlators and their OPE 57
   4.2 Chiral partners of open charm mesons 63
      4.2.1 The case of P–S 63
      4.2.2 The case of V–A 65
      4.2.3 Heavy-quark symmetry 67
      4.2.4 Numerical examples 68

	4.3 Interim summary	68
5	**VOC scenario for the ρ meson**	**69**
	5.1 Chiral transformations and QCD condensates	71
	5.2 VOC scenario for the ρ meson	74
	5.3 Mass shift vs. broadening	79
	5.4 Notes on ω and axial-vector mesons	92
	5.5 Interim summary	93
6	**Introduction to Dyson-Schwinger and Bethe-Salpeter equations**	**95**
	6.1 Dyson-Schwinger equations	95
	6.1.1 Derivation	95
	6.1.2 Renormalization	98
	6.1.3 Rainbow approximation	99
	6.2 Bethe-Salpeter equations	100
	6.2.1 Derivation	100
	6.2.2 Ladder approximation	104
	6.2.3 Solution in ladder approximation	106
7	**Dyson-Schwinger and Bethe-Salpeter approach**	**109**
	7.1 DSE for the quark propagator	109
	7.1.1 Analytical angle integration	112
	7.1.2 Solution along the real axis	116
	7.1.3 Solution in the complex plane	118
	7.2 Bethe-Salpeter equation for mesons	126
	7.3 Wigner-Weyl solution	135
8	**Summary and outlook**	**141**
	Appendices	**145**
A	**Chiral symmetry, currents and order parameters**	**147**
	A.1 The formalism for classical fields	148
	A.2 The formalism for quantized fields	165
	A.3 Brief survey on Quantum Chromodynamics	176
B	**Correlation functions**	**181**
	B.1 Källén-Lehmann representation and analytic properties	182

	B.2 Symmetry constraints	186
	B.3 Decomposition	191
	B.4 Non-anomalous Ward identities	194
	B.5 Subtracted dispersion relations	200
	B.5.1 Vacuum dispersion relations	201
	B.5.2 In-medium dispersion relations	204
C	**In-medium OPE for heavy-light mesons**	**207**
	C.1 Fock-Schwinger gauge and background field method	207
	C.2 Borel transformed sum rules	214
	C.3 Projection of color, Dirac and Lorentz indices	221
	C.4 OPE for the D meson	224
	C.5 Absorption of divergences	231
	C.6 Recurrence relations	235
	C.7 Sum rule analysis for heavy-light mesons	242
	C.8 Condensates near the deconfinement transition	245
D	**Chiral partner QCD sum rules addendum**	**249**
	D.1 Dirac projection	249
	D.2 The perturbative quark propagator	250
	D.3 Cancellation of IR divergences	252

Acronyms   257

List of Figures   259

List of Tables   263

Bibliography   265

1 Introduction and motivation

Within the present standard model of particle physics the building blocks of matter are quarks and leptons. These fermions interact via gravity, as well as weak, electromagnetic and strong interaction - the four known forces in nature. Apart from gravity, these interactions may be described by virtue of gauge field theories, where the interaction is transferred by gauge bosons. One such theory is quantum chromodynamics (QCD) which is believed to describe the strong interaction. The gauge bosons of QCD are called gluons. At the time when this thesis has been written, the last missing piece of the standard model is the Higgs boson, which is predicted in order to explain the masses of the gauge bosons carrying the weak interaction force. Moreover, the interaction of the Higgs boson with the fermion fields within the standard model also gives rise to the bare masses of the quarks and some of the leptons. Recently, first results of the Higgs boson search from the ATLAS and ALICE experiments at the Large Hadron Collider at CERN have been presented [ATL, ALI], and physicists around the world are excitingly awaiting the final announcements. However, the main contribution to the mass of matter that surrounds us has a different origin. The mass of "usual matter" resides in nuclei composed of nucleons, i. e. protons and neutrons. Nucleons belong to the hadrons and the fundamental degrees of freedom within an environment below the deconfinement transition of QCD (to be explained later on) are hadrons. Hadrons are strongly interacting, subatomic states of quarks and gluons, which interact via and form bound states by virtue of the strong interaction. They may be cataloged by their spin. Bosonic hadrons are called mesons and consist of an even number of valence quarks. Fermionic hadrons are called baryons and consist of an odd number of valence quarks, see Fig. 1.0.1. Valence quarks are the quarks which determine the quantum numbers of the hadron. The bare masses of quarks are well known from several experiments and gluons are supposed to be massless due to the exact local $SU_c(3)$ gauge symmetry. But against naive intuition, summing up all constituent masses of a hadron, one can not explain its mass, which would be expected to be smaller than the sum of the constituent masses due to the binding energy. In contrast, the hadron masses are much larger

1 Introduction and motivation

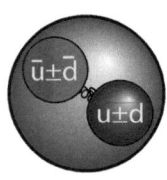 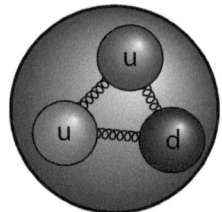 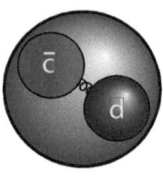

Figure 1.0.1: Illustrative picture depicting (from left to right) a ρ and ω meson (left), a proton (middle) and a D meson (right) in the constituent quark model. All hadrons are color-less, and the confining interactions carried by the gluons are depicted by curly lines.

than the sum of their constituents masses.

Let us consider, for example, the nucleon. In the naive quark model it consists of three light quarks with the sum of their bare current quark masses as obtained from several high energy experiments being less than 20 MeV, while the nucleon mass is about 938 MeV [Nak10]. Hence, the masses are even of different scales. In this special case, i. e. for nucleons, we might think of nearly massless quarks but nucleons would still have a mass of several hundreds MeV. Another example is given by the $D^{\pm}$ meson, which consists of a charm quark with a mass of $m_c = 1.25\ldots 1.45\,\text{GeV}$ and a down quark with a mass of $m_d = 4\ldots 8\,\text{MeV}$ [Nak10]. However, the mass of the $D^{\pm}$ meson is about $m_{D^{\pm}} = 1.87\,\text{GeV}$ [Nak10], which is again larger than the sum of its constituent masses. This mass difference is a direct result of the complicated structure of the strong interaction which binds quarks and gluons into hadrons. The given examples tell us that genuine quark masses, as supposed to be generated by the Higgs mechanism in the electro-weak sector of the standard model, are of minor importance in an explanation of the mass of the matter around us. It is the strong interaction which has to be understood in order to understand our "massive world".

One of the main goals of hadron physics is to understand the origin of hadron masses in terms of their underlying structure of quarks and gluons and the strong interaction, which binds them together to bound states. As already pointed out, QCD is the accepted theory which describes the strong interaction at quark-gluon level. It is a non-Abelian gauge field theory which provides basic features of the strong interaction such as "asymptotic freedom" (basic details of QCD which are needed throughout this work are reviewed in App. A.3). The phenomenon of "asymptotic freedom" means that in scattering experiments at large momentum transfer a projectile, say an electron, being scattered off a proton behaves as being scattered off almost free quarks. At low momenta, only the quarks are observed which constitute the quantum numbers of the hadron. The thus observed quarks are the above mentioned valence or current quarks and their masses, as measured in high energy experiments, are the so-called current-quark masses. Indeed, it can be shown [Gro73] that QCD entails this feature as it tells us that the coupling strength of strong interaction decreases with increasing momentum, such that in the limit of infinite energy the quarks behave as free (in the sense of non-interacting) quarks. This also enables one to apply perturbative techniques in the high momentum or small distance regime.

Another property of the interaction among quarks is that one can not observe isolated individual quarks. Looking up free quarks in the particle data booklet [Nak10] reveals this circumstance with the simple sentence: "All searches since 1977 have had negative results." This feature of the strong interaction is called confinement. In vacuum or in a strongly interacting medium, below a certain temperature or chemical potential (i. e. density), quarks are always bound into hadrons as color singlets. Only above this temperature and/or chemical potential quarks and gluons are the fundamental degrees of freedom. The transition from hadronic to quark-gluon degrees of freedom is called the deconfinement phase transition. May it be a strict law of nature that there are no free quarks or may it just be extremely seldom and difficult to observe free quarks, any theory of the strong interaction must provide a confining mechanism in order to explain this "non-observation." Unfortunately, a rigorous theoretical treatment of confinement within QCD is still missing. Indeed, the (mathematical) rigorous establishing of a Yang-Mills theory (non-Abelian gauge theory, e. g. QCD, without quarks) with mass gap and confinement is one of the seven Millennium Prize problems stated by the Clay Mathematics Institute.

One of the most common and well-established techniques of quantum field theory is that of perturbation series – an expansion in the coupling constant. If the coupling is sufficiently small the expansion is hoped to be convergent and may be applied to evaluate quantities within the theory or to draw conclusions from experiments. In case of the strong interaction the thus derived coupling increases with increasing distance. This is in line with confinement as the attraction between quarks therefore increases with increasing separations until the energy of the quarks is large enough to create quark–anti-quark pairs, such that it is not possible to isolate a single quark. QCD reflects this by its non-Abelian character, which accounts for the self-interaction of the gluons and causes the growing coupling. Along these lines the interaction of the valence quarks with the sea-quarks (virtual quark–anti-quark pairs) and gluons account for the heavily increased constituent quark mass addressed before, which is

1 Introduction and motivation

known as antiscreening or mass dressing. However, since the coupling strength of the strong interaction grows with decreasing momentum (or increasing distance), perturbative techniques are not applicable in a scenario of two quarks separating from each other. Also for the task of calculating bound state properties one has to look for different methods in order to perform calculations in the low momentum regime.

1.1 Dynamical chiral symmetry breaking

Apart from "asymptotic freedom", which governs the high-energy domain, or color-confinement, which is a low energy phenomenon, the dynamical (or spontaneous) breakdown of chiral symmetry (DCSB) is another important low-energy property of the strong interaction. It is always noteworthy that symmetries play a significant role in physics. By Noether's theorem [Noe18] every differentiable symmetry guarantees the existence of a conservation law. For example, energy-momentum conservation can be traced back to translational invariance in space and time. A system where the symmetry is spontaneously broken therefore strongly deviates from a system where the symmetry is realized.

For vanishing quark masses $M \to 0$, the classical, i.e. not quantized, chromodynamical Lagrangian

$$\mathscr{L} = \bar{\psi}\left(i\hat{D} - M\right)\psi - \frac{1}{4}G^A_{\mu\nu}G_A^{\mu\nu}, \qquad (1.1.1)$$

with N_f flavors, is invariant with respect to the global chiral $SU_R(N_f) \times SU_L(N_f)$ transformations.[1] The quantity $\psi = (q_1, \ldots, q_{N_f})$ collects the Dirac spinors of N_f quark flavors, $\hat{D} = \gamma^\mu D_\mu$ denotes the covariant derivative contracted with the Dirac matrix γ_μ, $G^A_{\mu\nu}$ is the gluon field-strength tensor, μ, ν, are Lorentz indices, and A is a color index in the adjoint representation.

Focusing for the time being on the $N_f = 2$ light (massless) quark sector, the corresponding left-handed transformations read for the left-handed quark field $\psi_L = \frac{1}{2}(1-\gamma_5)\psi$ and the right-handed quark field $\psi_R = \frac{1}{2}(1+\gamma_5)\psi$

$$\psi_L \to e^{-i\vec{\theta}_L \cdot \frac{\vec{\sigma}}{2}}\psi_L, \quad \psi_R \to \psi_R, \qquad (1.1.2a)$$

---

[1] In Eq. (1.1.1) the classical Lagrangian is given. For the canonically quantized Lagrangian see App. A.3. Canonical quantization necessitates the introduction of gauge fixing and Faddeev-Popov ghost terms [Pas84]. These, however, do not alter the transformation properties w.r.t. the chiral transformations discussed here.

while the right-handed transformations are

$$\psi_R \to e^{-i\vec{\theta}_L \cdot \frac{\vec{\sigma}}{2}} \psi_R, \quad \psi_L \to \psi_L, \tag{1.1.2b}$$

where $\vec{\sigma}$ are the isospin Pauli matrices, and $\psi = \begin{pmatrix} u \\ d \end{pmatrix}$ denotes the quark iso-doublet. Equation (1.1.2) represents isospin transformations acting separately on the right-handed and left-handed parts of the quark field operator $\psi = \psi_L + \psi_R$, i.e. the three-component vectors $\vec{\theta}_R$ and $\vec{\theta}_L$ contain arbitrary real numbers. Gluons and heavier quarks remain unchanged with respect to the transformations (1.1.2).

Consider, for example, a quark current which has the quantum numbers of the ρ meson,[2] i.e. is given by the vector–isospin-vector current

$$\vec{j}^\mu = \frac{1}{2} \bar{\psi} \gamma^\mu \frac{\vec{\sigma}}{2} \psi. \tag{1.1.3a}$$

If a chiral transformation according to (1.1.2) is applied to $\bar{\psi}$ and ψ, it becomes mixed with the axial-vector–isospin-vector current[3]

$$\vec{j}_5^\mu = \frac{1}{2} \bar{\psi} \gamma^\mu \gamma_5 \vec{\tau} \psi \tag{1.1.3b}$$

which has the quantum numbers of the a_1 meson. Indeed, experiments show that the vector current (1.1.3a) couples strongly to the ρ meson, while the axial-vector current (1.1.3b) couples to the a_1 meson [Sch05]. Therefore, ρ and a_1 are called chiral partners. As the Lagrangian (1.1.1) in the limit $M \to 0$ is invariant w.r.t. the transformations (1.1.2) on the one hand and, on the other hand, the currents (1.1.3) are mixed by these transformations, the spectra of ρ and a_1 meson, in particular their masses, should be degenerate if the chiral symmetry was realized in nature. The observed mass of the ρ meson is $m_\rho = 775.5\,\text{MeV}$ while the a_1 meson mass is $m_{a_1} = 1260\,\text{MeV}$ [Nak10]. Such a significant mass splitting is observed for many chiral partner mesons. Therefore, the symmetry must be spontaneously broken.

In a quantized theory the realization of a symmetry is manifest in the invariance of its ground state, i.e. the lowest energy state of the theory $|\Omega\rangle$, under the corresponding symmetry transformation. As a consequence $Q|\Omega\rangle = 0$, where Q is the generator of the transformation. This case is called the Wigner-Weyl phase or realization. In contrast, the spontaneous breakdown of a symmetry is reflected in the non-invariance of its ground state, i.e. $Q|\Omega\rangle \neq 0$. This case is called the Nambu-Goldstone phase or

---

[2] This means that the current has parity, charge, isospin and spin of the ρ meson.
[3] The choice $|\vec{\theta}_L| = 2|\vec{\theta}_R| = 2\pi$ explicitly reveals this mixing. See App. A.1.

1 Introduction and motivation

realization. The latter property is used to prove the so-called Goldstone theorem (see [Bur00, Kug97] for details of the proof and its consequences). It states that in the case of a spontaneously broken symmetry massless bosons must exist. These bosons where first found by Nambu [Nam60] within in the treatment of superconductivity. Goldstone generalized this phenomenon to quantum field theories [Gol61, Gol62]. The Goldstone bosons of QCD are the pions, which are pseudo-scalar mesons consisting of the two lightest (i. e. almost massless) quark flavors up and down. Pions belong to the isospin multiplet. The isospin singlet and the multiplet are degenerate due to the explicit breaking of the $U_A(1)$ symmetry by the axial anomaly (cf. App. A.2). While constituent quark models with quark masses of $\approx 300\ldots400\,\text{MeV}$ for up and down quarks may be employed to explain hadron masses, e. g. for the ρ meson, on a phenomenological ground, the pions do not fit into such a scheme due to the spontaneous chiral symmetry breaking.

Because the current quarks are not massless, which is referred to as explicit chiral symmetry breaking (in contrast to the dynamical or spontaneous chiral symmetry breaking), the pions are neither. Nevertheless, their masses are $m_{\pi^0} = 135.0\,\text{MeV}$ and $m_{\pi^\pm} = 139.6\,\text{MeV}$ [Nak10] are small as compared to the other hadrons.

The addressed property that the ground state of a theory is not invariant under a symmetry transformation of the Lagrangian may also be used to show that certain ground state expectation values are nonzero. Regarding the example given above, this is true for the chiral condensate $\langle\bar\psi\psi\rangle$. The Wigner-Weyl phase, in contrast, would require $\langle\bar\psi\psi\rangle = 0$.

1.2 Hadrons in the medium

As we have already seen, the ground state of QCD has a complicated non-trivial dynamical structure due to color-confinement and dynamical chiral symmetry breaking, which can not be explained perturbatively. Indeed, the perturbative ground state $|0\rangle$, defined such that it is annihilated by all annihilation operators, is invariant w. r. t. chiral transformations $Q|0\rangle = 0$. Consequently, it is not possible to describe chiral symmetry breaking effects on a perturbative basis. However, hadrons may be regarded as excitations of the QCD ground state, and hadron properties are therefore directly linked to it. The non-degeneracy of chiral partners and the small mass of the pions indicate that the dynamical breaking of the chiral symmetry is responsible for the observed hadron mass spectrum. Any theoretical attempt to determine hadron masses must incorporate this effect. Vice versa, comprehending this effect is a crucial

1.2 Hadrons in the medium

vacuum

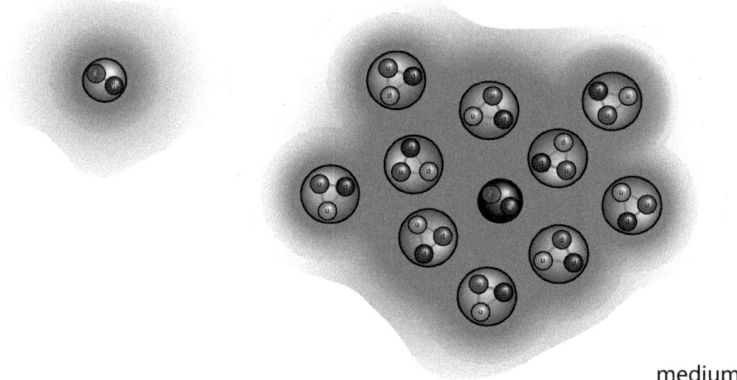

medium

Figure 1.2.1: Illustrative picture of a D^- meson in vacuum (left) and placed in nuclear matter (right). Gluons are depicted by curly lines and "clouds" symbolize the non-perturbative ground state illustrating creation and annihilation of quark–anti-quark pairs.

step towards an understanding of the origin of mass.

Similar to the deconfinement phase transition, the chiral symmetry is expected to be restored at higher temperatures and/or densities, which is called the chiral transition. Thus, hadronic properties may be modified if the hadrons are embedded in a strongly interacting medium. Such environments may be found in the early universe, in compact stars or if the hadron is embedded in a nucleus (zero temperature and low densities as compared to conditions at the chiral or deconfinement transition). Such scenarios are experimentally investigated in heavy-ion collisions. Understanding of high-temperature and high-density physics requires the understanding of medium modifications of hadrons.

Modifications of spectral properties are well known from the splitting of atomic or molecular spectral lines when external magnetic (Zeeman effect [Zee97b, Zee97a, Zee97c]) or electric (Stark effect [Sta14]) fields are present. An example of a spontaneously broken symmetry which is restored at higher temperatures due to a phase transition can be found in material science. For a ferromagnetic, antiferromagnetic or ferrimagnetic[4] material the rotational invariance is spontaneously broken due to

---

[4] In a ferrimagnetic material the antiparallel aligned magnetic moments are unequal.

1 Introduction and motivation

the spontaneous magnetization of the Weiss domains, i.e. the parallel or anti-parallel alignment of neighboring spins. Above a critical temperature, called the Curie temperature, the spontaneous polarization is lost and the material becomes paramagnetic. The spins are disordered and the rotational symmetry is restored, cf. [Kit86]. The strongly interacting medium in hadron physics acts as an external field which modifies the quark-gluon interaction and, therefore, the ground state of QCD. This is reflected in the modification of hadronic properties such as masses and/or widths, see Fig. 1.2.1. Hence, restoration or partial restoration of a spontaneously broken symmetry may be signaled by the change of hadronic spectral functions. However, considering the analog of a ferro- or ferrimagnetic material one may ask about other properties apart from the magnetization of the material above the critical temperature.

Correspondingly, if the quarks where massless, the spectra of chiral partners where degenerate in case of chiral symmetry restoration. In addition, one may ask about other hadronic properties such as interactions among hadrons or, more simply, the spectral function of a single hadron. Furthermore, the explicit symmetry breaking should be accounted for. In this sense, one may ask about the mass of the pion or any other hadron in the Wigner-Weyl phase. As already discussed above, the pion is the (almost massless) Goldstone boson and has a nonzero mass only due to the explicit breaking of the chiral symmetry, whereas its valence quarks acquire large dressed masses of several hundred MeV due to the DCSB.

1.3 QCD sum rules

As hadrons are composite objects consisting of confined quarks and gluons, hadronic properties should be related to the quark and gluon structure within the hadron, thus, to the non-perturbative properties of the theory. Instead of calculating directly hadronic properties from first principles, a tool to relate low-momentum properties of the hadron to its large momentum structure is provided by the method of QCD sum rules (QSRs). But, its major draw-back is that it can not be derived rigorously from first principles. Instead, heuristic arguments are used to set up the method which gains its reliability from the successful description of hadronic properties. Indeed, QSRs have been applied very successfully to a variety of mesons and baryons in vacuum and extended to in-medium situations.

The method was first developed in [Shi79d, Shi79b] and successfully applied to charmonium states (mesons consisting of charm quarks only), as wells as to ρ, ω, ϕ and K^* mesons [Shi79a]. The $\rho - \omega$ mixing was also addressed in [Shi79c], and a

1.3 QCD sum rules

Figure 1.3.1: Representation of some QCD condensates by Feynman diagrams. From left to right: quark, gluon and mixed quark-gluon condensate. Crosses symbolize condensation, i.e. interaction with the ground state.

technical guide has been published in [Nov84a]. Baryons have first been dealt with in [Iof81], see also [Dos89]. Apart from hadron masses, couplings, magnetic moments and many other observables have successfully been obtained within this method (cf. [Col00] for a modern perspective). Over the years, a variety of more or less didactic introductions have been accumulated, such as [Rei84, Wey85, Raf89]. Many results of vacuum QSRs and references to the corresponding publications are collected in [Nar89, Nar02]. A detailed and explicit evaluation for ρ mesons is given in [Pas84].

The method of QSRs is based on dispersion relations, which are used to relate different energy regimes. It introduces phenomenological parameters, called condensates (see Fig. 1.3.1), in order to parametrize non-perturbative physics and to understand and to reproduce hadronic properties, e. g. hadron masses. These parameters enter the theory as ground state expectation values of quantum field operators by virtue of the operator product expansion (OPE) and, hence, are directly connected to the fundamental properties of QCD and its ground state. The most famous among these condensates is the chiral condensate $\langle \Omega | \bar{q} q | \Omega \rangle$, where q is the field operator of a massless quark, introduced earlier. Some condensates are order parameters of the chiral symmetry, meaning that they are zero if the symmetry is restored and, if the condensates are nonzero, the symmetry must be broken. As a consequence, the difference of the spectra of chiral partners should only depend on those condensates which are order parameters of chiral symmetry.

Loosely speaking, a condensate can be read as a measure for the probability of a particle being annihilated by a virtual particle from the ground state and, therefore, creating another real particle somewhere else. A non-vanishing condensate indicates a particle interaction with the ground state. Hence, the hadron appears much heavier than its QCD constituents, because not only interactions among quarks transmitted by gluons or among gluons have to be considered but also interactions of these particles with the ground state. The hadrons acquire an effective mass by the interaction of their constituents with the ground state of QCD. This can be taken as a very simplified illustration of the mass effects mentioned at the beginning of this chapter. Such a concept might be known from many-body physics and indeed many concepts, such

1 Introduction and motivation

as the Lehmann representation or dispersion relations, are also subject to this field. Furthermore, continuing in this illustrative language, the probability of a virtual light-quark pair creation from the ground state must be much larger than the probability of a heavy-quark pair creation. Therefore, the mass increase via the coupling of a quark to a condensate shall be much lower for a heavy quark than for a light quark. The heavy quark may nevertheless emit gluons which in turn interact with virtual particles of the ground state. This explains why the mass of the nucleon compared to its constituent quarks is of a different scale, while the D meson is only somewhat heavier than its constituents. The charm quark does not contribute to the mass acquirement. Moreover, we see that the appearance of the heavy quark must require another treatment than the light-quark systems, because it is less affected by non-perturbative effects. In this respect it also seems reasonable that D and $\bar{D}$ behave different when embedded in nuclear matter. The nuclear medium breaks the charge conjugation symmetry. Therefore, a $c\bar{q}$ state, where q denotes a light quark, is expected to be differently affected than a $q\bar{c}$ state.

QSRs play a twofold role. On the one hand, they are used to determine the condensates in order to get valuable information about the structure of QCD and the strong interaction. On the other hand, they are used to evaluate hadronic properties, like masses, in terms of the condensates. In the latter case one does of course not attempt to derive non-perturbative effects, such as confinement and dynamical chiral symmetry breaking, from first principles, but instead takes them as a matter of fact (nonzero condensates) and relates them to hadron properties.

QSRs have been extended to finite temperatures [Boc84, Boc86] and densities [Hat92] for mesons and nucleons [Fur90]. A comprehensive but nevertheless detailed treatise about applying QSRs to nucleons in the medium can be found in [Fur92, Jin93, Jin94]. At small nonzero temperatures and chemical potentials the medium dependence of the system under consideration is encoded in the condensates. This enables one to perform in-medium investigations comparably easy and to systematically relate the medium dependence of the condensates to the change of hadronic properties. On the other side one may hope to relate medium modifications of hadrons to changes of QCD condensates, in particular to order parameters of the chiral symmetry such as the chiral condensate. Therefore it is desired to find systems which are sensitive to changes of the chiral condensate.

The Brown-Rho scaling [Bro91] suggested the direct relation of the light vector meson masses in the medium to the pion decay constant and, thus, via the Gell-Mann–Oakes–Renner relation with the chiral condensate. Similar, Joffe's formula for the nucleon [Iof81] relates the nucleon mass to squares of the chiral condensate. Thus, light-quark systems seem to be appropriate probes either to determine the medium dependence of the chiral condensate or to predict their masses in the medium using the medium dependence of the chiral condensate only. Taking both scaling laws rigorously, a strict consequence would also be the vanishing, or at least decreasing, of the hadron masses near the chiral transition. Vice versa, decreasing masses are often considered as signals for chiral symmetry restoration. As already pointed out, QSRs have been applied very successfully to light-quark hadrons such as the ρ and ω mesons. But as is already known since the early days of QSRs, the chiral condensate is actually numerically suppressed, by the light quark mass, for these mesons [Rei85]. Instead, a subtle balance of gluon and so-called four-quark condensates determine the ρ meson spectral density. Assuming vacuum saturation, the four-quark condensates in turn may be factorized into squares of the chiral condensate, which arranges for an enhanced impact of the chiral condensate. However, this assumption has already been questioned in [Shi79a] even in vacuum and, therefore, also in the medium; at the chiral transition point it is of course even more questionable. Not all four-quark condensates are order parameters of chiral symmetry breaking and it is not necessarily consistent that these are factorized in terms of an order parameter. Nevertheless, certain combinations of four-quark condensates may indeed serve as order parameters and the medium modifications may be investigated w. r. t. changes of them. Such an investigation is part of this thesis, with first results published in [Hil12b].

For mesons consisting of a heavy quark, such as the charm quark, and a light quark, e. g. the pseudo-scalar D mesons, the chiral condensate is amplified by the heavy charm-quark mass. In vacuum this has also been known for a long time [Rei85]. A delicate step thereby is the consistent separation of large-momentum and low-momentum scales, which, in case of heavy-light quark mesons in contrast to light-light or heavy-heavy quark mesons, is complicated due to the two different mass scales. The resulting infrared divergences have to be absorbed. Thereby, the condensates mix under renormalization.[5] While the vacuum treatment was finalized in [Jam93], the necessary in-medium relations where first derived in [Zsc06], generalized in [Hil08] and published in its final form in [Zsc11]. First results for heavy-light quark mesons have been published in [Rap11] and a concise treatment in [Hil09]. Further investigations were also published in [Hil10a, Hil10c, Käm10, Hil10b].

A natural way to isolate order parameters is to investigate chiral partners. In the

---

[5]This mixing is not to be confused with the mixing of condensates due to applications of the renormalization group equation [Mut87].

light-quark vector–axial-vector channel the first chiral partner sum rules have been derived in [Wei67], where first and second moment of the difference between the spectral densities of ρ and a_1 mesons have been related to the pion decay constant as another order parameter of chiral symmetry. These so-called Weinberg sum rules have been derived in the scope of current algebra in vacuum. Current algebra represents a general tool to investigate the structure of a field theory in terms of its symmetries. It was developed prior to the advent of QCD. Later on, the Weinberg sum rules have been generalized and extended in [Kap94] to nonzero temperatures and/or densities including a third sum rule. These so-called Weinberg-Kapusta-Shuryak sum rules have been derived within QSRs utilizing the OPE technique. It is part of this thesis to derive chiral partner sum rules in the vector–axial-vector and scalar–pseudo-scalar channel for mesons consisting of a heavy and a light quark at nonzero densities and/or temperatures. Early results of this work have been published in [Hil10a, Hil10c], and improvements in [Hil11]; an additional investigation has been initiated in [Hil12a]. Although the main results refer to heavy-light quark systems, the formalism may be applied to any other hadron and general conclusions have been made.

1.4 Bethe-Salpeter–Dyson-Schwinger approach

Another non-perturbative method to evaluate hadron properties is based on the Poincaré invariant Bethe-Salpeter equation (BSE). The equation was first proposed in [Sal51] and derived from field theory in [GM51], cf. [Nak69] for a general survey. It is an exact equation for two-body problems in a relativistic quantum field theory. As it does not directly involve potentials but rather relies on the propagators and vertices of the theory, it naturally incorporates recoil effects which would not be possible otherwise. To the best of our knowledge it has, up to now, not been possible to rigorously prove that the non-relativistic limit is a Schrödinger equation. Its non-perturbative character is manifested in being an integral equation and, in case of diquark bound states, maintaining dynamical chiral symmetry breaking and dynamical quark dressing. Thereby, an infinite number of interactions between the two particles may take place. In its homogeneous form it describes bound states, where an infinite number of particle interactions is mandatory. Otherwise, it is a scattering problem which can be addressed by using the inhomogeneous BSE. As perturbative techniques rely on an expansion in the coupling and thus in the number of interactions, it is also clear from this point of view that bound states are not treatable in perturbation theory. The homogeneous BSE is an integral equation for the Bethe-Salpeter amplitude

1.4 Bethe–Salpeter–Dyson-Schwinger approach

(BSA). It can not be solved analytically for realistic interaction kernels. Due to its high dimensionality (16 dimensional in coordinate space, 4 dimensional in momentum space in the bound state rest frame) there is no direct numerical solution at hand nowadays. Approximations have to be applied. Restricting the exchange of interaction particles to one at a time corresponds to the ladder approximation. But even for the ladder approximation there is only one model that can be solved analytically, the so-called Cutkosky model [Cut54], see also [Nak69]. It is often referred to as a toy model, because it entails unphysical states and gives the wrong limit for the ratio of the masses tending to infinity, i. e. one mass being much larger than the other. Employing the bare interaction vertex is referred to as rainbow approximation. In momentum space, the kernel depends parametrically on the bound state momentum and, in the rest frame of the bound state, the equation has solutions only at discrete values of the bound state mass. The BSA has no direct physical interpretation. Nevertheless it is a full description of the bound state and observables, such as form factors and decay constants, may be evaluated by virtue of the BSA. Indeed, in [Man55] it has been shown that, once the BSA is known, any dynamical variable which describes the transition between two bound states is calculable. The equation has been applied successfully to many different problems such as the hydrogen atom [New55], positronium, excitons and the deuteron [Dor08].

Apart from the propagator of the particle which mediates the interaction between the constituents, also the propagators of the constituents and the interaction vertex enter. These are two- and three-point functions which, in a quantum field theory, are determined by their Dyson-Schwinger equation (DSE). The DSEs where first given in [Dys49] by summing an infinite number of Feynman diagrams and generalized to arbitrary quantum field theories in [Sch51a, Sch51b]. They represent an infinite coupled system of non-linear integral equations for the n-point functions of a quantum field theory and represent the equations of motion (EoM) for the Green's functions. Therefore, they are often referred to as the Euler-Lagrange equations of a quantum field theory. Indeed, the BSE can also be derived from the DSE of a four-point function. Note, however, that the BSA itself is not an n-point function.

First applications of the relativistic BSE to diquark bound states have been made in [Mun92] and reviews may be found in e. g. [Kug91, Rob00, Rob94, Rob07, Mar03]. While for many applications of the BSE it is sufficient to employ the free constituent propagators, as for example in case of the deuteron, the low-momentum properties of the strong interaction demand to employ the exact quark propagator. Only the exact quark propagators (or appropriate models) are capable of generating the large dressed

1 Introduction and motivation

quark masses that are required to explain the observed hadron masses, as well as the correct ultraviolet limit of the quark masses as determined in high-energy scattering. As quarks obey confinement, it is reasonable that the free quark propagator with bare current quark masses is not appropriate in describing the low energy interaction among quarks which dominates the formation of bound states. Thus, the quark DSE has to be solved, and the resulting propagators serve as an input for the bound state BSE. It turns out, however, that it is not necessary to solve the gluon and ghost DSE or the DSE for the interaction vertices. Instead, a phenomenological gluon propagator and the bare quark-gluon vertex may be employed [Mar99]. Using the bare vertex and a phenomenological gluon propagator technically means to work in quenched approximation, i.e. there are no dynamically generated quark–antiquark pairs. However, adjusting the parameters of the phenomenological propagator effectively includes these effects up to a certain extend and provides therefore an appropriate tool to study hadron properties. Attempts to include other DSEs and to go beyond rainbow-ladder truncation are published in [Fis08, Fis09b].

In [Alk02] a very simple but nevertheless successful model was proposed which is capable of describing the masses of a variety of mesons in the vector, axial-vector, scalar and pseudo-scalar channels for equal quarks up to a bare quark mass of ≈ 1 GeV. It turns out, however, that solving the BSE is inferred by the analytic structure of the quark propagators in the complex plane. In order to solve the BSE for mesons, the quark propagators have to be known in the complex plane, because of the not negligible quark dressing. Indeed, such a complication does not exist for, e.g., the deuteron [Dor08]. Their evaluation in the complex plane is possible by various methods, but singularities are found in any case. For the ground state of equal quark bound states, the singularities do not infer the solution of the BSE, but for excited states they do. This problem has been known for a long time without resolution [Mar92, Sta92]. Unfortunately, neither the physical nor the mathematical nature of these singularities is known. In particular for heavy-light quark bound states, these poles infer the solution of the BSE. As a part of this thesis we will investigate the analytic structure in detail and test a method to circumvent this issue. Some of the results given in the course of this thesis have been published in [Dor11, Dor10].

1.5 Experimental and theoretical status and perspectives

Chiral symmetry and its breaking pattern represent important features of strong interaction physics. As mentioned above, the non-degeneracy of chiral partners of hadrons is considered to be a direct hint to the spontaneous chiral symmetry breaking in nature which characterizes the QCD vacuum. In fact, the distinct difference of vector–isospin-vector and axial-vector–isospin-vector spectral functions deduced from τ decays [Sch05, Ack99] gives one of the empirical and precise evidences for the breaking of chiral symmetry. The low-energy strengths of the mentioned spectral functions, concentrated in the resonances $\rho(770)$ and $a_1(1260)$, deviate strongly from each other and from perturbative QCD predictions. (For a dynamical interpretation of the two spectra see [Wag08a, Leu09a].) This clearly exposes the strong non-perturbative effects governing the low-energy part of the hadron spectrum. The spontaneous symmetry breaking is quantified by the chiral condensate $\langle \bar{q}q \rangle$, which plays an important role in the Gell-Mann–Oakes–Renner relation connecting hadronic quantities and quark degrees of freedom (cf. [GM68, Col01]).

Apart from the briefly described non-perturbative methods, other tools to explore the low-energy regime of hadron physics and QCD are, e.g., lattice QCD, chiral perturbation theory, instanton models and conformal field theory. In lattice QCD the theory is described on a discretized space-time. Chiral perturbation theory relies upon the effective description of hadrons utilizing phenomenological parameters. Of particular importance here is the vector meson dominance model. Instantons are classical, localized soliton solutions of the EoM.

As pointed out QSRs offer a link from hadronic properties, encoded in spectral functions, to QCD related quantities, like condensates, in the non-perturbative domain. A particularly valuable aspect of QSRs is, therefore, the possibility to predict in-medium modifications of hadrons, supposed the density and temperature dependence of the relevant condensates is known. Taking the attitude that this is the case, one arrives at testable predictions for changes of hadronic properties in an ambient strongly interacting medium. There is a vast amount of literature on the in-medium changes of light vector mesons, cf. [Hat95, Leu98a, Ste06, Leu01, Rup06, Pet98b, Pet98a, Pos01, Mue06, Kli96, Kli97, Tho08b, Tho05, Rap00, Lut02, Rap09, Tse09, Hay10, Met08, Leu10] and further references therein. Vector mesons are of special interest as their spectral functions determine, e.g., the dilepton emissivity of hot and compressed nuclear matter. Via the direct decays $V \to l^+l^-$, where V stands for a vector meson

1 Introduction and motivation

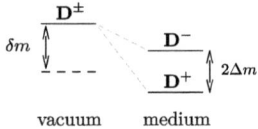

Figure 1.5.1: Expected spectral changes of D mesons in a cold nuclear medium.

with $\bar{q}q$ structure and l^+l^- for a dilepton, the spectral distribution of V can be probed experimentally. Accordingly, heavy-ion experiments are often accompanied by special devices for measurements of $l^+l^- = e^+e^-$ or $\mu^+\mu^-$. Addressed questions concern in particular signals for chiral restoration [Rap00]. Clearly, besides the QSRs, also purely hadronic models have been employed to understand the behavior of vector mesons in nuclear matter, cf. [Pet98b, Pet98a, Pos01, Mue06, Rap00, Lut02] for example.

Such hadronic models are also used in the strangeness sector [Waa96, Waa97, Kai01, Bor05, Tol06a]. Here, the distinct behavior of kaons and anti-kaons attracted much attention, cf. [Sch06, För07, Cro00] for experimental aspects. Similar to the K-$\bar{K}$-pattern one expects a D-$\bar{D}$-pattern as depicted in Fig. 1.5.1. The upcoming accelerator complex FAIR [FAI] at GSI/Darmstadt offers the opportunity to extend the experimental studies into the charm sector [Fri11]. The CBM collaboration [CBM] intends to study the near-threshold production of D and J/ψ mesons in heavy-ion-collisions, while the PANDA collaboration [PAN] will focus on charm spectroscopy, as well as on charmed mesons produced by anti-proton annihilation at nuclei. In the CBM experiments, charm degrees of freedom will serve as probes of nuclear matter at the maximum compression achievable in the laboratory, at moderate temperatures. Despite of this interest in D mesons and their behavior in nuclear matter, the literature on in-medium D mesons is fairly scarce. While there is a variety of calculations within a hadronic basis, e. g. [Tol08, Tol06b, Tol05, Tol04, Miz06, Lut06], or within the quark-meson coupling model, e. g. [Sai07], the use of QSRs is fairly seldom and resides in unpublished form [Hay00, Mor01, Mor99, Zsc06]. In contrast, the treatment of vacuum D (and D_s) ground states is performed in a concise manner [Hay04, Ali83, Pfa06]. A re-evaluation of the QSR for D and $\bar{D}$ mesons is thus in order and part of this thesis.

Weinberg type chiral sum rules for differences of moments between light vector and axial-vector spectral functions have been developed for vacuum [Wei67, Das67] and for a strongly interacting medium at finite temperature [Kap94]. As a part of this work the framework of chiral partner sum rules is extended to the heavy-light quark meson sector in the medium for spin-0 and spin-1 mesons.

Medium modifications of mesons may be observed in, e. g., photo-nuclear reactions, cf. [Met11] for a recent review. One distinguishes between measurements which are sensitive to the production point and those which are sensitive to the decay point. Short living mesons such as the ρ meson (with a lifetime of 4.5×10^{-24} s) may be investigated by measurements which are sensitive to the decay point. The meson line shape analysis determines the in-medium mass from the four-momenta of the decay products in the limit of zero three-momentum. Such an approach requires that the decay products leave the medium undistorted, i. e. without final state interactions with the nuclear medium. Therefore, as already discussed above, decays into dileptons are the preferred decay channels. In momentum distribution measurements, the momentum distribution with effective mass-shift is compared to those without such a mass shift. When a meson leaves the medium, it must be on its vacuum mass shell. If the in-medium mass is lower than its vacuum mass, the momentum distribution must be shifted downward due to energy-momentum conservation. Hence, the momentum distribution in case of a lowered in-medium mass differs from a scenario without mass-shift. In contrast, long living mesons, such as the D meson (with a lifetime of 10^{-12} s) most probably leave the medium if they are produced at high momenta. In this case, methods which are sensitive to the production point are preferable. Transparency ratio measurements determine the absorption of mesons by a nucleus by comparing the cross section of the meson production per nucleon with that of free nucleons. The effective reduction of the meson lifetime corresponds to an increase of its width. Alternatively, the excitation functions of mesons should be altered in case of changing effective masses. Consequently, if the effective mass of a meson decreases in the medium, the production threshold should be lowered as well.

1.6 Structure of the thesis

This thesis is organized as follows. The method of QSRs is introduced in Sec. 2. DSEs and BSEs are introduced in Sec. 6. The starting point for the introduction to QSRs is the current-current correlator. Its analytic structure is investigated using a Lehmann representation. Subtracted dispersion relations in vacuum and in medium are derived, relating large momenta to low momenta. The OPE is introduced as an asymptotic expansion. DSEs are derived following [Rom69] by introducing external classical, i. e. not quantized, sources. These allow to probe the response of the system w. r. t. external changes, i. e. variations of the source fields. The resulting equation for the quark propagator is renormalized. The rainbow approximation is introduced

1 Introduction and motivation

by approximating the quark-gluon vertex by its lowest order perturbative contribution. BSEs are discussed by introducing two-particle irreducibility. By restricting the interactions to one at a time, the ladder approximation is introduced.

The application of QSRs to pseudo-scalar mesons consisting of a heavy and a light quark is presented in Sec. 3, and is close to [Hil09, Hil10a, Hil10b, Käm10, Hil10c], where the main results have been published. The extension of Weinberg-Kapusta-Shuryak sum rules to the in-medium heavy-light sector in the spin-1 and spin-0 channel is demonstrated in Sec. 4 similarly to [Hil11, Hil12a]. The impact of chirally odd condensates on the ρ meson is investigated in Sec. 5; the main results have been published in [Hil12b]. In Sec. 7 the coupled Dyson-Schwinger–Bethe-Salpeter approach to heavy-light quark meson masses and to light quark mesons in the Wigner-Weyl mode is given.

In App. A an introduction to Schwinger's action principle in conjunction with symmetries and Noether's theorem in a classical, i. e. not quantized, setting is given, which serves as the basis for symmetry considerations throughout this thesis. Poisson brackets are introduced and generalized from constant time surfaces to arbitrary space-like surfaces. These are used to derive a classical algebra of currents for infinitesimal internal transformations. Furthermore, the particular case of fermions and chiral transformations is discussed. Order parameters are introduced in the scope of quantizing the theory.

App. B is an essay about Källén-Lehmann representations, symmetries and non-anomalous Ward identities of current-current correlators and dispersion relations.

Technical details of QSRs are discussed in App. C. These include the Fock-Schwinger gauge in App. C.1, which is used to introduce the background field method as a tool for calculating an OPE. Furthermore, Borel transformations, used to improve the convergence of the OPE and to enhance the weighting of the lowest resonance of the spectral function, are reviewed and applied to in-medium sum rules in App. C.2. A detailed evaluation of the OPE for heavy-light quark currents and the associated renormalization of condensates by introducing non-normal ordered condensates can be found in Apps. C.4 and C.5. Recurrence relations for the Wilson coefficients are derived in App. C.6. The strategy of analyzing the sum rule is explained in App. C.7. Finally, a perspective on gluon condensates at temperatures and densities close to the critical temperature is given.

In App. D explicit calculations of the in-medium OPE for chiral partner sum rules are exposed.

2 Introduction to QCD sum rules

As known from quantum field theory, n-point functions play an important role in evaluating observables. Many quantities can be calculated directly if an appropriate n-point function is known. The main object of investigation within a QSR analysis is the current-current correlation function $\Pi(q)$ defined in vacuum as the following two-point function

$$\Pi(q) = i \int d^4x \, e^{iqx} \langle \Omega | T \left[j(x) j^\dagger(0) \right] | \Omega \rangle, \tag{2.0.1a}$$

being the Fourier transform of the expectation value of the time-ordered product of two currents. The quantity defined by Eq. (2.0.1a) is called the causal correlator. An extensive essay about correlators, their analytic properties, symmetry constraints, non-anomalous Ward identities and subtracted dispersion relations can be found in App. B. We will merely summarize the main results here.

For contour integrations in the medium, it is more convenient to use the retarded or advanced correlator instead [Fet71]. Retarded (R) and advanced (A) correlators are defined as

$$R(q) = i \int d^4x \, e^{iqx} \langle \Omega | \Theta(x_0) \left[j(x), j^\dagger(0) \right]_- | \Omega \rangle, \tag{2.0.1b}$$

$$A(q) = i \int d^4x \, e^{iqx} \langle \Omega | \Theta(-x_0) \left[j^\dagger(0), j(x) \right]_- | \Omega \rangle, \tag{2.0.1c}$$

where $\Theta(x_0)$ is the Heaviside function, which obey the appropriate analytic structure. The spectral density is defined as

$$\rho(q) = -\int d^4x \, e^{iqx} \langle \Omega | \left[j(x), j^\dagger(0) \right]_- | \Omega \rangle. \tag{2.0.1d}$$

It may be regarded as a density of states which allows to express the correlator of an interacting particle by a sum of (integral over) free correlators. For illustrative purposes a spectral function, chosen similarly to the experimentally observed ρ

2 Introduction to QCD sum rules

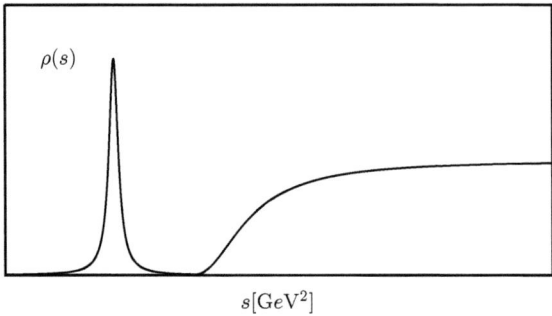

Figure 2.0.1: Schematic view of a spectral density. A relativistic Breit-Wigner curve $\rho(s) \approx 1/\left[(s-m^2)^2 + m^2\Gamma^2\right]$ has been used to model the resonance. The high energy tail is the continuum contribution to the spectral density.

spectral density, is given in Fig. 2.0.1. Only the causal correlator can be evaluated in perturbation theory. On the other hand, the quantities defined in Eqs. (2.0.1b) and (2.0.1c) are related to the causal correlator via analytic continuation in the complex plane, which can be used in conjunction with dispersion relations (to be introduced in Sec. 2.1) to determine the OPE. Furthermore, the ground state expectation value has to be replaced at nonzero temperatures by the Gibbs average:

$$\langle\Omega|\ldots|\Omega\rangle \to \langle\ldots\rangle \equiv \frac{1}{Z}\text{Tr}\left[e^{-\beta(H-\mu N)}\ldots\right]. \tag{2.0.2}$$

Here, Z is the grand canonical partition function, H stands for the Hamiltonian, β is the inverse temperature and N denotes some additive quantum number such as the particle number and μ the corresponding chemical potential. The ground state expectation value of an operator is the zero temperature and density limit of (2.0.2).

The physical ground state $|\Omega\rangle$ at zero temperature satisfies

$$H|\Omega\rangle = E_\Omega|\Omega\rangle, \tag{2.0.3a}$$
$$\langle\Omega|\Omega\rangle = 1, \tag{2.0.3b}$$
$$a^-|\Omega\rangle \neq 0, \tag{2.0.3c}$$

with H being the full Hamiltonian of the theory and $|\Omega\rangle$ the lowest eigenstate. The operator a^- denotes an arbitrary annihilation operator which annihilates the canonical ground state $|0\rangle$ of free particles, $a^-|0\rangle = 0$, used in perturbation theory. The latter one is, in contrast to the physical ground state $|\Omega\rangle$, the lowest eigenstate of the free

Hamiltonian. The state $|0\rangle$ is often referred to as the vacuum state, but in order to prevent confusions we will only refer to it as perturbative ground state. Equation (2.0.3) reflects the non-perturbative physics of the strong interaction, hence, it is referred to as non-perturbative, physical ground state or simply ground state. As a simple example of a ground state which is not annihilated by an annihilation operator recall the shifted harmonic oscillator. The sets of creation and annihilation operators of shifted and unshifted harmonic oscillator are related to each other and an operator of one set can be expressed by a linear combination of operators from the other set. The lowest energy state of the shifted harmonic oscillator, which is the shifted lowest energy state of the unshifted harmonic oscillator, is of course annihilated by its annihilation operators. It is however not annihilated by annihilation operators of the unshifted harmonic oscillator. Throughout the literature, $|\Omega\rangle$ is often simply referred to as vacuum state, with the meaning of zero temperature and density. In the course of this thesis the zero density, i. e. the vacuum ground state, and the finite density ground state are both denoted by $|\Omega\rangle$. It is understood that the nonzero density ground state is a many-particle state. Although the vacuum ground state must be understood as non-perturbative ground state it is defined to satisfy $E_\Omega = 0$ and hadrons as excitations of the ground state thus have an energy equal to their mass in the rest frame. The vacuum ground state is Lorentz invariant and invariant under time-reversal and parity transformations. Note that weak interaction processes violate invariance w. r. t. parity transversal [Wu57]. In contrast, the in-medium ground state is only assumed to be invariant under time reversal and parity in its local rest frame. It is not invariant under all Lorentz transformations, but expectation values calculated in this state, e. g. the above current-current correlator $\Pi(q)$, indeed transform covariantly [Jin93]. As a result of (2.0.3a), the ground state is not translational invariant but obeys

$$e^{iPx}|\Omega\rangle = e^{ip_\Omega x}|\Omega\rangle, \qquad (2.0.4)$$

where P is the momentum operator and p_Ω denotes the related momentum of the finite density ground state. The latter one is a function of the medium four-velocity v_μ. A brief discussion of the problem of defining the medium four-velocity is given in App. B.3. Hence, application of the translation operator results in a phase factor. However, expectation values w. r. t. $|\Omega\rangle$ are translationally invariant because the phase factors cancel each other. Analogously, we also assume translational invariance of

2 Introduction to QCD sum rules

the finite temperature medium and mean

$$\langle \mathcal{O}(x) \rangle = \langle \mathcal{O}(x+a) \rangle \tag{2.0.5}$$

for the Gibbs average (the zero temperature case is included in this notation).

The analytic structure of causal, retarded and advanced correlators and their relation to the spectral density is investigated in some detail in App. B.1 utilizing their Källén-Lehmann representation [Käl52, Leh54]. We generalize the canonical treatment of spin-0 current operators, as in Eq. (2.0.1a), or traces of correlators with tensorial rank-2 in Minkowski space, i. e. contracted Lorentz indices of two spin-1 currents in Eq. (2.0.1a), to the general case of non-contracted Lorentz indices of spin-1 currents. As a result, we find that simple dispersion relations which relate the imaginary part of the correlator to its real part can only be given if the correlator is symmetric w. r. t. its Lorentz indices. As a consequence the causal correlators can not be expressed by advanced and retarded correlator only. Instead, one has

$$\Pi_{\mu\nu}(q) = \frac{R_{\mu\nu}(q)}{1-e^{-\beta q_0}} + \frac{A_{\mu\nu}(q)}{1-e^{\beta q_0}} - \tanh^{-1}\left(\frac{\beta q_0}{2}\right)\mathrm{Im}\rho_{\mu\nu}(q), \tag{2.0.6}$$

at nonzero temperature and density. In App. B.1 we show that the spectral density is real if and only if it is symmetric w. r. t. Lorentz indices. Contracting the Lorentz indices gives the well-known representation of the causal correlator at nonzero temperatures in terms of retarded and advanced correlators only. Note that R and A have overlapping poles along the real energy axis for $\beta \neq 0$ (see App. B.1). At zero temperature Eq. (2.0.6) becomes

$$\Pi_{\mu\nu}(q) = \Theta(q_0)R_{\mu\nu}(q) + \Theta(-q_0)A_{\mu\nu}(q) - \mathrm{sign}(q_0)\mathrm{Im}\rho_{\mu\nu}(q), \tag{2.0.7}$$

clearly exposing non-overlapping pole contributions.

Furthermore, in App. B.2 we give a comprehensive analysis of symmetries and transformation properties of the correlators assuming different combinations of translational invariance, invariance under parity and time reversal or charge conjugation. These are used to decompose the tensor structure of the correlators in Minkowski space. It turns out that retarded and advanced correlators are transformed into each other if the ground state or the medium is translational invariant or invariant w. r. t. time reversal. Assuming translational invariance and invariance w. r. t. time inversion and parity reversal, the spectral density and the retarded, advanced and causal correlators are symmetric w. r. t. their Lorentz indices. This gives rise to the well-known transversal and longitudinal projections. In case of parity violation, an antisymmetric part contributes and has to be investigated separately. This important result allows a decomposition into symmetric tensors of rank 2 in Minkowski space, as demonstrated in App. B.3. Finally, in App. B.4 we derive non-anomalous Ward identities to express the longitudinal part in terms of the trace of the correlator (in Minkowski space), and spin-0 correlators of the according parity and condensates. These interrelations between correlators of different spins are required in order to separate the spin-0 contribution from the spin-1 correlator when analyzing vector or axial-vector spectra. As a byproduct we show under what conditions the longitudinal part is zero and demonstrate that current conservation is not sufficient. In particular, for different quark flavors entering the currents, the Gibbs average must be symmetric w. r. t. to these flavors. The results obtained in App. B are extensively used in Sec. 4.

The current $j(x)$ entering Eq. (2.0.1a), which describes a hadron in the low-energy region, is assumed to be represented by a composite operator consisting of quark field operators in the Heisenberg picture. It has to reflect the quantum numbers and valence quark content of the particle under consideration. The correlator $\Pi(q)$ can be understood as a function describing the propagation of a particle from 0 to x. In the high-momentum regime, i. e. at small distances, it must reflect the quark structure and valence quark content of the particle, while at low momentum, i. e. large distances, it is determined by hadronic properties of the respective particle.

The theorem of Gell-Mann–Low relates matrix elements of interacting Heisenberg field operators ("sandwiched" between interacting states, e. g. the ground state of the interacting theory $|\Omega\rangle$) to expectation values of interaction picture field operators, i. e. free fields, in non-interacting states (e. g. the perturbative ground state $|0\rangle$) [GM51, Pas84]:

$$\Pi(q) = i \int d^4x\, e^{iqx} \frac{\langle 0|T\left[j(x)j^\dagger(0)e^{i\int d^4x\, \mathscr{L}_{\text{int}}^{(0)}(y)}\right]|0\rangle}{\langle 0|T\left[e^{i\int d^4x\, \mathscr{L}_{\text{int}}^{(0)}(y)}\right]|0\rangle}, \qquad (2.0.8)$$

where $\mathscr{L}_{\text{int}}^{(0)}(x)$ is the interaction Lagrange density and the superscript (0) indicates that all fields have to be taken as free fields. If the coupling strength g in $\mathscr{L}_{\text{int}}^{(0)}$ is small, the exponential may be expanded into a series which is hoped to converge and may serve as a starting point for a perturbative treatment taking into account a finite number of terms. Noticeable, the BSE has also been proven in [GM51].

The basic idea of the QSR method is to relate large-momentum properties of the

2 Introduction to QCD sum rules

correlation function to properties in the region of small momenta [Shi79b]. Instead of deriving properties of hadrons which are results of the non-perturbative character of the interaction from first principles of the theory, one introduces non-vanishing ground state expectation values of quantum field operators as power corrections to a perturbative expansion of Eq. (2.0.8) in each order of $\alpha_s = g^2/4\pi$, called condensates, in order to reproduce the hadronic properties of the current-current correlator. In this sense, by introducing condensates, a possibility is offered to implement non-perturbative physics. A clear separation of long and short distances is mandatory. All the non-perturbative physics, i.e. large distance dynamics, must be contained in the condensates. This separation is achieved within the scope of a rigorous OPE, whereas the short distance dynamics is accessed via dispersion relations.

Either hadronic phenomenology is used to determine the condensates or known condensates are used to predict and to understand hadronic properties. The condensates as expectation values of QCD operators are accessible, e.g., in lattice QCD.

2.1 Subtracted dispersion relations

We will now relate the different energy regimes of the current-current correlator (2.0.1a) (in vacuum) or (2.0.1b) (in the medium) to each other by using the analytic properties only. Thereby, as distinguished from the treatment in the original articles of Källén [Käl52] and Lehmann [Leh54], an approach is used which exhibits the role of subtractions in detail and follows the treatment of [Sug61]. The treatment of all correlators proceeds along the same line of arguments. In particular, only general analytic properties which are common to the correlators are used. Therefore we restrict the discussion without loss of generality to the causal correlator. However, a distinction between vacuum and medium is necessary, as we carefully calculate explicitly subtracted dispersion relations for the in-medium case. All derivations can be found in App. B.5.

Vacuum dispersion relation

In vacuum, the correlator is a function of q^2, $\Pi(q) \rightarrow \Pi(q^2)$, and, thus, has poles which correspond to the resonances and a cut which stems from the continuum contribution starting from a threshold s_+ and extends to infinity on the positive real axis. Therefore, one may use the contour given in the left panel of Fig. 2.1.1 in conjunction with Cauchy's integral formula to relate the correlator at q^2 off the

2.1 Subtracted dispersion relations

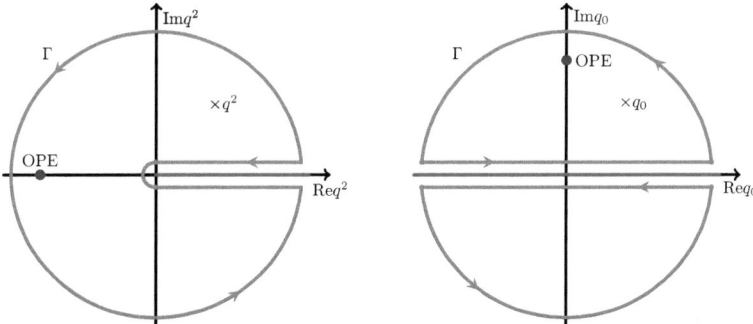

Figure 2.1.1: Integration contours Γ in the complex q^2-plane (left panel) and in the complex q_0-plane (right panel). The parallel integration paths tend to the real axis (but do not coincide with it), while the outer circle or half-circles tend to infinity.

positive real axis to its discontinuities along the positive real axis. Indeed, for any contour Γ and q^2 within the contour, one may write

$$\Pi(q^2) = \frac{1}{2\pi i} \int_\Gamma \frac{\Pi(s)}{s-q^2}\, ds\,. \tag{2.1.1}$$

Assuming $|\Pi(q^2)| \leq |q^2|^N$ for $|q^2| \to \infty$, where $N \in \mathbb{N}$ is a finite and fixed number, and letting the outer circle tend to infinity, we arrive at the N-fold subtracted dispersion relation in vacuum

$$\Pi(q^2) - \sum_{n=0}^{N-1} \frac{\Pi^{(n)}(0)}{n!}(q^2)^n = \frac{1}{\pi}\int_{s_0}^{\infty}\left(\frac{q^2}{s}\right)^N \frac{\Delta\Pi^{\text{vac}}(s)}{s-q^2}\, ds\,. \tag{2.1.2}$$

We have defined the discontinuities as

$$\Delta\Pi^{\text{vac}}(q^2) = \frac{1}{2i}\lim_{\epsilon \to 0}\left[\Pi(q^2 + i\epsilon) - \Pi(q^2 - i\epsilon)\right]. \tag{2.1.3}$$

The polynomial in Eq. (2.1.2) is called subtraction and is introduced to suppress the contribution of the outer circle, which is polynomial as well, if the correlator does not vanish fast enough for $|q^2| \to \infty$.

2 Introduction to QCD sum rules

Medium dispersion relation

In the medium the correlator is a function of all possible scalar products of the medium four-velocity v_μ and the momentum q_μ, i. e. $\Pi(q) = \Pi(q,v) \to \Pi(q^2, v^2, qv)$. For a fixed medium four-velocity, it is a function of q_0. In contrast to the vacuum case, it has poles along the entire real energy axis and one may therefore give a dispersion relation in the complex energy plane rather than in the energy-squared plane. Applying Cauchy's integral formula to the contour depicted in the right panel of Fig. 2.1.1 gives

$$\Pi(q_0, \vec{q}) = \frac{1}{2\pi i} \int_\Gamma \frac{\Pi(\omega, \vec{q})}{\omega - q_0} d\omega . \tag{2.1.4}$$

The main difference to the vacuum case is that the correlator may be split into a part which is even (e) and a part which is odd (o) in the energy q_0

$$\Pi(q_0, \vec{q}) = \Pi^e(q_0, \vec{q}) + q_0 \Pi^o(q_0, \vec{q}) . \tag{2.1.5}$$

Defining the discontinuities along the real axis as

$$\Delta\Pi(\omega, \vec{q}) = \frac{1}{2i} \lim_{\epsilon \to 0} \left[\Pi(\omega + i\epsilon, \vec{q}) - \Pi(\omega - i\epsilon, \vec{q}) \right] , \tag{2.1.6}$$

the N-fold subtracted dispersion relation in the medium for the even part read

$$\Pi^e(q_0, \vec{q}) - \frac{1}{2} \sum_{n=0}^{N-1} \frac{\Pi^{(n)}(0, \vec{q})}{n!} (q_0)^n (1 + (-1)^n)$$
$$= \frac{1}{2\pi} \int_{-\infty}^{+\infty} d\omega \, \Delta\Pi(\omega, \vec{q}) \frac{q_0^N}{\omega^{N-1}} \frac{\left(1 + (-1)^N\right) + \frac{q_0}{\omega}\left(1 - (-1)^N\right)}{\omega^2 - q_0^2} \tag{2.1.7a}$$

and for the odd part

$$\Pi^o(q_0, \vec{q}) - \frac{1}{2} \sum_{n=0}^{N-1} \frac{\Pi^{(n)}(0, \vec{q})}{n!} (q_0)^{n-1} (1 - (-1)^n)$$
$$= \frac{1}{2\pi} \int_{-\infty}^{+\infty} d\omega \, \Delta\Pi(\omega, \vec{q}) \frac{q_0^{N-1}}{\omega^{N-1}} \frac{\left(1 - (-1)^N\right) + \frac{q_0}{\omega}\left(1 + (-1)^N\right)}{\omega^2 - q_0^2} . \tag{2.1.7b}$$

As can be seen from both equations, even and odd parts only depend on q_0^2.

Dispersion relations are exact relations between a current-current correlation function at arbitrary (complex) values of q_0 or q^2, respectively, off the real axis (positive real axis) and its values at the real axis (positive real axis). This enables us to relate properties of the current-current correlation function at real (physical) values of the energy q_0 for the in-medium case and positive real values of the momentum squared q^2 for the vacuum case to the hadronic properties of the correlation function encoded in the discontinuities along the real axis.

2.2 Operator product expansion

Applying the OPE [Wil69] relates the current-current correlator to the quark degrees of freedom encoded in the respective currents. Intuitively, it is clear that such a relation can only be valid at large external momenta q^2. At low momentum the fundamental degrees of freedom are hadrons. In fact, the OPE has only been proved in perturbation theory [Zim73]. Therefore, application of the OPE is restricted to the high-momentum regime.

Consider the product of two local field operators $j(x)j^\dagger(y)$. In the limit $x \to y$ this product is not well defined. Instead, it is singular [Mut87]. This also holds true for the time-ordered product of two local field operators, $\mathrm{T}\left[j(x)j^\dagger(y)\right]$, as can be seen by applying Wick's theorem to the time-ordered product[6]

$$\mathrm{T}\left[j(x)j^\dagger(y)\right] =: j(x)j^\dagger(y) : + \contraction{}{j}{(x)}{j} j(x)j^\dagger(y), \qquad (2.2.1)$$

introducing the free-field propagators and setting $x = y$. The singularities are contained in the free-field propagators. In general, it can be shown that the square of a local free field operator diverges if the operator has non-vanishing matrix elements between the vacuum and one particle states [Zim73]. Thereby the divergence is carried by the ground state expectation value. For matrix elements of two interacting local field operators this can also be seen by investigating the Källén-Lehmann representation (see Sec. B.1) in coordinate space for $x \to y$. The same holds true for operator products which are not time ordered. This is intimately connected with the canonical equal-time commutators (ETCs) of the theory.

Wilson proposed [Wil69] that an operator product can be written as a sum of

---

[6] $: \ldots :$ means normal ordering and $\contraction{}{}{\;\;}{}$ indicates a Wick contraction. Note that normal ordering requires a unique splitting of an operator into positive and negative frequency parts. For free fields, the EoM guarantee that this is possible. In general this is not true for interacting fields [Rom69]. Normal ordering is therefore defined w. r. t. the perturbative ground state $|0\rangle$.

2 Introduction to QCD sum rules

c-number functions $C_\mathcal{O}(x-y)$, which are singular for $x \to y$, and non-singular operators $\mathcal{O}$. For the time-ordered product, this expansion therefore reads

$$\mathrm{T}\left[j(x)j^\dagger(y)\right] = \sum_\mathcal{O} C_\mathcal{O}(x-y)\mathcal{O}, \tag{2.2.2}$$

where $C_\mathcal{O}(x-y)$ are the so-called Wilson coefficients or coefficient functions, being singular, and $\mathcal{O}$ being finite in the limit $x \to y$. Wilson's proposal was strongly motivated by current algebra considerations and Schwinger terms (see Sec. A.2). Indeed, the ETC of two local currents A and B is expected to be of the form

$$\left[A(x_0,\vec{x}), B(x_0,\vec{y})\right]_- = \sum_n C_n(\vec{x}-\vec{y})\mathcal{O}_n(x), \tag{2.2.3}$$

where $\mathcal{O}_n$ is a local field and C_n denotes Dirac's delta distribution and derivatives thereof. Wilson proposed that a similar expansion must hold for any operator product. In App. A.2 a brief discussion of the interrelation between divergent operator products, Schwinger terms and OPEs is given. Approximate scale invariance at small distances [Pok00, Col84] dictates the nature of the singularities, i.e. the behavior of the functions $C_n(\vec{x}-\vec{y})$ at small distances [Pas84], and relates it to the mass dimension. Accordingly, the operators $\mathcal{O}_n$ can be ordered with increasing mass dimension $\dim_m$ (which can be determined for free field operators via their canonical commutation relations [Pok00]) and the leading contribution has the lowest mass dimension. We list here all operators and their mass dimension which are considered throughout this thesis:

$$D_\mu : \quad \dim_m = 1, \tag{2.2.4a}$$

$$q : \quad \dim_m = \frac{3}{2}, \tag{2.2.4b}$$

$$\mathcal{G}_{\mu\nu} : \quad \dim_m = 2, \tag{2.2.4c}$$

and the operator products

$$\bar{q}q : \quad \dim_m = 3, \tag{2.2.5a}$$

$$\bar{q}D_\mu q : \quad \dim_m = 4, \tag{2.2.5b}$$

$$G^2 : \quad \dim_m = 4, \tag{2.2.5c}$$

$$\bar{q}D_\mu D_\nu q : \quad \dim_m = 5, \tag{2.2.5d}$$

$$\bar{q}\sigma^{\mu\nu}\mathcal{G}_{\mu\nu}q : \quad \dim_m = 5, \tag{2.2.5e}$$

$$\bar{q}\Lambda T\Gamma q \bar{q}\Lambda' T'\Gamma' q : \quad \dim_m = 6, \tag{2.2.5f}$$

where the last term denotes, quite generally, four-quark condensates with flavor matrices Λ, Λ', color matrices T, T' and Dirac structures Γ, Γ'. Possible Lorentz indices have been suppressed in the last term.

One can show that, if non-perturbative effects dominate the dynamics, a consistent separation of large distance and short distance physics is necessary [Gen84]. This means that the non-perturbative effects must be contained in expectation values of the operators $\mathcal{O}$, while the coefficients are completely determined by perturbative physics.

It is important to note that (2.2.2) can only be understood as an asymptotic expansion, i.e. it diverges. The series is an approximation to the operator product, if truncated and only a finite number of terms is taken into account [Shi79b]. A divergent series which is asymptotic to the operator product for $x \to y$, i.e. an asymptotic expansion of the divergent operator product, has to fulfill [Itz80]

$$\lim_{x \to y} \frac{\mathrm{T}\left[j(x)j^\dagger(y)\right] - \sum_{\mathcal{O}}^{\mathcal{O}_{max}} C_\mathcal{O}(x-y)\mathcal{O}}{C_{\mathcal{O}_{max}}(x-y)} = 0 \quad \forall\, \mathcal{O}_{max}, \tag{2.2.6}$$

where the l.h.s. converges weakly. In contrast to a Taylor expansion of a function $f(x)$ at $x = x_0$, where the quality of the approximation increases with increasing number of terms that have been taken into account, the asymptotic series of a, possibly divergent, function $g(x)$ at $x = x_0$ has to be truncated at some finite order $N = N(x - x_0)$ to give the best approximation that can be achieved.[7] Taking more or less terms into account, the approximation gets worse. We will use this property of asymptotic expansions for the analysis of our sum rule.

As the OPE is an expansion at operator level, the coefficient functions are state independent. They are completely determined by the structure of the operator product. This means that choosing a certain current to express the quark structure of the considered particle completely determines the coefficient functions. On the other hand, there might be different interpolating currents which describe the same particle and, hence, lead to different OPEs. The nucleon, for example, may be

---

[7] A Laurent expansion would only be a possible expansion if the function $g(x)$ has isolated singularities. Moreover, such an expansion would mean that the series is convergent inside some ring around the singularity. Instead, an asymptotic expansion can also be used to approximate functions $g(x)$ which are well defined in some region, e.g. $x > 0$, have a divergence at $x = 0$ and are ill defined for $x < 0$. In such a case it is impossible to find a Laurent series for the function $g(x)$ at $x = 0$ with maximum radius of convergence being $R > 0$ [Win06].

2 Introduction to QCD sum rules

described by various interpolating currents such as the two ones given by Ioffe [Iof81]. Many other choices may be useful for more sophisticated problems, e. g. [Bra93, Fur96, Lei97, Kon06]. The question whether a current with specified quantum numbers couples to a certain particle or not, can not be answered a priori and has to checked by explicit calculations.

Calculating matrix elements of (2.2.2) by using the non-perturbative vacuum ground state $|\Omega\rangle$ or the in-medium Gibbs average, one obtains

$$\langle T[j(x)j^\dagger(y)]\rangle = \sum_{\mathcal{O}} C_{\mathcal{O}}(x-y)\langle \mathcal{O}\rangle . \qquad (2.2.7)$$

The matrix elements of the operators $\langle \mathcal{O}\rangle$ on the r. h. s. of (2.2.7) are called condensates. Using the perturbative ground state $|0\rangle$, only the unity operator gives a contribution to the sum, because all the other matrix elements vanish. In regimes where perturbative QCD applies, e. g. for heavy quarks, the condensates are comparably small and it may be a reasonable approximation to set them to zero. Whenever non-perturbative effects are not negligible, perturbative techniques fail and, by (2.0.3), the condensates acquire nonzero values. In this sense, by introducing non-vanishing matrix elements one is able to deal with non-perturbative physics. From (2.2.7) we also see that the medium dependence must be completely contained in the condensates, because the Wilson coefficients are state independent.[8] In Fig. 2.2.1, the graphical representation of an OPE for a current-current correlator is shown. Although the medium dependence shall be contained in the condensates, the matrix elements of the OPE differ in vacuum and medium. However, this is not a result of vanishing Wilson coefficients. Rather, it is a result of vanishing matrix elements of certain operators on the r. h. s. of Eq. (2.2.7) in vacuum, which do not vanish in the medium. Therefore, the complete medium dependence is indeed contained in the condensates, although the OPEs in form of Eq. (2.2.7) differ. Note that due to the different mass scales in heavy-light quark mesons standard OPE techniques fail to properly separate long and short distance dynamics as addressed in the discussion after Eq. (2.2.4). Additional renormalization of the condensates is required and results in the famous operator mixing, see Sec. C.4.

For our purposes the condensates can be considered at this stage as phenomenological parameters used to reproduce the properties of the correlation function in the

---

[8]This is true for low densities or temperatures. A full thermal field theoretic approach leads to a medium dependence of the Wilson coefficients [Hat92]. This partly contradicts the separation of scales demanded by the OPE.

non-perturbative, i.e. hadronic, region. Knowledge of the current-current correlation function in the low momentum regime, i.e. the hadronic properties of the particle under investigation, together with the dispersion relations reviewed in App. B will result in restrictions for the correlator in the large-momentum regime. By virtue of the OPE, this gives us valuable information about the condensates.[9]

Moreover, instead of reproducing hadronic properties, e.g. hadron masses, by explicitly deriving non-perturbative dynamical effects of the theory, such as confinement or DCSB from first principles of QCD, the existence of these effects is implicitly taken as granted by introducing condensates and relating them to measurable quantities. As the condensates are expectation values of quark and gluon operators in the ground state, they give us valuable information about the dynamical structure of the theory. They can be read as particles, propagating in the ground state of QCD or the strongly interacting medium, being annihilated by virtual particles and creating other real particles (the remaining virtual particles become real) somewhere else. Although Feynman diagram techniques are applicable, throughout this thesis they are merely used to visualize the formulas which are evaluated or as an intuitive notation in order to indicate the terms that have to be inserted for further calculations.

However, because the operators $\mathcal{O}$ appearing in the sum (2.2.2) are completely determined by the theory, their matrix elements do not depend on the specific currents of the operator product. Of course, a certain operator could be absent from the sum, if the corresponding Wilson coefficient vanishes due to the structure of the currents, but the condensate itself is independent. Hence, the condensates may be considered as universal parameters in the sense that their knowledge is not restricted to a specific sum rule. Instead, parameters obtained from a certain sum rule or any other method are universal and valid for any other sum rule.

In order to calculate the OPE, one may use the circumstance that it is an expansion at operator level. Applying the Gell-Mann–Low equation (2.0.8) in conjunction with Wick's theorem [Wic50, Gre92] gives all orders of the strong coupling in a perturbative expansion but, strictly speaking, eliminates all condensates due to the non-perturbative vacuum. However, as we are only interested in the Wilson coefficients of the OPE, Eq. (2.0.8) may be applied without making use of the defining property of the perturbative ground state, i.e. retaining matrix elements of normal ordered operators, and hence delivering coefficient functions and non-singular operators. Formally, Eq. (2.0.8) serves to obtain the expansion at operator level. Once at our disposal, the ground state expectation value or Gibbs' average of this expansion

---

[9]Strictly speaking, a QSR gives us information about certain combinations of condensates.

2 Introduction to QCD sum rules

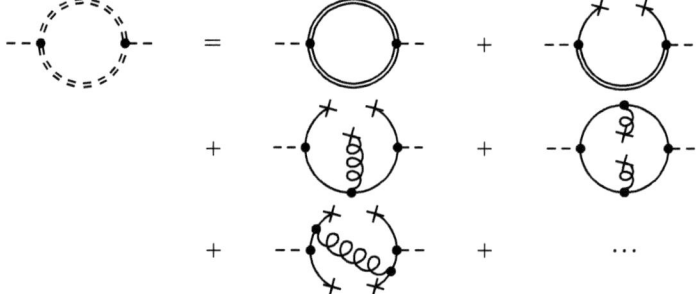

Figure 2.2.1: Graphical visualization of an OPE for a current-current correlation function including (from top left to bottom right) the perturbative contribution, the chiral condensate, the mixed quark-gluon condensate, the digluon condensate and a four-quark contribution which stems from power corrections to the $\mathcal{O}(\alpha_s)$ perturbative contribution. In the latter case, the gluon carries hard momenta, while the gluons carry soft momenta in case of the digluon and mixed quark-gluon condensate. Apart from the last condensate, all condensates are power corrections to the lowest order perturbative term. In this figure, double dashed lines correspond to exact propagators, double lines to perturbative correlators and single lines denote free propagators. Crosses symbolize condensates.

is evaluated, introducing the condensates as power corrections to the perturbative expansion

$$\Pi_{\text{OPE}}(q) = \sum_{\mathcal{O}} C_{\mathcal{O}}(q)\langle\mathcal{O}\rangle. \qquad (2.2.8)$$

The unit operator $\mathbb{1}$, which is the operator of the lowest mass dimension, thereby corresponds to the genuine perturbative series. This approach is the basis for the background field method [Nov84b], where the interaction with the ground state or the medium is modeled by a weak gluonic background field. Strictly speaking, this interaction is therefore limited to soft gluons, whereas hard gluon contributions arise due to gluon fields introduced by higher orders of the interaction. Details of this method are reviewed in App. C.1. Other techniques are, e. g., the plane wave method [Nar02] and a recently developed method based on canonical commutation relations [Hay12].[10]

Recapitulatory we recall that the singular behavior of operator products like $j(x)j^\dagger(y)$ for $x \to y$ gives rise to the OPE (2.2.2), while the non-perturbative large distance behavior gives rise to the introduction of nonzero condensates $\langle\mathcal{O}\rangle$. The condensates are introduced in order to reproduce or to predict hadronic properties, which can not be understood from a perturbative point of view. A graphical representation can be found in Fig. 2.2.1.

2.3 Sum rules

The sum rules are set up by inserting the OPE into the dispersion relation. For the vacuum case the sum rules read

$$\Pi_{\text{OPE}}(q^2) - \sum_{n=0}^{N-1} \frac{\Pi_{\text{ph}}^{(n)}(0)}{n!}(q^2)^n$$
$$= \frac{1}{\pi}\int_0^{s_0}\left(\frac{q^2}{s}\right)^N \frac{\Delta\Pi_{\text{ph}}(s)}{s-q^2}\,ds + \frac{1}{\pi}\int_{s_0}^{+\infty}\left(\frac{q^2}{s}\right)^N \frac{\Delta\Pi_{\text{OPE}}(s)}{s-q^2}\,ds, \qquad (2.3.1)$$

---

[10]Note first that this method actually circumvents an OPE in QSRs. And second, it originates from solid state physics, cf. e. g. [Nol05].

2 Introduction to QCD sum rules

where we introduced the threshold parameter s_0.[11] Similarly, for the in-medium case the sum rule for the even part read

$$\Pi^e_{\text{OPE}}(q_0^2, \vec{q}\,) - \frac{1}{2}\sum_{n=0}^{N-1} \frac{\Pi^{(n)}_{\text{ph}}(0, \vec{q}\,)}{n!}(q_0)^n (1+(-1)^n)$$

$$= \frac{1}{2\pi}\int_{s_0^-}^{s_0^+} d\omega\, \Delta\Pi_{\text{ph}}(\omega, \vec{q}\,) \frac{q_0^N}{\omega^{N-1}} \frac{\left(1+(-1)^N\right) + \frac{q_0}{\omega}\left(1-(-1)^N\right)}{\omega^2 - q_0^2}$$

$$+ \frac{1}{2\pi}\left(\int_{-\infty}^{s_0^-} + \int_{s_0^+}^{+\infty}\right) d\omega\, \Delta\Pi_{\text{OPE}}(\omega, \vec{q}\,) \frac{q_0^N}{\omega^{N-1}}$$

$$\times \frac{\left(1+(-1)^N\right) + \frac{q_0}{\omega}\left(1-(-1)^N\right)}{\omega^2 - q_0^2}, \quad (2.3.2a)$$

and for the odd part

$$\Pi^o_{\text{OPE}}(q_0^2, \vec{q}\,) - \frac{1}{2q_0}\sum_{n=0}^{N-1} \frac{\Pi^{(n)}_{\text{ph}}(0, \vec{q}\,)}{n!}(q_0)^n (1-(-1)^n)$$

$$= \frac{1}{2\pi q_0}\int_{s_0^-}^{s_0^+} d\omega\, \Delta\Pi_{\text{ph}}(\omega, \vec{q}\,) \frac{q_0^N}{\omega^{N-1}} \frac{\left(1-(-1)^N\right) + \frac{q_0}{\omega}\left(1+(-1)^N\right)}{\omega^2 - q_0^2}$$

$$+ \frac{1}{2\pi q_0}\left(\int_{-\infty}^{s_0^-} + \int_{s_0^+}^{+\infty}\right) d\omega\, \Delta\Pi_{\text{OPE}}(\omega, \vec{q}\,) \frac{q_0^N}{\omega^{N-1}}$$

$$\times \frac{\left(1-(-1)^N\right) + \frac{q_0}{\omega}\left(1+(-1)^N\right)}{\omega^2 - q_0^2}. \quad (2.3.2b)$$

The subscript "ph" indicates that a phenomenological model is used in order to relate condensates to hadronic properties. The threshold parameters s_0 ($s_0^\pm$) have been introduced to separate the continuum contribution from the lowest lying excitations of the spectrum. Thereby we assumed semi-local quark-hadron duality, which means that integrating the spectral strength above the threshold is well approximated by an integral over Π_{OPE}. It turns out that the perturbative contribution approximates the continuum very well, cf. e. g. [Kwo08].

The major drawbacks of the QSR method are that, on the one hand, only integrated spectral densities are probed and, on the other hand, only combinations of condensates enter. Possibilities of extracting certain parameters of the spectral density

---
[11] The quantity s_0 differs from the threshold parameter introduced in App. B.5 to characterize the regions of analyticity.

or isolating particular condensates, e. g. a particular order parameter, are therefore limited. In order to enhance the weight of the lowest excitation of the spectrum a Borel transformation may be applied. It transforms the spectral integral as given in Eqs. (2.3.1) and (2.3.2) into the Laplace transform of the spectral density, i. e. the spectral density is integrated with an exponential weight. As a byproduct the subtractions are eliminated. On one side, the Borel transform improves the convergence of the OPE, as the result is nothing else but the Borel sum of the OPE. The Borel summation is a technique to sum an actually divergent series. Details about the Borel transformation and the resulting expressions are given in App. C.2. In some cases it might be useful to introduce a different weighting which enhances other energy domains. An example is given by the Gaußian sum rule [Raf97], cf. [Oht11a, Oht11b] for the nucleon. Throughout this thesis, models for the spectral density, e. g. a Breit-Wigner curve, have been employed to extract bound state masses and other parameters. During the last years the maximum entropy method has been applied to QSRs [Gub10]. However, although the method was also successfully applied to nucleons [Oht11a] and charmonium states [Gub11], it fails to give the correct width for the ρ meson [Gub10] and is not considered in this thesis.

3 QCD sum rules for heavy-light spin-0 mesons in the medium*

As argued in Sec. 1.3 a careful re-evaluation of the D and $\bar{D}$ OPE is mandatory. The aim of the present chapter is the analysis of these sum rules in cold nuclear matter. Extensions to B and $\bar{B}$, D_s and $\bar{D}_s$ and the scalar D mesons are included. Even for the OPE up to mass dimension 5, there are conflicting results in the literature concerning the open charm sector [Hay00, Mor01, Mor99, Zsc06, Ali83, Neu92, Nar89, Nar01]. While in [Hay00] only the even part of the in-medium OPE up to mass dimension 4 has been used, we present here the even as well as the odd in-medium OPE up to mass dimension 5. Moreover, a term $\propto \langle \bar{q} g \sigma \mathcal{G} q \rangle$, i.e. the lowest-order quark-gluon condensate, can be found in the literature with various factors and signs already for the vacuum. As the subtle $D - \bar{D}$ mass splitting is of paramount experimental interest, a safe basis is mandatory.

3.1 Operator product expansion

Employing the current operators $j_{D^+} = i\bar{d}\gamma_5 c$, $j_{D^-} = j_{D^+}^\dagger = i\bar{c}\gamma_5 d$ we obtain the Borel transformed (see App. C.2) OPE for the correlator given in Eq. (2.0.1a) up to mass dimension 5, in the rest frame of nuclear matter $v = (1, \vec{0})$ (v stands for the medium four-velocity), in the limit $m_d \to 0$ and sufficiently large charm-quark pole mass m_c,

$$\mathcal{B}\left[\Pi^e_{\text{OPE}}(\omega^2, \vec{q} = 0)\right](M^2)$$
$$= \frac{1}{\pi} \int_{m_c^2}^{\infty} ds\, e^{-s/M^2} \text{Im}\Pi^{D^+}_{\text{per}}(s, \vec{q} = 0)$$
$$+ e^{-m_c^2/M^2} \left(-m_c \langle \bar{d}d \rangle + \frac{1}{2}\left(\frac{m_c^3}{2M^4} - \frac{m_c}{M^2}\right) \langle \bar{d} g \sigma \mathcal{G} d \rangle + \frac{1}{12}\langle \frac{\alpha_s}{\pi} G^2 \rangle \right)$$

---

*The presentation is based on [Hil09, Hil10a, Käm10, Hil10c, Hil10b].

$$+ \left[\left(\frac{7}{18} + \frac{1}{3} \ln \frac{\mu^2 m_c^2}{M^4} - \frac{2\gamma_E}{3} \right) \left(\frac{m_c^2}{M^2} - 1 \right) - \frac{2}{3} \frac{m_c^2}{M^2} \right] \langle \frac{\alpha_s}{\pi} \left(\frac{(vG)^2}{v^2} - \frac{G^2}{4} \right) \rangle$$
$$+ 2 \left(\frac{m_c^2}{M^2} - 1 \right) \langle d^\dagger i D_0 d \rangle + 4 \left(\frac{m_c^3}{2M^4} - \frac{m_c}{M^2} \right) \left[\langle \bar{d} D_0^2 d \rangle - \frac{1}{8} \langle \bar{d} g \sigma \mathcal{G} d \rangle \right] \Bigg),$$
(3.1.1a)

$$\mathcal{B}\left[\Pi^o_{\text{OPE}}(\omega^2, \vec{q} = 0) \right](M^2)$$
$$= e^{-m_c^2/M^2} \left(\langle d^\dagger d \rangle - 4 \left(\frac{m_c^2}{2M^4} - \frac{1}{M^2} \right) \langle d^\dagger D_0^2 d \rangle - \frac{1}{M^2} \langle d^\dagger g \sigma \mathcal{G} d \rangle \right), \quad (3.1.1b)$$

where $\alpha_s = g^2/4\pi$. (Analog relations hold for $j_{D^0}(x) = i\bar{u}\gamma_5 c$ with $j_{\bar{D}^0}(x) = j^\dagger_{D^0}(x) = i\bar{c}\gamma_5 u$.) The calculational details are documented in App. C.4. Even (e) and odd (o) correlators are defined in Eq. (B.5.18). The sum rules are set up according to the N-fold subtracted dispersion relations given in Eqs. (2.3.2) and, for the Borel transformed relations, (C.2.27). While the perturbative spectral function $\text{Im}\Pi^{D^+}_{\text{per}}(s)$ (see [Ali83, Nar89] for an explicit representation in terms of the pole mass) is known for a long time, discrepancies especially for Wilson coefficients of medium specific condensates exist. An important intermediate step is the careful consideration of the operator mixing, which occurs due to the introduction of non-normal ordered condensates and the corresponding cancellation of infrared divergent terms $\propto m_q^{-2}$ and $\log m_q$ (m_q is the light-quark mass) at zero and nonzero densities [Hil08]. This is not to be confused with the operator mixing within renormalization group methods. In vacuum, our expression differs from [Hay00] in the coefficient of $\langle (\alpha_s/\pi) G^2 \rangle$; [Ali83] reports an opposite sign; [Mor01, Mor99] finds the same result. For the medium case [Mor01, Mor99] does not give explicit results, while terms $\propto \langle \bar{d}d \rangle$, $\propto \langle (\alpha_s/\pi)G^2 \rangle$, $\propto \langle (\alpha_s/\pi)((vG)^2/v^2 - G^2/4) \rangle$ have different coefficients compared to [Hay00]. Higher order terms are partially considered in [Mor01, Mor99] and are found to be numerically not important.

We stress the occurrence of the term $m_c \langle \bar{d}d \rangle$. In the pure light quark sector, say for vector mesons, it would read $m_d \langle \bar{d}d \rangle$, i.e., the small down-quark mass strongly suppresses the numerical impact of the chiral condensate $\langle \bar{d}d \rangle$. In fact, only within the doubtful factorization of four-quark condensates into the squared chiral condensate it would become important [Tho05]. Here, the large charm-quark mass acts as an amplifier of the genuine chiral condensate entering the QSRs for the D^+ meson.

3.2 Parameterizing the spectral function

Especially in vacuum the spectral strength of the vector–isospin-scalar excitation exhibits a well-defined sharp peak (the ω meson) and a well-separated flat continuum. Assuming the same features for the ω meson in a medium gives rise to the often exploited "pole + continuum" ansatz. One way to avoid partially such a strong assumption is to introduce certain moments of the spectral function, thus replacing the assumed pole mass by a centroid of the distribution [Tho05, Kwo08].

For D mesons the sum rule includes an integral which arises from the dispersion relation over positive and negative energies, see Eq. (C.2.27). Similar to baryons [Coh95, Tho07], one may try to suppress the antiparticle contribution corresponding here to D^-. This, however, is not completely possible [Tho08b]. Nevertheless, one can identify with the ansatz $\Delta\Pi(s) = \pi F_+ \delta(s - m_+) - \pi F_- \delta(s + m_-)$, motivated by the Lehmann representation of the correlation function (cf. App. B.1), the meaning of the even and odd sum rules (C.2.27) with (3.1.1):

$$e \equiv \int_{s_0^-}^{s_0^+} d\omega\, \omega\, \Delta\Pi e^{-\omega^2/M^2} = m_+ F_+ e^{-m_+^2/M^2} + m_- F_- e^{-m_-^2/M^2}, \quad (3.2.1a)$$

$$o \equiv \int_{s_0^-}^{s_0^+} d\omega\, \Delta\Pi e^{-\omega^2/M^2} = F_+ e^{-m_+^2/M^2} - F_- e^{-m_-^2/M^2}. \quad (3.2.1b)$$

With the decomposition $m_\pm = m \pm \Delta m$ and $F_\pm = F \pm \Delta F$ the leading order terms of an expansion in Δm for the first and second lines become $\propto Fme^{-m^2/M^2}$ and $\propto (\Delta F - 2\Delta m\, F \frac{m}{M^2})e^{-m^2/M^2}$ meaning that (3.2.1a) is related to the average $D + \bar{D}$ properties, while (3.2.1b) refers to the $D - \bar{D}$ splitting. If one assumes for the moment being $m_\pm$ and $F_\pm$ to be independent of the Borel mass M, Eq. (3.2.1) can be rewritten as

$$\Delta m = \frac{1}{2} \frac{oe' - eo'}{e^2 + oo'}, \quad (3.2.2a)$$

$$m = \sqrt{\Delta m^2 - \frac{ee' + (o')^2}{e^2 + oo'}}, \quad (3.2.2b)$$

where a prime denotes the derivative w. r. t. $1/M^2$. Having fixed Δm and m, ΔF and

3 QCD sum rules for heavy-light mesons

F are given as

$$\Delta F = \frac{1}{2} \frac{e^{(m^2 + \Delta m^2)/M^2}}{m} \left[(e - o\Delta m)\sinh\left(\frac{2m\Delta m}{M^2}\right) + om\cosh\left(\frac{2m\Delta m}{M^2}\right) \right], \quad (3.2.3a)$$

$$F = \frac{1}{2} \frac{e^{(m^2 + \Delta m^2)/M^2}}{m} \left[(e - o\Delta m)\cosh\left(\frac{2m\Delta m}{M^2}\right) + om\sinh\left(\frac{2m\Delta m}{M^2}\right) \right]. \quad (3.2.3b)$$

In order to gain further insight into the dependencies of Δm and m on the different OPE contributions, we expand (3.2.2) up to first order in the density n employing $e(n) \approx e(0) + n \, de/dn|_{n=0}$ and $o(n) \approx n \, do/dn|_{n=0}$, since $o(0)$ must vanish to reproduce the vacuum sum rules where $\Delta m(n = 0) = 0$ holds. We remark that these expansions are exact for a linear density dependence of the condensates and if $s_0^2 = ((s_0^+)^2 + (s_0^-)^2)/2$ as well as $\Delta s_0^2 = ((s_0^+)^2 - (s_0^-)^2)/2$ are density independent and the Borel mass M is kept fixed. This implies $\Delta s_0^2 = 0$ for all densities, because otherwise $o(0) = 0$ cannot be fulfilled. For small densities we get accordingly

$$\Delta m(n) \approx \frac{1}{2} \frac{\frac{do}{dn}\big|_0 e'(0) - e(0) \frac{do'}{dn}\big|_0}{e(0)^2} n, \quad (3.2.4a)$$

$$m(n) \approx \sqrt{-\frac{e'(0)}{e(0)} + \frac{1}{2} \sqrt{-\frac{e(0) \frac{de}{dn}\big|_0 e'(0) - e(0) \frac{de'}{dn}\big|_0}{e'(0)}} \frac{}{e(0)^2} n, \quad (3.2.4b)$$

which can be written as

$$\Delta m(n) \approx -\frac{1}{2} \frac{\frac{do}{dn}\big|_0 m^2(0) + \frac{do'}{dn}\big|_0}{e(0)} n \equiv \alpha_{\Delta m} n, \quad (3.2.5a)$$

$$m(n) \approx m(0) - \frac{1}{2m(0)} \frac{\frac{de}{dn}\big|_0 m^2(0) + \frac{de'}{dn}\big|_0}{e(0)} n. \quad (3.2.5b)$$

Equation (3.2.2) and the approximations in Eq. (3.2.5) offer a transparent interpretation. In vacuum ($n = 0$), there is no mass splitting, of course; the mass parameter $m(0)$ is determined merely by the even part of the OPE. In first order of n, the mass splitting Δm depends on both the even and odd parts of the OPE, whereas only the even part of the OPE determines the mass parameter m, having the meaning of the centroid of the doublet D^+, D^-. If one is only interested in the mass shift of the doublet as a whole, for small densities it is sufficient to consider the even OPE part alone, as was done in [Hay00]. However, for the mass splitting the odd part of the

3.2 Parameterizing the spectral function

OPE is of paramount importance. In particular, it is the density dependence of the odd part of the OPE alone which drives the mass splitting in first order of n. Interestingly, the density dependent part of the chiral condensate, which belongs to the even part of the OPE, enters the mass splitting in order n^2. The chiral condensate comes about in the combination $m_c \langle \bar{d}d \rangle$. The large charm mass amplifies the numerical impact, as stressed above.

We remark that Eqs. (3.2.2) or (3.2.5) are a consequence of using a pole-ansatz for the first excitation. The OPE and the special form of the continuum contribution to the spectral integral are encoded in e and o. Likewise, the arguments following Eq. (3.2.2) merely use $o(0) = 0$. The last requirement must always be fulfilled in any sum rule and/or dispersion relation because, at zero density, the current-current correlation function (2.0.1a) only depends on q^2 and, hence, the odd part (B.5.18b) vanishes. This can also be confirmed directly from Eq. (3.2.1b), where $s_0^+ = -s_0^-$, due to particle anti-particle symmetry, and $\Delta\Pi(s) \to \Delta\Pi(s^2)$, meaning that the spectral density in vacuum merely depends on the squared energy, on account for $o(0) = 0$.

To arrive at a more general result, one may seek for a relation of $m_\pm$ to certain normalized moments of $\Pi(s)$ (or ratios thereof) independent of a special ansatz, as can be done in the case of vector mesons [Tho05, Kwo08]. In this spirit one would be tempted to define $\int_0^{s_0^+} d\omega\, \omega \Delta\Pi e^{-s^2/M^2} \to m_+ F_+ e^{-m_+^2/M^2}$ and $\int_0^{s_0^+} d\omega\, \Delta\Pi e^{-\omega^2/M^2} \to F_+ e^{-m_+^2/M^2}$ and analogously for m_- and F_-. However, such a separation of positive and negative frequency parts leads to multiple but different expressions for $m_\pm$ which can be fulfilled consistently only for special cases of $\Pi(s)$, as for the above pole ansatz. (This can be seen by combining these relations with derivatives according to M^{-2}.) Therefore, one is left with either the somewhat vague statement that Eq. (3.2.1) refers to $D + \bar{D}$ and $D - \bar{D}$ properties or one has to employ another explicit ansatz for the function $\Pi(s)$.

Alternatively, one can define moments which correspond to the integrals defined in Eq. (3.2.1)

$$S_n(M) \equiv \int_{s_0^-}^{s_0^+} d\omega\, \omega^n \Delta\Pi(\omega) e^{-\omega^2/M^2}. \tag{3.2.6}$$

The odd and even OPE, $o = S_0(M)$ and $e = -S_1(M)$, and their derivatives with respect to M^{-2}, $o' = -S_3(M)$ and $e' = S_4(M)$, can then be related via Eq. (3.2.2) to these moments. Thereby, new quantities $\overline{\Delta m}$ and $\overline{m}$ may be defined which encode the

3 QCD sum rules for heavy-light mesons

combined mass–width properties of the particles under consideration:

$$\overline{\Delta m} \equiv \frac{1}{2}\frac{S_1 S_2 - S_0 S_3}{S_1^2 - S_0 S_2}, \tag{3.2.7a}$$

$$\overline{m_+ m_-} \equiv -\frac{S_2^2 - S_1 S_3}{S_1^2 - S_0 S_2} \tag{3.2.7b}$$

and $\overline{m}^2 \equiv \overline{\Delta m}^2 + \overline{m_+ m_-}$. For the above pole ansatz, these quantities become $\Delta m = \overline{\Delta m}$ and $m = \overline{m}$, i.e., they allow for an interpretation as mass splitting and mass centroid. The relations (3.2.6) and (3.2.7) avoid the use of a special ansatz of the spectral function, but prevent a direct physical and obvious interpretation.

3.3 Evaluation for D and $\bar{D}$ mesons

3.3.1 The pseudo-scalar case

We proceed with the above pole ansatz and evaluate the behavior of $m_\pm$ having in mind that these parameters characterize the combined $D, \bar{D}$ spectral functions, but need not necessarily describe the pole positions in general. According to the above defined current operators, D stands either for D^+ or D^0 and $\bar{D}$ for D^- or $\bar{D}^0$.

Because $dm_\pm/dM = 0$ has been used to derive Eq. (3.2.2) we have to look for the extrema of $m_\pm(M)$. Furthermore, in order to solve consistently the system of equations defined by (3.2.1), the values taken for $m_\pm$ must be fixed at the same Borel mass M. Therefore, we evaluate the sum rules using two threshold parameters

$$(s_0^\pm)^2 = s_0^2 \pm \Delta s_0^2 \tag{3.3.1}$$

and demand that the minima of the respective Borel curves $m_+(M)$ and $m_-(M)$ must be at a common Borel mass M. Hence, the thresholds are prescribed and offer the possibility to give a consistent solution to Eq. (3.2.1). It is important to note that the consistent evaluation of $F_\pm$ within this method is only possible because $F_\pm$ and $m_\pm$ have extrema at the same Borel mass, as also $dF_\pm/dM = 0$ has been used. The consistency is shown in App. C.7.

Analogously to the analysis in [Hay00], we chose the threshold parameter $s_0^2 = 6.0\,\text{GeV}^2$, which approximately reproduces the vacuum case. At zero density we obtain for $m_\pm$ a value of 1.863 GeV, representing a reasonable reproduction of the experimental value of the D mass. The employed condensates are parametrized

3.3 Evaluation for D mesons

Table 3.3.1: List of employed condensate parameters. The condensates are given in linear density approximation $\langle\ldots\rangle = \langle\ldots\rangle_{\text{vac}} + \langle\ldots\rangle_{\text{med}} n$. A discussion of these numerical values can be found in [Coh95]; further remarks on $\langle q^\dagger g \sigma \mathcal{G} q \rangle$ are given in [Mor01, Mor99]. For the strong coupling we utilize $\alpha_s = 4\pi / \left[((11 - 2N_f/3) \ln(\mu^2/\Lambda_{\text{QCD}}^2)) \right]$ with μ being the renormalization scale, taken to be of the order of the largest quark mass in the system, and N_f being the number of quark flavors with mass smaller than μ; $\Lambda_{\text{QCD}}^2 = 0.25\,\text{GeV}^2$ is the dimensional QCD parameter. The employed quark pole masses are $m_c = 1.5\,\text{GeV}$ and $m_b = 4.7\,\text{GeV}$ [Nar01].

condensate	vacuum value $\langle\cdots\rangle_{\text{vac}}$	density dependent part $\langle\cdots\rangle_{\text{med}}$
$\langle \bar{q} q \rangle$	$(-0.245\,\text{GeV})^3$	$45/11$
$\langle \frac{\alpha_s}{\pi} G^2 \rangle$	$(0.33\,\text{GeV})^4$	$-0.65\,\text{GeV}$
$\langle \bar{q} g \sigma \mathcal{G} q \rangle$	$0.8\,\text{GeV}^2 \times (-0.245\,\text{GeV})^3$	$3\,\text{GeV}^2$
$\langle q^\dagger q \rangle$	0	1.5
$\langle \frac{\alpha_s}{\pi} \left(\frac{(vG)^2}{v^2} - \frac{G^2}{4} \right) \rangle$	0	$-0.05\,\text{GeV}$
$\langle q^\dagger i D_0 q \rangle$	0	$0.18\,\text{GeV}$
$\langle \bar{q} \left[D_0^2 - \frac{1}{8} g \sigma \mathcal{G} \right] q \rangle$	0	$-0.3\,\text{GeV}^2$
$\langle q^\dagger D_0^2 q \rangle$	0	$-0.0035\,\text{GeV}^2$
$\langle q^\dagger g \sigma \mathcal{G} q \rangle$	0	$0.33\,\text{GeV}^2$

in linear density approximation as $\langle\ldots\rangle = \langle\ldots\rangle_{\text{vac}} + \langle\ldots\rangle_{\text{med}} n$ with values listed in Tab. 3.3.1.

The density dependence of the mass splitting parameter Δm and the $D + \bar{D}$ doublet mass centroid m are exhibited in Fig. 3.3.1. We observe an almost linear behavior of the mass splitting with increasing density. At $n = 0.15\,\text{fm}^{-3}$ a mass splitting of $2\Delta m \approx -60\,\text{MeV}$ is obtained. The mass splitting has negative values, i.e. $m_- > m_+$ or $m_{\bar{D}} > m_D$ in line with previous estimates in [Mor01, Mor99]. For the mass centroid m our result differs from the one in [Hay00], where a mass shift of the order of $-50\,\text{MeV}$ is obtained, while we find about $+45\,\text{MeV}$. At $n = 0.15\,\text{fm}^{-3}$ the splitting of the threshold parameters is $\Delta s_0^2 \approx -0.3\,\text{GeV}^2$ for the used set of parameters, and the minima of the Borel curves are located at $M \approx 0.95\,\text{GeV}$ being slightly shifted upwards with increasing density.

While the mass splitting is fairly robust, we find a sensitivity of the centroid mass shift under variation of the continuum threshold parameter s_0^2. The above reported value of the mass centroid changes towards zero when lowering s_0^2. In Fig. 3.3.1 we therefore also use a density dependent prescription for the threshold $s_0^2(n) = s_0^2(0) \pm n/n_0\,\text{GeV}^2$, where $n_0 = 0.15\,\text{fm}^{-3}$ is the nuclear saturation density; $\pm 1/n_0\,\text{GeV}^2$

3 QCD sum rules for heavy-light mesons

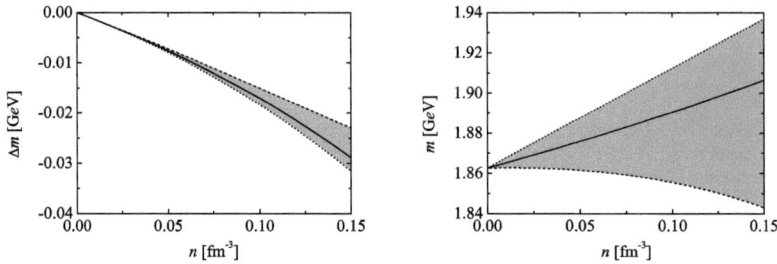

Figure 3.3.1: Mass splitting parameter Δm (left) and mass centroid m (right) for $D, \bar{D}$ mesons for density independent threshold (solid line) and a density dependent threshold $s_0^2(n) = s_0^2(0) \pm n/n_0\,\text{GeV}^2$, where the dotted (dashed) curve is for the positive (negative) sign. Note that the $D - \bar{D}$ mass splitting is $2\Delta m$.

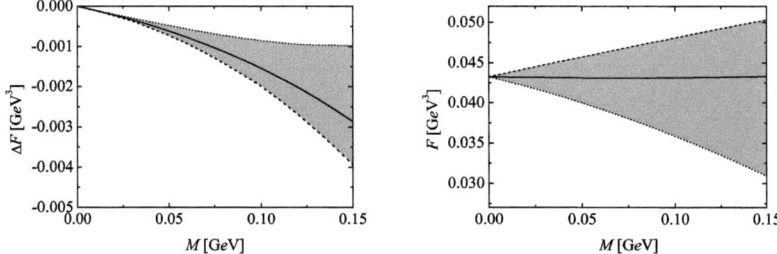

Figure 3.3.2: ΔF (left) and F (right) as evaluated from the QSR analysis. Line codes as in Fig. 3.3.1.

corresponds to the first Taylor coefficient $ds_0^2/dn(0)$. This simple choice enables us to identify the uncertainties which might emerge due to the introduction of a density independent threshold. As can be seen, the average mass shift may change in sign. In contrast, the result for Δm shows only a weak dependence on s_0^2.

The density dependence of ΔF and F are exhibited in Fig. 3.3.2. Although the uncertainties due to variations of the density dependence of the continuum threshold $s_0^2(n)$ are rather large for both quantities, the overall conclusion is the same as for the mass parameters. While F ranges from $\approx -0.030\,\text{GeV}^3$ to $\approx 0.050\,\text{GeV}^3$, ΔF can be determined to be negative, ranging from $\approx -0.004\,\text{GeV}^3$ to $\approx -0.001\,\text{GeV}^3$. However, the overall dependence on the threshold behavior seems rather large.

At this point a comment concerning the sign of $\langle q^\dagger g \sigma \mathcal{G} q \rangle$ is in order. If one would use $\langle q^\dagger g \sigma \mathcal{G} q \rangle = -0.33\,\text{GeV}^2\,n$ instead (this option is also discussed in [Coh95],

3.3 Evaluation for D mesons

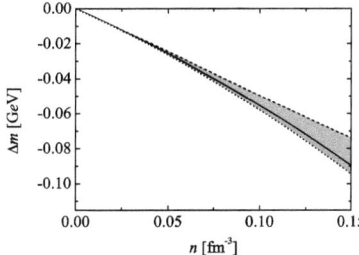 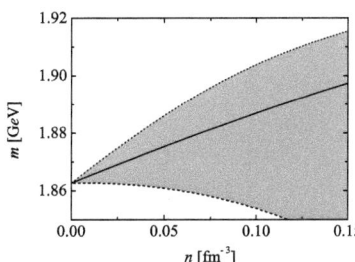

Figure 3.3.3: Evaluation of the QSR for the mass splitting Δm (left) and the centroid $\bar{m}$ (right) of pseudo-scalar D and $\bar{D}$ mesons for $\langle q^\dagger g\sigma\mathcal{G}q\rangle = -0.33\,\text{GeV}^2\,n$. Line codes as in Fig. 3.3.1.

$\langle q^\dagger D_0^2 q\rangle$ would acquire a value of $-0.0585\,\text{GeV}^2\,n$ accordingly) one would get a much larger mass splitting of about $-180\,\text{MeV}$, which is far beyond the estimates obtained in [Tol08, Tol06b, Tol05, Tol04, Miz06, Lut06]. Hence, we favor the positive sign of $\langle q^\dagger g\sigma\mathcal{G}q\rangle$ as advocated in [Mor01, Mor99], too. The density dependancies of mass splitting and mass centroid for this set of parameters are exhibited in Fig. 3.3.3. Clearly, further correlators should be studied to investigate the role of the condensate $\langle q^\dagger g\sigma\mathcal{G}q\rangle$. In view of the strong influence of this poorly known condensate, the $D-\bar{D}$ mass splitting may be considered as an indicator for its actual value.

We emphasize the special evaluation strategy employed so far. Another possibility is, e. g., variation of s_0^2 and Δs^2 so that $m_\pm(M)$ develop a section of maximum flatness. Interestingly, this method leads to a rather low threshold $s_0^2 \approx 4\,\text{GeV}^2$ and a low vacuum mass of about $m \approx 1.6\,\text{GeV}$. In contrast, averaging over the Borel curves in the interval $[0.9M_0, 1.2M_0]$, around the minimum M_0, we find the values for the mass splitting $\Delta m \approx -40\,\text{MeV}$ and the average mass shift to be of the same order as quoted above, whereas the absolute value of the vacuum mass becomes $m = 1.877\,\text{GeV}$.

Let us now further consider the impact of various condensates. The result for the mass splitting Δm strongly depends on the quark density $\langle q^\dagger q\rangle$, whose density dependence is uniquely fixed. The odd mixed quark-gluon condensate $\langle q^\dagger g\sigma Gq\rangle$ and the chiral condensate $\langle \bar{q}q\rangle$ are the next influential ones for the mass splitting. The density dependent part of the chiral condensate enters in order $\mathcal{O}(n^2)$ gaining its influence from the heavy quark mass amplification factor. The influence of the chiral condensate is illustrated in Fig. 3.3.4. In a strictly linearized sum rule evaluation, the density dependent part of $m_c\langle \bar{q}q\rangle$ would be omitted for the mass splitting. However, numerically the influence of the chiral condensate is of the same order as (but still

3 QCD sum rules for heavy-light mesons

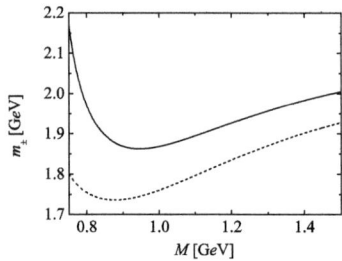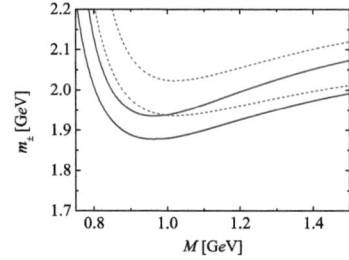

Figure 3.3.4: Borel curves $m_\pm(M)$ for the D meson for two densities $n = 0$ (left), $n = 0.15\,\text{fm}^{-3}$ (right) and two values of the chiral condensate. Solid curves: chiral condensate from Tab. 3.3.1, dotted curves: (left) doubling the chiral condensate or (right) doubling the density dependent part of the chiral condensate; lower (upper) curves in the right panel are for m_+ (m_-), while $m_+ = m_-$ for the vacuum case in the left panel.

smaller than) the above discussed condensate $\langle q^\dagger g \sigma \mathcal{G} q \rangle$, which enters the odd part of the OPE. As expected, the density dependence of the mass centroid is basically determined by the even part of the OPE. The density dependent parts of the other even condensates are of minor importance for the mass splitting. The shift of the centroid's mass is anyhow fragile.

Digression: D_s mesons

Within the given formulation and with the first evaluation strategy, one may also consider D_s and $\bar{D}_s$ mesons with the replacements

$$
\begin{aligned}
m_q &\longrightarrow m_s = 150\,\text{MeV}\,, \\
\langle \bar{q} q \rangle &\longrightarrow \langle \bar{s} s \rangle = 0.8 \langle \bar{q} q \rangle_{\text{vac}} + y \langle \bar{q} q \rangle_{\text{med}}\,, \\
\langle \bar{q} g \sigma \mathcal{G} q \rangle &\longrightarrow \langle \bar{s} g \sigma \mathcal{G} s \rangle = 0.8\,\text{GeV}^2 \langle \bar{s} s \rangle\,, \\
\langle q^\dagger q \rangle &\longrightarrow \langle s^\dagger s \rangle = 0\,, \\
\langle q^\dagger g \sigma \mathcal{G} q \rangle &\longrightarrow \langle s^\dagger g \sigma \mathcal{G} s \rangle = y \langle q^\dagger g \sigma \mathcal{G} q \rangle\,, \\
\langle q^\dagger i D_0 q \rangle &\longrightarrow \langle s^\dagger i D_0 s \rangle = 0.018\,\text{GeV}\, n\,, \\
\langle \bar{q} \left[D_0^2 - \frac{1}{8} g \sigma \mathcal{G} \right] q \rangle &\longrightarrow \langle \bar{s} \left[D_0^2 - \frac{1}{8} g \sigma \mathcal{G} \right] s \rangle = y \langle \bar{q} \left[D_0^2 - \frac{1}{8} g \sigma \mathcal{G} \right] q \rangle\,, \\
\langle q^\dagger D_0^2 q \rangle &\longrightarrow \langle s^\dagger D_0^2 s \rangle = y \langle q^\dagger D_0^2 q \rangle\,.
\end{aligned}
$$

3.3 Evaluation for D mesons

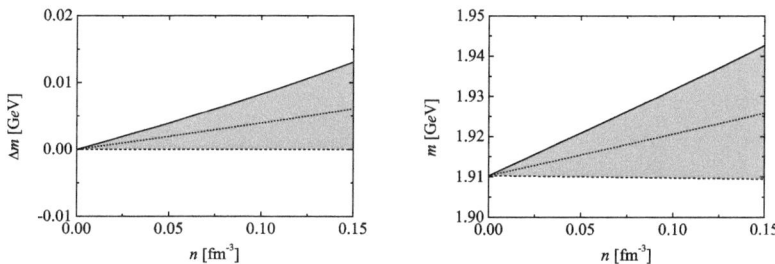

Figure 3.3.5: Δm (left) and m (right) for D_s and $\bar{D}_s$ at $s_0^2 = 8.0\,\text{GeV}^2$ and for $y = 0.5$ (solid), $y = 0.25$ (dotted), $y = 0$ (dashed).

The anomalous strangeness content of the nucleon is varied as $0 \leq y \leq 0.5$ [Nav05]; lattice calculations, for example, point to $y = 0.36$ [Don96]. The results are exhibited in Fig. 3.3.5. At $n = 0.15\,\text{fm}^{-3}$ and for $y = 0.5$ we observe a mass splitting of $2\Delta m \approx +25\,\text{MeV}$ and a shift of the mass centroid of about $+30\,\text{MeV}$. The splitting of the thresholds becomes $\Delta s_0^2 \approx 0.83\,\text{GeV}^2$, and the minima of the Borel curves are located at $M \approx 0.89\,\text{GeV}$ and are slightly shifted upwards with increasing density. The main reason for the positive sign of the mass splitting is the vanishing strange quark net density $\langle s^\dagger s \rangle$. The mass splitting acquires positive values for $\langle s^\dagger s \rangle \lesssim 0.4\,n$ (at $y = 0.5$). Mass splitting and the average mass shift tend to zero for $y \to 0$. In this case only the pure gluonic condensates, which enter the even OPE and are numerically suppressed compared to other condensates, have a density dependence. Note that these evaluations are, at best, for a rough orientation, as mass terms $\propto m_s$ have been neglected. The too low vacuum mass of $1.91\,\text{GeV}$ compared to the experimental value $m_{D_s} = 1.968\,\text{GeV}$ is an indication for some importance of strange quark mass terms. Such mass terms $\propto m_s$ have been accounted for in [Hay04] for the vacuum case. The complete in-medium OPE and sum rule evaluation deserves separate investigations, as m_s introduces a second mass scale.

Digression: Finite width analysis

Before proceeding with the QSR analysis for scalar D mesons we will comment on the finite width analysis for pseudo-scalar D mesons. Investigating finite width effects within QSRs always meets the problem of additional parameters which have to be fixed by the same set of equations. Hence, one only is able to give the mass of the particle (which may be defined as, e. g., the center of gravity or the peak position of

3 QCD sum rules for heavy-light mesons

the spectral function) as a function of its width (cf. Sec. 5 for an analog study w. r. t. the ρ meson). In the case of D mesons, the widths of particle (+) and anti-particle (−) enter, and the mass–width correlation of the particle is locked with the width of the antiparticle as well. Employing the following Breit-Wigner ansatz for the spectral function

$$\Delta\Pi(s) = \frac{F_+}{\pi} \frac{s\Gamma_+}{(s^2 - m_+^2)^2 + s^2 \Gamma_+^2} \Theta(s) + \frac{F_-}{\pi} \frac{s\Gamma_-}{(s^2 - m_-^2)^2 + s^2 \Gamma_-^2} \Theta(-s) \qquad (3.3.2)$$

and determining $m_\pm$, $\Gamma_\pm$ such that they fulfill Eq. (3.2.7) reveals this effect. A detailed analysis, as was done for the VOC scenario of the ρ meson in Sec. 5, requires the coupled solution of Eq. (3.2.1) for $m_\pm$ for a given pair $\Gamma_\pm$ over the whole range of Borel masses, finding the minima of the thus obtained curves $m_\pm(M)$ and varying the splitting of thresholds in order to align the minima. Variation of the pair $\Gamma_\pm$ finally delivers the desired mass–width curves. As only the demonstration of the mass–width coupling between particle and anti-particle is intended, we refrain from performing the full mass–width analysis. Instead, we choose $\overline{\Delta m}$ and $\overline{m}$ as determined from the zero-width analysis and solve Eq. (3.2.1) for $m_\pm$ for all pairs of $\Gamma_\pm$ at the Borel mass which corresponds to the minima of the Borel curves for $m_\pm$ in the zero-width analysis. For $\overline{\Delta m} = -30$ MeV and $\overline{m} = 1.915$ GeV one obtains the results depicted in Fig. 3.3.6.

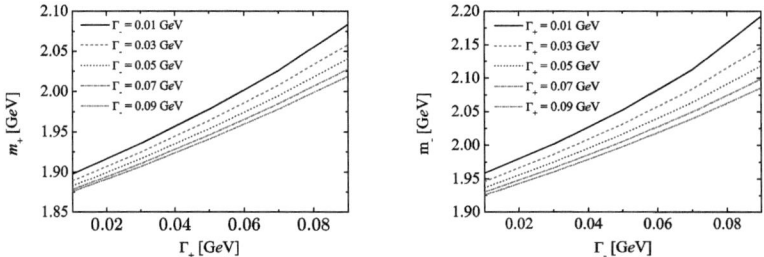

Figure 3.3.6: Mass–width correlations for m_+ (left) and m_- (right) for different assumptions of widths of the respective other particle. $\overline{\Delta m} = -30$ MeV, $\overline{m} = 1.915$ GeV and $F_+ = F_-$ are assumed. For simplicity, the integration limits in Eq. (3.2.1) have been extended to $\pm\infty$ without including the continuum contribution, to avoid the influence of possible threshold effects.

3.3.2 The scalar case

The OPE needed here can be obtained by combining the OPE for pseudo-scalar D mesons and the OPE for difference sum rules in Sec. 4:

$$\mathcal{B}_{Q^2 \to M^2}\left[\Pi^e_{\text{OPE}}(Q^2, \vec{q} = 0)\right](M^2)$$
$$= \frac{1}{\pi}\int_{m_c^2}^{\infty} ds\, e^{-s/M^2} \text{Im}\Pi^{D^*}_{\text{per}}(s, \vec{q} = 0)$$
$$+ e^{-m_c^2/M^2}\left(m_c\langle \bar{d}d\rangle - \frac{1}{2}\left(\frac{m_c^3}{2M^4} - \frac{m_c}{M^2}\right)\langle \bar{d}g\sigma\mathcal{G}d\rangle + \frac{1}{12}\langle\frac{\alpha_s}{\pi}G^2\rangle\right.$$
$$+ \left[\left(\frac{7}{18} + \frac{1}{3}\ln\frac{\mu^2 m_c^2}{M^4} - \frac{2\gamma_E}{3}\right)\left(\frac{m_c^2}{M^2} - 1\right) - \frac{2}{3}\frac{m_c^2}{M^2}\right]\langle\frac{\alpha_s}{\pi}\left(\frac{(vG)^2}{v^2} - \frac{G^2}{4}\right)\rangle$$
$$\left. + 2\left(\frac{m_c^2}{M^2} - 1\right)\langle d^\dagger i D_0 d\rangle - 4\left(\frac{m_c^3}{2M^4} - \frac{m_c}{M^2}\right)\left[\langle \bar{d}D_0^2 d\rangle - \frac{1}{8}\langle \bar{d}g\sigma\mathcal{G}d\rangle\right]\right),$$

(3.3.3a)

$$\mathcal{B}_{Q^2 \to M^2}\left[\Pi^o_{\text{OPE}}(Q^2, \vec{q} = 0)\right](M^2)$$
$$= e^{-m_c^2/M^2}\left(\langle d^\dagger d\rangle - 4\left(\frac{m_c^2}{2M^4} - \frac{1}{M^2}\right)\langle d^\dagger D_0^2 d\rangle - \frac{1}{M^2}\langle d^\dagger g\sigma\mathcal{G}d\rangle\right)$$
$$\equiv e^{-m_c^2/M^2}\langle K(M)\rangle n,$$

(3.3.3b)

where $\text{Im}\Pi^{D^*}_{\text{per}}$ is given in [Nar02] and $m_c = 1.3$ GeV is the charm quark mass. Note that we use a different charm quark mass here. We remark again that our main goal is to predict medium modifications. Therefore, parameters are chosen such that they reproduce the vacuum case.

Low-density expansion:

The primary goal of the present sum rule analysis is to find the dependence of Δm and m on changes of the condensates entering Eq. (3.3.3). However, also the continuum thresholds $s_0^\pm$ can depend on the density, as discussed above for the pseudo-scalar case. The employed evaluation strategy, cf. App. C.7, naturally provides a density dependent splitting of the threshold parameters, but the average of both thresholds is still density independent. We will now include its medium dependence.

To study this influence, we consider Eq. (3.2.5) in detail using an asymmetric

3 QCD sum rules for heavy-light mesons

splitting of the continuum thresholds $\Delta s_0^2 = ((s_0^+)^2 - (s_0^-)^2)/2$ and parameterize its density dependence by $\Delta s_0^2(n) = \alpha_{\Delta s} n + \mathcal{O}(n^2)$. Up to order n, only $s_0^2(n=0)$ and $\left.\frac{d\Delta s_0^2}{dn}\right|_0$ enter Δm (i.e. neither $\left.\frac{ds_0^2}{dn}\right|_0$ nor $\left.\frac{dM}{dn}\right|_0$), whereas $s_0^2(n=0)$, $\left.\frac{ds_0^2}{dn}\right|_0$ and $\left.\frac{dM}{dn}\right|_0$ enter m (not $\left.\frac{d\Delta s_0^2}{dn}\right|_0$) as can be seen from the derivatives needed to calculate m and Δm from Eq. (3.2.4):

$$\left.\frac{do}{dn}\right|_0 = \left(\frac{e^{-s_0^2/M^2}}{\pi s_0}\mathrm{Im}\Pi_{\mathrm{per}}(s_0^2)\frac{d\Delta s_0^2}{dn}\right)_{n=0} + e^{-m_c^2/M^2}\langle K(M)\rangle, \qquad (3.3.4\mathrm{a})$$

$$\left.\frac{do'}{dn}\right|_0 = \left(-\frac{e^{-s_0^2/M^2}}{\pi}s_0\mathrm{Im}\Pi_{\mathrm{per}}(s_0^2)\frac{d\Delta s_0^2}{dn}\right)_{n=0} + e^{-m_c^2/M^2}\langle K'(M) - m_c^2 K(M)\rangle, \qquad (3.3.4\mathrm{b})$$

$$\left.\frac{de}{dn}\right|_0 = \left(\frac{e^{-s_0^2/M^2}}{\pi}\mathrm{Im}\Pi_{\mathrm{per}}(s_0^2)\frac{ds_0^2}{dn} - \frac{1}{\pi}\int_{m_c^2}^{s_0^2} ds\,\mathrm{Im}\Pi_{\mathrm{per}}(s)s e^{-s/M^2}\frac{dM^{-2}}{dn}\right)_{n=0}$$
$$+ \frac{d}{dn}\mathcal{B}_{Q^2\to M^2}\left[\Pi^e_{\mathrm{OPE}}(Q^2,\vec{q}=0)\right](M^2)\bigg|_{n=0}. \qquad (3.3.4\mathrm{c})$$

While Eq. (3.2.5) suggests that one can independently adjust $m(0)$ to the respective vacuum value, Eq. (3.3.4) evidences that further vacuum parameters (such as M, $\left.\frac{dM}{dn}\right|_0$, s_0^2, Δs_0^2, $\left.\frac{ds_0^2}{dn}\right|_0$ and $\left.\frac{d\Delta s_0^2}{dn}\right|_0$) enter the density dependence and have to be chosen consistently to the vacuum mass. That means, one has to evaluate the complete sum rule, including consistently the vacuum limit.

Note that the Borel mass, at which the sum rule is analyzed, changes with the density, i.e.

$$\frac{d}{dn}\mathcal{B}\left[\Pi^o_{\mathrm{OPE}}\right](M^2) = \left(\frac{\partial}{\partial n} + \frac{dM^{-2}}{dn}\frac{\partial}{\partial M^{-2}}\right)\mathcal{B}\left[\Pi^o_{\mathrm{OPE}}\right](M^2). \qquad (3.3.5)$$

However, the derivative w.r.t. M^{-2} of the perturbative term is still an integral with range of integration being the threshold splitting Δs_0^2. Likewise, the derivative of the non-perturbative terms w.r.t. M^{-2} is still proportional to the density. Hence, both contributions vanish at zero density, and the density dependence of the Borel mass drops out.

In linear density approximation of the condensates, $\alpha_{\Delta m}$ is thus given as

$$\alpha_{\Delta m} = -\frac{1}{2e(0)}\left(e^{-m_c^2/M^2}\left(\left[m^2(0) - m_c^2\right]\langle K(M)\rangle + \langle K'(M)\rangle\right)\right.$$

50

3.3 Evaluation for D mesons

$$+\frac{e^{-s_0^2/M^2}}{\pi s_0}\mathrm{Im}\Pi_{\mathrm{per}}(s_0^2)\left[m^2(0)-s_0^2\right]\alpha_{\Delta s}\Bigg), \quad (3.3.6)$$

which is dominated by the non-perturbative terms. We choose the Borel mass range and the thresholds according to [Nar05].

As an estimate for the order of $\alpha_{\Delta s}$ we rely on the splitting of the thresholds for the pseudo-scalar channel and obtain $\alpha_{\Delta s}\approx 0.25\cdot 10^3\,\mathrm{GeV}^{-1}$. It is an overestimation of the pseudo-scalar $\mathcal{O}(n)$ threshold splitting as it would correspond to a linear interpolation of Δs_0^2 from the vacuum to nuclear saturation density and, hence, includes higher order terms in the density. We choose $\alpha_{\Delta s}\approx 10^2\ldots 10^3\,\mathrm{GeV}^{-1}$.

The results are depicted in Fig. 3.3.7 for $\alpha_{\Delta s}=\pm 10^2\,\mathrm{GeV}^{-1}$ (left panel) and for $\alpha_{\Delta s}=\pm 10^3\,\mathrm{GeV}^{-1}$ (right panel). In Fig. 3.3.8, $\alpha_{\Delta m}$ as a function of $\alpha_{\Delta s}$ is displayed for $M=1.37\,\mathrm{GeV}$, the minimum of the vacuum Borel curve for the scalar D meson.

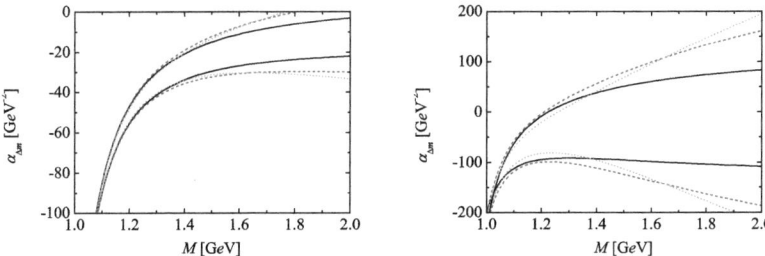

Figure 3.3.7: Low density approximation of the mass splitting $\alpha_{\Delta m}$ as a function of the Borel mass. Left: for $\alpha_{\Delta s}$ from $-10^2\,\mathrm{GeV}^{-1}$ (lower bundle of curves) to $+10^2\,\mathrm{GeV}^{-1}$ (upper bundle of curves). Right: for $\alpha_{\Delta s}$ from $-10^3\,\mathrm{GeV}^{-1}$ (lower bundle of curves) to $+10^3\,\mathrm{GeV}^{-1}$ (upper bundle of curves). The threshold values are $s_0^2=6.0\,\mathrm{GeV}^2$ (solid black), $7.5\,\mathrm{GeV}^2$ (dashed red) and $9.0\,\mathrm{GeV}^2$ (dotted blue).

Considering the results for the mass splitting of heavy-light pseudo-scalar mesons, e.g. D and B, one could raise the question if the splitting is mainly caused by a splitting of the thresholds and, hence, might be an artifact of the method which determines Δs_0^2. From the above study we find that $\alpha_{\Delta s}$ indeed influences the mass splitting. A direct correlation in the sense of a correlation in sign can not be confirmed. Furthermore, the results for D_s mesons allow for a positive mass splitting if the net strange quark density falls below a critical value. As the strange quark density enters through the vector quark condensate, this already points to a suppressed influence of the threshold splitting on the mass splitting.

3 QCD sum rules for heavy-light mesons

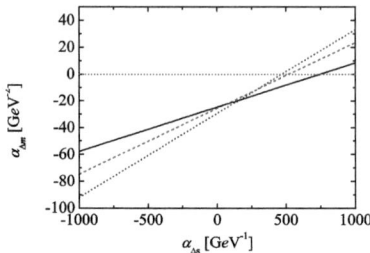

Figure 3.3.8: $\alpha_{\Delta m}$ as a function of $\alpha_{\Delta s}$ for $M = 1.37$ GeV. For line code see Fig. 3.3.7.

Beyond low-density approximation

In the sum rule analysis of this chapter, the threshold splitting is not considered as a free parameter but determined by the requirement that the minima of the Borel curves for particle and antiparticle are at the same Borel mass.

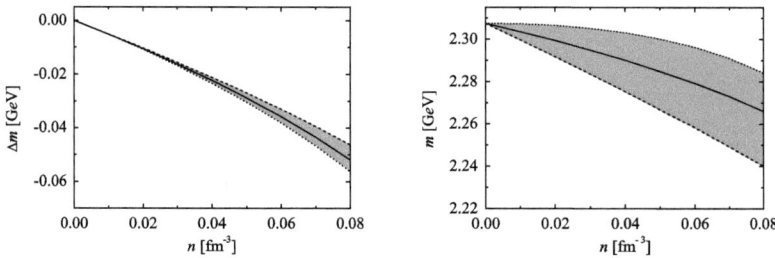

Figure 3.3.9: Mass splitting parameter Δm (left) and mean mass m (right) of scalar $D^* - \bar{D}^*$ mesons with the pole + continuum ansatz as a function of density at zero temperature. For a charm quark mass parameter of $m_c = 1.3$ GeV and mean threshold value $s_0^2(0) = 7.5$ GeV^2. The curves are for $s_0^2(n) = s_0^2(0) + \xi n/n_0$ with $\xi = 0$ (solid), $\xi = 1$ GeV^2 (dotted) and $\xi = -1$ GeV^2 (dashed).

Employing the condensates listed in Tab. 3.3.1, one obtains the results exhibited in Fig. 3.3.9. As in Sec. 3.3, this analysis goes beyond the strict linear density expansion of m and Δm and uses the full solution (3.2.2) of the sum rule. The medium dependent part of the chiral condensate (the density dependence of which is only in linear density approximation, as the other condensates too) enters the mass splitting next to leading order of the density. The determination of the mass center m depends strongly on the chosen center of continuum thresholds $s_0^2 = ((s_0^+)^2 + (s_0^-)^2)/2$ as indicated by the broad range in the right panel of Fig. 3.3.9 when varying their medium dependence. In contrast, the splitting is again fairly robust as evidenced by Fig. 3.3.7, where the difference between curves of different thresholds is negligible.

A linear interpolation of the threshold splitting from vacuum to a density of $n = 0.01\,\text{fm}^{-3}$ gives an estimate for the $\mathcal{O}(n)$ term $\alpha_{\Delta s} \approx 7 \cdot 10^2\,\text{GeV}^{-1}$, which justifies the range chosen in the previous section.

3.4 Evaluation for B and $\bar{B}$ mesons

We turn now to B and $\bar{B}$ mesons. The corresponding current operators are $j_{B^+} = i\bar{b}\gamma_5 u$ or $j_{B^0} = i\bar{b}\gamma_5 d$. The antiparticles correspond to $j_{B^-} = j_{B^+}^\dagger = i\bar{u}\gamma_5 b$ or $j_{\bar{B}^0} = j_{B^0}^\dagger = i\bar{d}\gamma_5 b$. The above equations and, in particular, the OPE are applied with the replacements $m_c \to m_b$ and $m_{B^\pm} \to m_{\mp}$ in order to take into account the distinct heavy-light structure compared to the D meson case. The Borel curves $m_\pm(M)$ display, analogously to the case of open charm, pronounced minima at a Borel mass of about 1.7 GeV. We utilize again the first evaluation strategy outlined in App. C.7. Numerical results are exhibited in Fig. 3.4.1. We employ $s_0^2 = 40\,\text{GeV}^2$ and obtain $m \approx 5.33\,\text{GeV}$ for the vacuum mass. One observes a mass splitting of $2\Delta m \approx -130\,\text{MeV}$ at $n = 0.15\,\text{fm}^{-3}$. The centroid is shifted upwards by about 60 MeV. The splitting of the threshold parameters becomes $\Delta s_0^2 \approx -3.4\,\text{GeV}^2$ and the minima of the Borel curves $m_\pm(M)$ are shifted from $M \approx 1.67\,\text{GeV}$ in vacuum to $M \approx 1.71\,\text{GeV}$ at $n = 0.15\,\text{fm}^{-3}$. In case of $\bar{B}, B$ mesons, the combination $m_b \langle \bar{d}d \rangle$ is expected to have numerically an even stronger impact than the term $m_c \langle \bar{d}d \rangle$ in the charm sector. Indeed, the influence of the chiral condensate becomes even larger than that of the odd mixed quark-gluon condensate $\langle q^\dagger g \sigma \mathcal{G} q \rangle$ at higher densities. The overall pattern resembles the results exhibited in Fig. 3.3.4, but with shifted mass scale for m. The other evaluation strategies yield the same results. Setting $\langle q^\dagger g \sigma \mathcal{G} q \rangle = -0.33\,\text{GeV}^2 n$, and, hence, $\langle q^\dagger D_0^2 q \rangle = -0.0585\,\text{GeV}^2 n$, a mass splitting of $2\Delta m \approx -220\,\text{MeV}$ and an average mass shift $\approx 45\,\text{MeV}$ would be obtained.

3.5 Interim summary

To summarize the results of this chapter, we have evaluated the Borel transformed QSRs for pseudo-scalar mesons composed of a combination of a light and heavy quark. The heavy quark mass introduces a new scale compared to QSRs in the light

3 QCD sum rules for heavy-light mesons

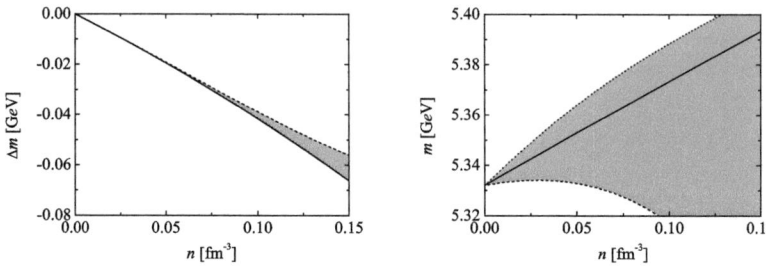

Figure 3.4.1: Δm (left) and m (right) for $\bar{B}$ and B at $s_0^2 = 40\,\text{GeV}^2$ and $m_b = 4.7\,\text{GeV}$. For line codes see Fig. 3.3.1. For density dependent thresholds, $s_0^2 = s_0^2(0) \pm 7n/n_0\,\text{GeV}^2$ has been used.

quark sector. The evaluation of the sum rules, complete up to mass dimension 5, has been performed for D, $\bar{D}$ and $\bar{B}$, B mesons with a glimpse on D_s, $\bar{D}_s$ and the scalar D^*, $\bar{D}^*$ mesons as well. Our analysis relies on the often employed pole + continuum ansatz for the hadronic spectral function. Complications concerning a finite-width ansatz have been briefly discussed. This is a severe restriction of the generality of the practical use of sum rules. In this respect, the extracted parameters refer to this special ansatz and should be considered as indicators for changes of the true spectral functions of hadrons embedded in cold nuclear matter. Particles and antiparticles are coupled – a problem which is faced also for hadrons with conserved quantum numbers in the light quark sector [Tho08b, Coh95, Tho07].

We presented a transparent approximation to highlight the role of the even and odd parts of the OPE. Numerically, we find fairly robust mass splittings (for the employed set of condensates), while an assignment of a possible mass shift of the centroids is not yet on firm ground. The impact of various condensates is discussed, and $\langle q^\dagger q \rangle$, $\langle q^\dagger g \sigma \mathcal{G} q \rangle$ and $\langle \bar{q} q \rangle$ are identified to drive essentially the mass splitting. While $\langle \bar{q} q \rangle$ is amplified by the heavy quark mass, it enters nevertheless the sum rules beyond the linear density dependence. A concern is the sign of the condensate $\langle q^\dagger g \sigma \mathcal{G} q \rangle$, vanishing in vacuum but with poorly known medium dependence, which determines the size of the $D - \bar{D}$ mass splitting. These findings, in particular for D, $\bar{D}$, D_s, $\bar{D}_s$, are of relevance for the planned experiments at FAIR [FAI].

4 Chiral partner QCD sum rules*

It is the aim of the present chapter to present in-medium sum rules for the scalar and pseudo-scalar as well as vector and axial-vector mesons composed essentially of a heavy quark and a light quark, e. g. $u\bar{c}$ and $d\bar{c}$ realized in vacuum as $D^{\pm}(1867)$, $J^P = 0^-$, $D^0(1865)$, $J^P = 0^-$, $D_0^*(2400)^0$, $J^P = 0^+$, $D^*(2007)^0$, $J^P = 1^-$, $D^*(2010)^{\pm}$, $J^P = 1^-$, $D_1(2420)^0$, $J^P = 1^+$ (there is no confirmed charged scalar or axial-vector state in the open charm sector) [Nak10]. As mentioned in Sec. 1.5, such open charm degrees of freedom will be addressed in near future by the CBM [CBM] and PANDA [PAN] collaborations at FAIR [FAI] in proton-nucleus and anti-proton–nucleus reactions [Fri11]. Accordingly we are going to analyze the chiral sum rules in nuclear matter.

Chiral symmetry is explicitly broken by nonzero quark masses (cf. Sec. 1.1 and App. A), but QCD is still approximately invariant under transformations which are restricted to the light quark sector. Although currents, related to mesons which are represented in the quark model as composed of a light quark and a heavy quark, are neither conserved nor associated with a symmetry transformation, vector and axial-vector currents still mix under a transformation which is restricted to the light quark sector. Consider for example the infinitesimal rotation in flavor space $\psi \to e^{-it^a \Theta^a \Gamma} \psi \approx (1 - it^a \Theta^a \Gamma) \psi$, where $t^a \in SU(N_f)$, $\Gamma = \mathbb{1}$ for the vector transformations and $\Gamma = \gamma_5$ for the axial transformations, and Θ^a denote a set of infinitesimal rotation parameters (rotation angles). The vector current transforms as

$$j_\mu^{V,\tau} = \bar{\psi} \gamma_\mu \tau \psi \to j_\mu^{V,\tau} \pm i \bar{\psi} \Gamma \gamma_\mu \Theta^a [t^a, \tau]_- \psi, \tag{4.0.1}$$

where $+$ ($-$) refers to the vector (axial) transformations and τ denotes a matrix in flavor space. In the 3-flavor sector $\psi = (\psi_1, \psi_2, \psi_3)$, where the first two quarks are light, the choice $\vec{\Theta} = (\Theta_1, \Theta_2, \Theta_3, 0, \dots, 0)$ and $2t^a = \lambda^a$, the well-known Gell-Mann matrices, clearly leaves the QCD Lagrangian invariant even if the third quark is heavy. To be specific, let us consider $\tau = (\lambda^4 + i\lambda^5)/2$, where the corresponding current

---
*The presentation is based on [Hil11, Hil12a].

4 Chiral partner QCD sum rules

transforms as

$$j_\mu^{V,\tau} = \bar{\psi}_1 \gamma_\mu \psi_3 \rightarrow j_\mu^{V,\tau'} = j_\mu^{V,\tau} - \frac{i}{2}\left(\bar{\psi}_1 \Gamma \gamma_\mu \psi_3 \Theta_3 + (\Theta_1 + i\Theta_2)\bar{\psi}_2 \Gamma \gamma_\mu \psi_3\right). \quad (4.0.2)$$

Obviously, the heavy-light vector current mixes with heavy-light axial-vector currents if an axial transformation, with $\Gamma = \gamma_5$, is applied. Analog expressions hold for spin-0 mesons, and the result is the same as in the spin-1 case, but with the replacements $\gamma_\mu \rightarrow \mathbb{1}$ for the vector transformation and $i\gamma_\mu [t^a, \tau]_- \rightarrow [t^a, \tau]_+$ for the axial transformation. Hence, a symmetry and its spontaneous breakdown in the light quark sector must also be reflected in the spectrum of mesons composed of a heavy and a light quark. In particular, the splitting of the spectral densities between heavy-light parity partners must be driven by order parameters of spontaneous chiral symmetry breaking.

Probing the chiral symmetry restoration via the change of order parameters requires a reliable extraction of their medium dependence. As is known for the vacuum case, the chiral condensate is numerically suppressed in QSRs in the light quark sector due to the tiny light quark mass, but occurs amplified by the large heavy quark mass in QSRs involving a light and a heavy quark [Rei85]. Despite of the amplification in terms of $m_c \langle \bar{u}u \rangle$, emphasized in Sec. 3, with m_c and $\langle \bar{u}u \rangle$ to be evolved to the appropriate scale, in case of the D mesons, the dependence of the in-medium D meson spectrum on the chiral condensate is not as direct as anticipated. This is clear in so far as there are always particle and antiparticle contributions to the spectrum of pseudo-scalar states and one has to deal with both species. This accounts for inherent suppressions and amplifications of different types of condensates due to the generic structure of the sum rule in that case. Indeed, a precise analytic and numerical investigation points to competitive numerical impacts of various condensates for the mass splitting of particle and antiparticle, whereas, the determination of the mass center rather depends on the modeling of the continuum threshold (cf. Sec. 3).

In the difference of chiral partner spectra (Weinberg type sum rules) the dependence on chirally symmetric condensates drops out. At the same time, the amplification of the chiral condensate by the heavy quark mass is still present. This makes Weinberg type sum rules of mesons composed of heavy and light quarks an interesting object of investigation.

4.1 Differences of current-current correlators and their operator product expansion

We consider the currents

$$j^S(x) \equiv \bar{q}_1(x) q_2(x) \qquad \text{(scalar)}, \qquad (4.1.1a)$$

$$j^P(x) \equiv \bar{q}_1(x) i\gamma_5 q_2(x) \qquad \text{(pseudo-scalar)}, \qquad (4.1.1b)$$

$$j^V_\mu(x) \equiv \bar{q}_1(x) \gamma_\mu q_2(x) \qquad \text{(vector)}, \qquad (4.1.1c)$$

$$j^A_\mu(x) \equiv \bar{q}_1(x) \gamma_5 \gamma_\mu q_2(x) \qquad \text{(axial-vector)}, \qquad (4.1.1d)$$

which are self adjoint in the case of equal quark flavors, and the corresponding causal correlators (cf. App. B)

$$\Pi^{(S,P)}(q) = i \int d^4x \, e^{iqx} \langle T \left[j^{(S,P)}(x) j^{(S,P)\dagger}(0) \right] \rangle, \qquad (4.1.2a)$$

$$\Pi^{(V,A)}_{\mu\nu}(q) = i \int d^4x \, e^{iqx} \langle T \left[j^{(V,A)}_\mu(x) j^{(V,A)\dagger}_\nu(0) \right] \rangle. \qquad (4.1.2b)$$

In the rest frame of the nuclear medium, i.e. $v = (1, \vec{0})$, and for mesons at rest, i.e. $q = (1, \vec{0})$, the (axial-) vector correlator can be decomposed as (cf. App. B.3)

$$\Pi^{(V,A)}_{\mu\nu}(q) = \left(\frac{q_\mu q_\nu}{q^2} - g_{\mu\nu} \right) \Pi^{(V,A)}_T(q) + \frac{q_\mu q_\nu}{q^2} \Pi^{(V,A)}_L(q) \qquad (4.1.3)$$

with $\Pi^{(V,A)}_T(q) = \frac{1}{3} \left(\frac{q^\mu q^\nu}{q^2} - g^{\mu\nu} \right) \Pi^{(V,A)}_{\mu\nu}(q)$ and $\Pi^{(V,A)}_L(q) = \frac{1}{q^2} q^\mu q^\nu \Pi^{(V,A)}_{\mu\nu}(q)$ written explicitly covariant. Π_T contains the information about (axial-) vector degrees of freedom. Π_L refers to (pseudo-) scalar states and can be related to the (pseudo-) scalar correlator (cf. App. B.4 for details).

We now proceed with the OPEs for $\Pi^X \in \{\Pi^{(S,P)}, \Pi^{(V,A)} \equiv g^{\mu\nu} \Pi^{(V,A)}_{\mu\nu}\}$. We do not include the α_s corrections which would arise from inserting the next-to-leading order interaction term in the time-ordered product of the current-current correlator. Such terms account, e.g., for four-quark condensates. They are of mass dimension 6 and beyond the scope of this investigation. According to standard OPE techniques (cf. e.g. [Nov84b, Shi79a, Nar02] and App. C), the time-ordered product can be expanded into normal ordered products multiplied by Wick-contracted quark-field operators. Dirac indices can be projected onto elements of the Clifford algebra,

4 Chiral partner QCD sum rules

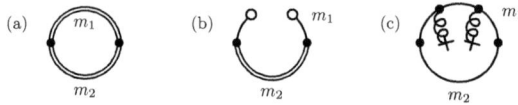

Figure 4.1.1: Feynman diagrams of (a) $\Pi^{(0)}$ and (b) $\Pi^{(2)}$. (c) is for $\Pi^{(0)}$ with two gluon lines attached to the free propagator of quark 1. Double lines stand for the complete perturbative series of the quark propagator (C.1.25), curly lines are for gluons, and circles denote non-local quark condensates, while crosses symbolize local quark or gluon condensates.

$\Gamma \in \{\mathbb{1}, \gamma_\mu, \sigma_{\mu\nu}, i\gamma_5\gamma_\mu, \gamma_5\}$, which provides an orthonormal basis in the space of 4×4 matrices with the scalar product $(A,B) \equiv \frac{1}{4}\text{Tr}_D[AB]$. Color indices can be projected onto an analogously appropriate basis. Thereby, color and Dirac traces of the quark propagator occur. Using the background field method in fixed-point gauge (cf. [Nov84b, Shi79a] and App. C.1) one arrives, after a Fourier transformation, at

$$\Pi^X(q) = -i^3 \int \frac{d^4p}{(2\pi)^4} \langle: \text{Tr}_{C,D}\left[S_1(p)\Gamma^X S_2(q+p)\Gamma^X\right]:\rangle$$
$$+ i^2 \sum_\Gamma \frac{1}{4} \sum_{n=0}^\infty \frac{(-i)^n}{n!} \partial_q^{\bar{a}_n} \langle: (-1)^n \bar{q}_1 \Gamma \text{Tr}_D\left[\Gamma\Gamma^X S_2(q)\Gamma^X\right] D_{\bar{a}_n} q_1$$
$$+ \bar{q}_2 \Gamma \text{Tr}_D\left[\Gamma\Gamma^X S_1(-q)\Gamma^X\right] D_{\bar{a}_n} q_2 :\rangle$$
$$= \Pi^{X(0)}(q) + \Pi^{X(2)}(q), \quad (4.1.4)$$

where $D_{\bar{a}_n} = D_{\alpha_1} \ldots D_{\alpha_n}$ (with an analog notation for the partial derivative) and quark fields and their derivatives are taken at $x = 0$. For X denoting vector and axial-vector states, $(\Gamma^X)_{ij}(\Gamma^X)_{kl} \equiv (\Gamma^X_\mu)_{ij}(\Gamma^{X,\mu})_{kl}$ is understood. $\Pi^{(0)}(q)$ denotes the fully contracted (depicted in Fig. 4.1.1 (a)) and $\Pi^{(2)}(q)$ the 2-quark term (see Fig. 4.1.1 (b) for $\langle \bar{\psi}\ldots\psi\rangle$), $\text{Tr}_{C,D}$ means trace w.r.t. color and Dirac indices. To have the same structures in both quantities $\Pi^{(0)}$ and $\Pi^{(2)}$, we also project the matrix product $\Gamma^X S \Gamma_X$ onto this basis (see App. D.1) which allows to consider the propagator properties in the space of Dirac matrices for each quark separately:

$$\Pi^{(P,S)(0)}(q) = (-1)^{(P,S)} i \int \frac{d^4p}{(2\pi)^4} \langle: \frac{1}{4}\text{Tr}_C\left[\text{Tr}_D[S_2(p+q)]\text{Tr}_D[S_1(p)]\right]$$
$$+ (-1)^{(P,S)} \text{Tr}_D[S_2(p+q)\gamma_\mu]\text{Tr}_D[S_1(p)\gamma^\mu]$$
$$+ \frac{1}{2}\text{Tr}_D[S_2(p+q)\sigma_{\mu\nu}]\text{Tr}_D[S_1(p)\sigma^{\mu\nu}]$$

4.1 Differences of current-current correlators and their OPE

$$+ (-1)^{(P,S)} \text{Tr}_D[S_2(p+q)\gamma_5\gamma_\mu]\text{Tr}_D[S_1(p)\gamma_5\gamma^\mu]$$
$$+ \text{Tr}_D[S_2(p+q)\gamma_5]\text{Tr}_D[S_1(p)\gamma_5]] :\rangle , \quad (4.1.5a)$$

$$\Pi^{(V,A)(0)}(q) = -(-1)^{(V,A)} i \int \frac{d^4p}{(2\pi)^4} \langle : \text{Tr}_C \left[\text{Tr}_D[S_2(p+q)]\text{Tr}_D[S_1(p)] \right.$$
$$+ (-1)^{(V,A)} \frac{1}{2} \text{Tr}_D[S_2(p+q)\gamma_\mu]\text{Tr}_D[S_1(p)\gamma^\mu]$$
$$- (-1)^{(V,A)} \frac{1}{2} \text{Tr}_D[S_2(p+q)\gamma_5\gamma_\mu]\text{Tr}_D[S_1(p)\gamma_5\gamma^\mu]$$
$$- \text{Tr}_D[S_2(p+q)\gamma_5]\text{Tr}_D[S_1(p)\gamma_5]] :\rangle , \quad (4.1.5b)$$

where $(-1)^{(P,V)} = -1$ for pseudo-scalar and vector mesons and $(-1)^{(S,A)} = 1$ for scalar and axial-vector mesons. Note that only the Dirac structure has been projected. Still the product of the propagators in color space must be accounted for, which is indicated by the overall trace in color space. Apart from the prefactors of each term, the result has the structure of a scalar product in the linear space spanned by the Clifford base. $\Pi^{X(2)}(q)$ in Eq. (4.1.4) may be simplified using Tab. D.1.2 in App. D.1. Also note that $\Pi^{(V,A)(0)}$ and $\Pi^{(V,A)(2)}$ have no $\sigma_{\mu\nu}$ part as can be seen from Tab. D.1.2. Later on, this expansion allows to identify and to separate certain parts in order to investigate their distinct properties, e.g. for $m_d \to 0$.

To obtain the OPEs for the difference of chiral partners $\Pi^{P-S} \equiv \Pi^P - \Pi^S$ and $\Pi^{V-A} \equiv \Pi^V - \Pi^A$, one can use (4.1.5) or directly project the occurring anticommutators. The result reads

$$\Pi^{P-S(0)}(q) = -i \int \frac{d^4p}{(2\pi)^4} \langle : \frac{1}{2} \text{Tr}_C \{ \text{Tr}_D[S_2(p+q)]\text{Tr}_D[S_1(p)]$$
$$+ \frac{1}{2} \text{Tr}_D[S_2(p+q)\sigma_{\mu\nu}]\text{Tr}_D[S_1(p)\sigma^{\mu\nu}] + \text{Tr}_D[S_2(p+q)\gamma_5]\text{Tr}_D[S_1(p)\gamma_5] \} :\rangle ,$$
$$(4.1.6a)$$

$$\Pi^{P-S(2)}(q) = \sum_n \frac{(-i)^n}{n!} \frac{1}{2} \sum_\Gamma^{\{1,\sigma_{\alpha<\beta},\gamma_5\}} \langle : \bar{q}_1 \overleftarrow{D}_{\bar{a}_n} \Gamma \partial^{\bar{a}_n} \left(\text{Tr}_D[\Gamma S_2(q)] \right) q_1$$
$$+ \bar{q}_2 \Gamma \partial^{\bar{a}_n} \left(\text{Tr}_D[\Gamma S_1(-q)] \right) \overrightarrow{D}_{\bar{a}_n} q_2 :\rangle , \quad (4.1.6b)$$

4 Chiral partner QCD sum rules

$$\Pi^{V-A(0)}(q) = i \int \frac{d^4 p}{(2\pi)^4} \langle :2\text{Tr}_C \{\text{Tr}_D[S_2(p+q)]\text{Tr}_D[S_1(p)]$$

$$- \text{Tr}_D[S_2(p+q)\gamma_5]\text{Tr}_D[S_1(p)\gamma_5]\}:\rangle, \quad (4.1.6c)$$

$$\Pi^{V-A(2)}(q) = -\sum_n \frac{(-i)^n}{n!} 2 \sum_\Gamma^{\{\mathbb{1},i\gamma_5\}} \langle :\bar{q}_1 \overleftarrow{D}_{\bar{a}_n} \Gamma \partial^{\bar{a}_n} \left(\text{Tr}_D[\Gamma S_2(q)]\right) q_1$$

$$+ \bar{q}_2 \Gamma \partial^{\bar{a}_n} \left(\text{Tr}_D[\Gamma S_1(-q)]\right) \overrightarrow{D}_{\bar{a}_n} q_2 :\rangle, \quad (4.1.6d)$$

where the Clifford basis is modified now by the imaginary unit in front of γ_5 for the vector–axial-vector difference. The advantage is that we are left with three different types of Dirac traces for the quark propagators in the P–S case and only two in the V–A case.

The perturbative series for a momentum-space quark propagator in a gluonic background field in fixed-point gauge is presented in App. C.1. Its derivative is given by the Ward identity $\partial^\mu S(q) = -S(q)\Gamma^\mu(q,q;0)S(q)$, where $\Gamma^\mu(q,q;0)$ denotes the exact quark-gluon vertex function at vanishing momentum transfer. $\Gamma^\mu(q,q;0) = \gamma^\mu$ holds for a classical background field meaning that the Ward identity for the complete perturbation series has the same form as for free quarks [Lan86].

For the limit of a massless quark flavor attributed to q_1, $m_1 \to 0$, one can show (see App. D.2) that $\text{Tr}_D[\Gamma S_1(q)] = 0$ for $\Gamma \in \{\mathbb{1}, \sigma_{\mu\nu}, \gamma_5\}$ and $q^2 \neq 0$. In this limit only the diagram in Fig. 4.1.1 (b) gives a contribution to $\Pi^{(2)}$. The corresponding diagram for $1 \longleftrightarrow 2$ vanishes, because the sum over Dirac matrices in (4.1.6) covers such elements where the corresponding traces vanish. Hence, for chiral partner sum rules the often used approximation of a static quark, which results in vanishing heavy-quark condensates, is not necessary since their Wilson coefficients vanish in the limit of the other quark being massless. This means that, if one is seeking the occurrence of quark condensates in lowest (zeroth) order of the strong coupling α_s, at least one quark must have a nonzero mass. (Higher order interaction term insertions cause the occurrence of further quark condensates proportional to powers of α_s, cf. [Kap94, Leu07, Leu06b] for examples.) Hence, the structure of the OPE side of the famous Weinberg sum rules for light quarks w. r. t. quark condensates is shown in all orders of the quark propagators and quark fields.

For the completely contracted term $\Pi^{(0)}$ the situation is somewhat more involved. A superficial view on Eq. (4.1.6) together with the trace theorem of App. D.2 may tempt to the conclusion that $\Pi^{(0)}$ is zero for chiral partner OPEs of heavy-light quark

4.1 Differences of current-current correlators and their OPE

meson currents. On the other hand it is clear that only matrix elements of chirally odd operators may enter the OPE, whereas gluon condensates are chirally even. In this sense, the cancellation of $\Pi^{(0)}$ is in line with naive expectations. But with the introduction of non-normal ordered condensates, also gluon condensates together with infrared divergences would be introduced. Taking this as a heuristic argument for a nonzero $\Pi^{(0)}$ raises the question of the precise cancellation of these terms and terms added by introducing non-normal ordered condensates. Indeed, from the in-medium OPE of D mesons (see App. C.4) it is known that the medium specific divergences are canceled because of the renormalization of $\langle : \bar{d} \gamma_\mu D_\nu d : \rangle$. But this term is chirally even (formally, it is the matrix element of an isospin-singlet vector current, cf. Eqs. (A.1.69) and (A.1.70c)) and does not enter Eq. (4.1.6). Moreover, it is known that introducing non-normal ordered condensates in order to cancel infrared divergent Wilson coefficients of gluon condensates leads to additional finite gluon contributions. Clearly, these have to cancel out in case of chiral partner sum rules. Again, this can be taken as a heuristic argument that only those mass divergences can remain in $\Pi^{(0)}$ which are cancelled by chirally odd condensates. Two questions have to be answered. Do all infrared divergences cancel out? And does the renormalization procedure introduce chirally even condensates? Thus, a careful analysis is mandatory to prove that the obtained results are infrared stable and that the renormalization procedure is consistent.

In case of two light quarks, i. e. $m_{1,2} \to 0$, the limiting procedure and the momentum integration commute and, hence, $\Pi^{(0)} = 0$ is obvious, because all non-vanishing terms drop out in the chiral difference. If one of the quarks has a nonzero mass, the limiting procedure and the momentum integration do not commute due to the occurrence of infrared divergences [Che82b, Tka83b, Gro95, Jam93, Hil08, Zsc11], e. g. for the term depicted in Fig. 4.1.1 (c) which is proportional to the gluon condensate. As the integration domains in Eqs. (4.1.5) and (4.1.6a), (4.1.6c) involve momenta $p = 0$, $\text{Tr}[\Gamma S(p, m)]$ does not converge uniformly for $m \to 0$. Hence, the integration and the limit $m \to 0$ cannot be interchanged. A careful treatment is in order. In fact, a direct calculation of the perturbative contribution to chiral partner OPEs, which is presented in App. D.3, shows that the infrared divergences are the only remaining terms.

All finite terms cancel out in the chiral difference. Hence, what remains in the chiral difference are infrared divergent terms stemming from the free propagation of the heavy quark and the light quark propagator with two gluon lines attached (see Fig. 4.1.1 (c)). These divergences have to be absorbed by introducing condensates

4 Chiral partner QCD sum rules

which are not normal ordered. In App. D.3 their cancellation is demonstrated in detail. The separation done in Eq. (4.1.5) now allows for an unambiguous identification of the singular terms. Up to mass dimension 5, the only product of traces which contributes in the limit $m_d \to 0$ to $\Pi^{(0)}$ is $\text{Tr}_D[S_1]\text{Tr}_D[S_2]$. Indeed, up to order α_s^1 one can show that the contribution $-\int d^4p\, \text{Tr}_D[S_2^{(0)}(p+q)] \langle :\text{Tr}_D[S_1^{(2)}(p)]: \rangle$, which is the remaining term of Eq. (4.1.5) after the limit $m_1 \to 0$ has been taken, is canceled by $\int d^4p\, \text{Tr}_D[S_2^{(0)}(q)] \langle \text{Tr}_D[S_1^{(2)}(p)] \rangle$. The latter quantity is introduced by the definition of non-normal ordered condensates (cf. App. C.5)

$$\langle \bar{q}\hat{O}[D_\mu]q \rangle = \langle :\bar{q}\hat{O}[D_\mu]q: \rangle - i \int \frac{d^4p}{(2\pi)^4} \langle \text{Tr}_{C,D}[\hat{O}[-ip_\mu - i\tilde{A}_\mu]S_q(p)] \rangle, \quad (4.1.7)$$

resulting in a factor $\int \langle \text{Tr}_D[S_1^{(2)}(p)] \rangle$ to the Wilson coefficient $\text{Tr}_D[S_2^{(0)}(q)]$ of the chiral condensate in Eq. (4.1.4). Hence, in case of heavy-light mesons, first the integration has to be performed, then one has to introduce non-normal ordered condensates according to Eq. (4.1.7), and afterwards the limit $m_1 \to 0$ can be taken. In case of equal quark masses, $m_1 = m_2$, the divergences cancel each other by virtue of

$$\int d^4p \left(\text{Tr}_D[S_1^{(2)}(p+q)]\text{Tr}_D[S_2^{(0)}(p)] + \text{Tr}_D[S_1^{(0)}(p+q)]\text{Tr}_D[S_2^{(2)}(p)] \right) = 0. \quad (4.1.8)$$

If two heavy (static) quarks are considered, only $\Pi^{(0)}$ gives a contribution to the chiral OPE, whereas for two massless quarks, both $\Pi^{(0)}$ and $\Pi^{(2)}$ vanish.

Putting everything together this means that for light quarks ($m_{1,2} \ll \Lambda_{\text{QCD}}$) in the chiral-difference OPE the corresponding traces and, therefore, the corresponding Wilson coefficients vanish, while for heavy quarks ($m_{1,2} \gg \Lambda_{\text{QCD}}$) the condensates vanish. To obtain quark condensates in order α_s^0 the two flavors must be of different mass scales, i.e. $q_1 \in \{u,d(,s)\}$ is a light quark and $q_2 \in \{c,b,t\}$ is a heavy quark. Hence, to seek for condensates which are connected to chiral symmetry breaking as possible order parameters in order α_s^0, a natural choice is to consider chiral partner mesons composed of a light and a heavy quark. Given the above mentioned experimentally envisaged research programs at FAIR [FAI, CBM, PAN, Fri11] we focus on open charm mesons. The presented formulas may be directly transferred to open bottom mesons by $m_c \to m_b$.

4.2 Chiral partners of open charm mesons

We now consider a light ($q_1 \equiv q$) and a heavy ($q_2 \equiv q_c$) quark entering the currents in Eq. (4.1.1), but the presented formalism can be applied to arbitrary flavor content. Thereby, traces w. r. t. flavor indices are introduced.

4.2.1 The case of P–S

For the P–S case we consider the pseudo-scalar $D(0^-)$ ($D^\pm$, D^0 and $\bar{D}^0$) and its scalar partners $D_0^*(0^+)$. Of course, all the results also account for other heavy–light (pseudo-) scalar mesons. (For open charm mesons D_s which contain a strange quark, however, the limit $m_s \to 0$ may not be a good approximation and terms $\propto m_s$ should be taken into account as well.)

Up to and including mass dimension 5, after absorbing the divergences in non-normal ordered condensates, the OPE gets the following compact form

$$\Pi^{P-S}(q) \equiv \Pi^P(q) - \Pi^S(q) = \Pi^{P(2)}(q) - \Pi^{S(2)}(q)$$
$$= \sum_n \frac{(-i)^n}{n!} \sum_\Gamma^{\{1,\sigma_{\alpha<\beta}\}} \langle \bar{q} \overleftarrow{D}_{\bar{a}_n} \Gamma \partial^{\bar{a}_n} \text{Tr}_D[\Gamma S_c(q)] q \rangle \,, \qquad (4.2.1)$$

where the sum over the elements of the Clifford algebra does not contain γ_5 up to this mass dimension anymore. To evaluate the condensates in Eq. (4.2.1) in terms of expectation values of scalar operators, Lorentz indices have to be projected onto $g_{\mu\nu}$ and $\epsilon_{\mu\nu\kappa\lambda}$ in vacuum, whereas the medium four-velocity v_μ provides an additional structure at finite densities and/or temperatures [Jin93]. Hence, new condensates must be introduced which vanish in the vacuum. Thereby, temperature and density dependencies stem from Gibbs averages of medium specific operators. The evaluation in the nuclear matter rest frame $v^\mu = (1, \vec{0})$ for mesons at rest yields

$$\Pi^{P-S}(q_0) = 2\langle \bar{q}q \rangle \frac{m_c}{q_0^2 - m_c^2} - \langle \bar{q} g \sigma \mathcal{G} q \rangle \frac{m_c q_0^2}{(q_0^2 - m_c^2)^3}$$
$$+ \left[\langle \bar{q} g \sigma \mathcal{G} q \rangle - 8 \langle \bar{q} D_0^2 q \rangle \right] \frac{m_c q_0^2}{(q_0^2 - m_c^2)^3} \,, \qquad (4.2.2)$$

where we separated a medium-specific term (last line, $\langle \bar{q} g \sigma \mathcal{G} q \rangle - 8 \langle \bar{q} D_0^2 q \rangle \equiv \langle \Delta \rangle$) vanishing in vacuum and q denotes either d or u quark field operators. Equation (4.2.2) reduces to the vacuum result [Nar05, Hay04, Rei85] at zero density and

4 Chiral partner QCD sum rules

temperature. The condensates $\langle \bar{q}q \rangle$, $\langle \bar{q}g\sigma\mathcal{G}q \rangle$ and $\langle \bar{q}D_0^2 q \rangle$ may have different medium dependencies.

An odd part of the OPE of chiral partner sum rules does not appear up to this mass dimension. Although there is no γ_μ projection of the condensates for difference OPEs of chiral partner, it may arise from an odd number of derivatives in Eq. (4.2.1).

The sum rule is set up according to Eq. (C.2.27) and after a Borel transformation (cf. App. C.2) the result is

$$\frac{1}{\pi}\int_{-\infty}^{+\infty} d\omega\, e^{-\omega^2/M^2}\, \omega\, \Delta\Pi^{\text{P-S}}(\omega) = e^{-m_c^2/M^2}\left[-2m_c\langle \bar{q}q \rangle \right.$$
$$\left. + \left(\frac{m_c^3}{2M^4} - \frac{m_c}{M^2}\right)\langle \bar{q}g\sigma\mathcal{G}q \rangle - \langle \Delta \rangle \left(\frac{m_c^3}{2M^4} - \frac{m_c}{M^2}\right)\right]. \quad (4.2.3)$$

It is instructive to cast Eq. (4.2.3) in the form of Weinberg type sum rules [Wei67, Kap94]. This can be accomplished by expanding the exponential on both sides and comparing the coefficients of inverse powers of the Borel mass. In such a way we can relate moments of the spectral P–S differences to condensates via

$$\frac{1}{\pi}\int_{-\infty}^{+\infty} d\omega\, \omega\, \Delta\Pi^{\text{P-S}}(\omega) = -2m_c\langle \bar{q}q \rangle, \quad (4.2.4\text{a})$$

$$\frac{1}{\pi}\int_{-\infty}^{+\infty} d\omega\, \omega^3\, \Delta\Pi^{\text{P-S}}(\omega) = -2m_c^3\langle \bar{q}q \rangle + m_c\langle \bar{q}g\sigma\mathcal{G}q \rangle - m_c\langle \Delta \rangle \quad (4.2.4\text{b})$$

$$\frac{1}{\pi}\int_{-\infty}^{+\infty} d\omega\, \omega^5\, \Delta\Pi^{\text{P-S}}(\omega) = -2m_c^5\langle \bar{q}q \rangle + 3m_c^3\langle \bar{q}g\sigma\mathcal{G}q \rangle - 3m_c^3\langle \Delta \rangle + \ldots, \quad (4.2.4\text{c})$$

where "..." denote a neglected contribution of mass dimension 7. Combining the recurrence relations derived in App. C.6 with the prescription to determine the required orders of the expansions within the background field technique for given mass dimension shows that such a contribution must exist in Eq. (4.2.4c). This generalizes the OPE side of Weinberg type sum rules to scalar and pseudo-scalar mesons in the heavy-light quark sector for the first time.

If one attributes chiral symmetry to the degeneracy of chiral partners (i. e. the l. h. s. of Eq. (4.2.4) vanishes) the vanishing of $\langle \bar{q}q \rangle$ and $8\langle \bar{q}D_0^2 q \rangle = \langle \bar{q}g\sigma\mathcal{G}q \rangle - \langle \Delta \rangle$ on the r. h. s. is required. In this spirit, these condensates may be considered as possible order parameters of chiral symmetry. Note that Eq. (4.2.4) also allows to consider the omitted mass dimension 7 condensate as an order parameter. Remarkably, the chiral condensate $\langle \bar{q}q \rangle$ of light quarks figures here in conjunction with the heavy quark mass as parameter for chiral symmetry breaking in each of the moments (for vacuum, cf. [Nar05]). Of next importance is $\langle \bar{q}D_0^2 q \rangle$, again in combination with the heavy quark mass. The r. h. s. quantities must be taken at a proper renormalization scale. Formally, in the chiral (i. e. strictly massless) limit for all quarks the r. h. s. of Eq. (4.2.4) would vanish.

4.2.2 The case of V–A

In the same manner we proceed in the V–A case. From Eqs. (4.1.6c) and (4.1.6d) we obtain up to and including mass dimension 5

$$\Pi^{V-A}(q) \equiv \Pi^{V}(q) - \Pi^{A}(q) = \Pi^{V(2)}(q) - \Pi^{A(2)}(q)$$
$$= -\sum_n \frac{(-i)^n}{n!} 2 \sum_{\Gamma}^{\{1, i\gamma_5\}} \langle \bar{q}_1 \overleftarrow{D}_{\bar{a}_n} \Gamma \partial^{\bar{a}_n} \mathrm{Tr}_{\mathrm{D}}[\Gamma S_2(q)] q_1 \rangle, \qquad (4.2.5)$$

where only the $\mathbb{1}$-projection survives. The in-medium evaluation results in

$$\Pi^{V-A}(q) = -8 \langle \bar{q}q \rangle \frac{m_c}{q_0^2 - m_c^2} + 4 \langle \bar{q}g\sigma\mathscr{G}q \rangle \frac{m_c^3}{(q_0^2 - m_c^2)^3} - 4\langle\Delta\rangle \frac{m_c q_0^2}{(q_0^2 - m_c^2)^3} \qquad (4.2.6)$$

and together with Eqs. (B.4.20) and (4.2.2) we obtain for the correlator containing the information about the vector and axial-vector degrees of freedom

$$\Pi_T^{V-A}(q) = \Pi^{P-S} + \frac{m_c}{(q_0^2 - m_c^2)^2} \langle \bar{q}g\sigma\mathscr{G}q \rangle + \frac{1}{3} \frac{m_c}{(q_0^2 - m_c^2)^2} \langle \Delta \rangle. \qquad (4.2.7)$$

For vacuum, the result of [Hay04, Rei85] is recovered. The Borel transformed sum rule is given by

$$\frac{1}{\pi} \int_{-\infty}^{+\infty} d\omega\, e^{-\omega^2/M^2} \omega\, \Delta\Pi_T^{V-A}(\omega) = e^{-m_c^2/M^2} \left[-2m_c \langle \bar{q}q \rangle \right.$$
$$\left. + \frac{m_c^3}{2M^4} \langle \bar{q}g\sigma\mathscr{G}q \rangle - \langle \Delta \rangle \left(\frac{m_c^3}{2M^4} - \frac{4}{3} \frac{m_c}{M^2} \right) \right]. \qquad (4.2.8)$$

4 Chiral partner QCD sum rules

The corresponding moments therefore are

$$\frac{1}{\pi}\int_{-\infty}^{+\infty} d\omega\, \omega\, \Delta\Pi_T^{V-A}(\omega) = -2m_c\langle\bar{q}q\rangle,\tag{4.2.9a}$$

$$\frac{1}{\pi}\int_{-\infty}^{+\infty} d\omega\, \omega^3\, \Delta\Pi_T^{V-A}(\omega) = -2m_c^3\langle\bar{q}q\rangle - \frac{4}{3}m_c\langle\Delta\rangle,\tag{4.2.9b}$$

$$\frac{1}{\pi}\int_{-\infty}^{+\infty} d\omega\, \omega^5\, \Delta\Pi_T^{V-A}(\omega) = -2m_c^5\langle\bar{q}q\rangle + m_c^3\langle\bar{q}g\sigma\mathcal{G}q\rangle - \frac{11}{3}m_c^3\langle\Delta\rangle + \ldots\tag{4.2.9c}$$

with the same meaning of "..." as above. This second set of Weinberg type sum rules contains the same condensates as the first set in Eq. (4.2.4) but in different combinations. Again, addressing chiral symmetry to the l.h.s. one may consider the chiral condensate $\langle\bar{q}q\rangle$ and $\langle\Delta\rangle$ as possible order parameters. As the P–S case allowed us to identify $\langle\bar{q}D_0^2q\rangle$ as order parameter, $\langle\bar{q}g\sigma\mathcal{G}q\rangle$ also qualifies as order parameter.

In addition,

$$\frac{1}{\pi}\int_{-\infty}^{+\infty} d\omega\, \omega\, \Delta\tilde{\Pi}_T^{V-A}(\omega) = 0\tag{4.2.10}$$

follows from Eq. (B.3.6) but for the decomposition $\Pi_{\mu\nu} = (q_\mu q_\nu - q^2 g_{\mu\nu})\tilde{\Pi}_T + q_\mu q_\nu \tilde{\Pi}_L$. This corresponds to Weinberg's first sum rule [Wei67]. Note that, in contrast to Weinberg's original sum rule [Wei67], no Goldstone boson properties appear on the right hand side of Eq. (4.2.10) because the heavy-light currents involved in our case are generally not conserved. The Borel transformed sum rule for $\tilde{\Pi}_T = \Pi_T/q^2$ reads

$$\frac{1}{\pi}\int_{-\infty}^{+\infty} d\omega\, e^{-\omega^2/M^2}\, \omega\, \Delta\tilde{\Pi}_T^{V-A}(\omega)$$
$$= \langle\bar{q}q\rangle \frac{2}{m_c}\left[1 - e^{-m_c^2/M^2}\right] - \langle\bar{q}g\sigma\mathcal{G}q\rangle \frac{1}{m_c^3}\left[1 - e^{-m_c^2/M^2}\left(1 + \frac{m_c^2}{M^2} + \frac{m_c^4}{2M^4}\right)\right]$$
$$+ \langle\Delta\rangle \frac{1}{m_c^3}\left[\frac{7}{3}\left[1 - e^{-m_c^2/M^2}\left(1 + \frac{m_c^2}{M^2}\right)\right] - e^{-m_c^2/M^2}\frac{m_c^4}{2M^4}\right]\tag{4.2.11}$$

and reproduces Eq. (4.2.9). Note that using Π_T, instead of $\tilde{\Pi}_T$, is more appropriate for the heavy-quark limit which is considered in the following.

4.2 Chiral partners of open charm mesons

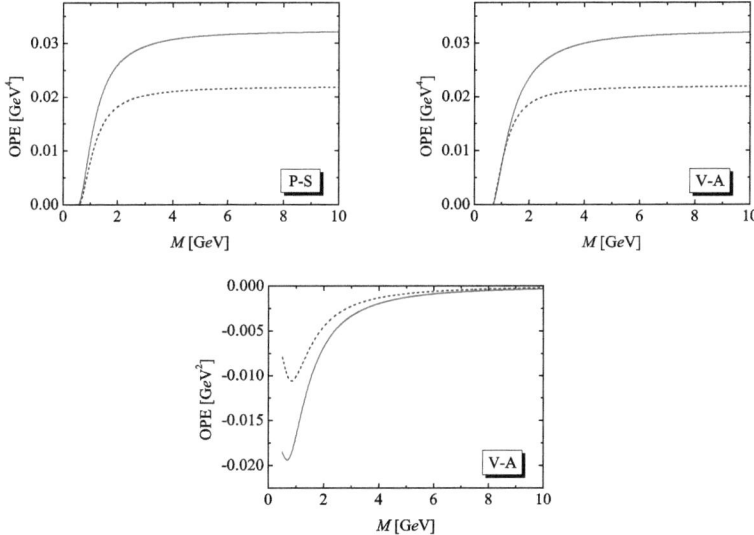

Figure 4.2.1: The OPE sides of Eqs. (4.2.2) (upper left panel, P–S), (4.2.8) (upper right panel, V–A) and (4.2.11) (lower panel V–A) as a function of the Borel mass in vacuum (solid red line) and at nuclear saturation density $n = 0.15$ fm^{-3} (dashed blue line) with $\langle \Delta \rangle = 8 \times 0.3\,n$ GeV^2.

4.2.3 Heavy-quark symmetry

In the heavy-quark limit, $m_2^2 \gg |q^2|$, the leading contributions in Eq. (4.2.7) are

$$\Pi_T^{V-A}(q)\Big|_{m_2^2 \gg |q^2|} \approx \Pi^{P-S}(q)\Big|_{m_2^2 \gg |q^2|} \approx -\frac{2}{m_2}\langle \bar{q}q \rangle, \qquad (4.2.12)$$

where $\frac{1}{q^2 - m_2^2} = -\frac{1}{m_2^2}\sum_{n=0}^{\infty}\left(\frac{q^2}{m_2^2}\right)^n$ has been exploited. This result is in agreement with the expected degeneracy of vector and pseudo-scalar mesons and of axial-vector and scalar mesons in the heavy-quark limit [Isg89]. Actually the $\langle \bar{q}q \rangle$ parts of Π^{P-S} and Π_T^{V-A} agree as one can check by comparing Eq. (4.2.3) with Eq. (4.2.8).

4 Chiral partner QCD sum rules

4.2.4 Numerical examples

Using the condensates from Tab. 3.3.1 the r.h.s of Eqs. (4.2.2), (4.2.8) and (4.2.11), i. e. the OPE sides, are exhibited in Fig. 4.2.1 as a function of the Borel mass for vacuum and cold nuclear matter. The Borel curves are significantly modified by changes of the entering condensates due to their density dependencies, i. e. their magnitudes are lowered by approximately 30% of their vacuum values. It should be emphasized, however, that the density dependence is estimated in linear approximation. For the chiral condensate $\langle \bar{q}q \rangle$ it is known [Lut00, Kai08] that the actual density dependence at zero temperature is not so strong. One may expect accordingly a somewhat weaker impact of the medium effects.

For further evaluations, the hadronic spectral functions or moments thereof must be specified, e. g. by using suitable moments as in [Zsc02, Tho05].

4.3 Interim summary

To summarize the results of this chapter, we present difference QSRs for chiral partners of mesons with a simple quark structure. Focussing on the OPE of the current-current correlator in lowest (zeroth) order of an interaction term insertion, only the combination of a light and a heavy (massive) quark yields a non-trivial result: Differences of spectral moments between pseudo-scalar and scalar as well as vector and axial-vector mesons for condensates up to mass dimension ≤ 5 are determined by the combinations $m_c \langle \bar{q}q \rangle$, $m_c \langle \bar{q} g \sigma \mathcal{G} q \rangle$, and $m_c (\langle \bar{q} g \sigma \mathcal{G} q \rangle - 8 \langle \bar{q} D_0^2 q \rangle)$ (to be taken at an appropriate scale) which may be considered as elements of "order parameters" of chiral symmetry breaking (see [Doi04] for a lattice evaluation of the mixed quark-gluon condensate at finite temperature). Vanishing of these condensates at high baryon density and/or temperature would mean chiral restoration, i. e. the degeneracy of spectral moments of the considered chiral partners. Chiral partners of mesons with light–light or heavy–heavy quark currents are non-degenerate in higher orders of α_s, as exemplified by the Kapusta-Shuryak sum rule. The famous Weinberg sum rules generalized to a hot medium in [Kap94] refer to lower mass dimension moments where the OPE side vanishes in the chiral limit.

Our results show a significant change of the OPE side when changing from vacuum to normal nuclear matter density. This implies that also the hadronic spectral functions may experience a significant reshaping in the medium.

5 The impact of chirally odd condensates on the ρ meson*

The impact of chiral symmetry restoration on the properties of hadrons, in particular of light vector mesons, is a much debated issue. As discussed in Sec. 1.3, in-medium modifications of hadrons made out of light quarks and especially their possible "dropping masses" are taken often synonymously for chiral restoration. However, experimentally the main observation of in-medium changes of light vector mesons via dilepton spectra is a significant broadening of the spectral shape [Arn06, Ada08]. Such a broadening can be obtained in hadronic many-body approaches, e. g., [Fri97, Rap97, Pet98c, Pos04, Hee06, Her93, Cha93], which at first sight are not related directly to chiral restoration in the above spirit. Pion dynamics and resonance formation, both fixed to vacuum data, provide the important input for such many-body calculations. Clearly the pion dynamics is closely linked to the vacuum phenomenon of chiral symmetry breaking, but the connection to chiral restoration is not so clear. For the physics of resonances the connection is even more loose. There are recent approaches which explain some hadronic resonances as dynamically generated from chiral dynamics [Kai95, Kol04, Sar05, Lut04, Roc05, Wag08a, Wag08b], but again this primarily points towards an intimate connection between hadron physics and chiral symmetry breaking whereas effects of the chiral restoration transition on hadron physics remain open. As suggested, e. g., in [Rap99, Leu09b] the link to chiral restoration might be indirect: The in-medium broadening could be understood as a step towards deconfinement. In the deconfined quark-gluon plasma also chiral symmetry is presumed to be restored. All these considerations suggest that the link between chiral restoration and in-medium changes of hadrons is more involved as one might have hoped.

Additional input could come from approaches which are closer to QCD than standard hadronic models. One such approach is the QSR method (cf. Sec. 2). A somewhat superficial view on QSRs for vector mesons seems to support the original picture of

---
*The presentation is based on [Hil12b].

5 VOC scenario for the ρ meson

an intimate connection between chiral restoration and in-medium changes. Here the previously popular chain of arguments goes as follows:

1. Four-quark condensates play an important role for the vacuum mass of the light vector mesons [Shi79b, Shi79a].

2. The four-quark condensates factorize into squares of the two-quark condensate [Nov84a].

3. The two-quark condensate decreases in the medium due to chiral restoration [Ger89, Dru91].

4. Thus the four-quark condensates decrease in the medium accordingly.

5. Therefore the masses of light vector mesons change (decrease) in the medium due to chiral restoration.

Before we critically assess this line of reasoning an additional remark concerning four-quark condensates is in order: In [Shi79b, Shi79a] it is shown that the vector meson mass emerges from a subtle balance between the gluon and four-quark condensates. In that sense four-quark condensates are important. There are, however, approaches (employing, e. g. finite energy sum rules) which deduce ρ meson properties without using the four-quark condensates, see, e. g., [Kwo08]. In this case, one needs additional input to determine the ρ meson properties (cf. the discussion in [Hat95]). In [Kwo08] this input is provided by the assumption that the continuum threshold is related to the scale of chiral symmetry breaking. In the following we use the original sum rule approach of [Shi79b, Shi79a]. Note that the different approaches of [Shi79b, Shi79a] and [Kwo08] are not mutually exclusive.

In the previous line of reasoning (items 1-5) one seems to have a connection between chiral restoration – descent of two- and four-quark condensates – and in-medium changes, no matter whether it is a mass shift or a broadening [Kli97, Leu98b] or a more complicated in-medium modification [Pos04, Ste06]. However, at least item 2 and, as its consequence item 4, are questionable: Whether the four-quark condensates factorize is discussed since the invention of QSRs, see, e. g., [Shi79b, Shi79a, Nar83, Lau84, Ber88, Dom88, Gim91, Lei97, Bor06] in vacuum and for in-medium situations [Ele93, Hat93, Bir96, Zsc03, Leu05a, Tho05]. With such doubts the seemingly clear connection between chiral restoration and in-medium changes gets lost.

Indeed, a closer look on the QSR for light vector mesons reveals that most of the condensates, whose in-medium change is translated into an in-medium modification for the respective hadron, are actually chirally symmetric (see below). Physically, it is of course possible that the same microscopic mechanism which causes the restoration of chiral symmetry is also responsible for changes of chirally symmetric condensates. For example, in the scenario [Par05] about half of the (chirally symmetric) gluon condensate vanishes together with the two-quark condensate. These considerations show that the connection between the mass of a light vector meson and chiral symmetry breaking is not as direct as often expected.

We take these considerations as a motivation to study in the present section a clear-cut scenario where we ask and answer the question: How large would the mass and/or the width of the ρ meson be in a world where the chiral symmetry breaking objects/condensates are zero? In the following we will call this scenario VOC (vanishing of chirally odd condensates). Note that we leave all chirally symmetric condensates untouched, i. e. they retain their respective vacuum values. We stress again that such a scenario may not reflect all the physics which is contained in QCD. There might be intricate interrelations between chirally symmetric and symmetry breaking objects. In that sense the VOC scenario shows for the first time the minimal impact that the restoration of chiral symmetry has on the properties of the ρ meson.

5.1 Chiral transformations and QCD condensates

For the ρ^0 meson we investigate the correlator

$$\Pi_{\mu\nu}(q) = i \int d^4x \, e^{iqx} \langle T \left[j_\mu^3(x) j_\nu^3(0) \right] \rangle, \tag{5.1.1}$$

where the current j_μ^3 stands for $j_\mu^{V,\tau}$ (cf. Eq. (A.1.42a)) with $\tau = \sigma^3/2$, σ the isospin Pauli matrices and $j_\mu^{3\dagger} = j_\mu^3$ has been used. We use the sum rule as given in [Leu98b]

$$\frac{1}{\pi} \int_0^\infty ds \, s^{-1} \Delta\Pi(s) e^{-s/M^2} = \tilde{\Pi}(M^2), \tag{5.1.2}$$

where we denoted the Borel transformed correlator for brevity as $\tilde{\Pi}(M^2)$. We consider a ρ meson at rest, therefore, and due to current conservation and flavor symmetry (cf. Apps. B.3 and B.4), the tensor structure of (5.1.1) reduces to a scalar $\Pi = \frac{1}{3}\Pi_\mu^\mu$.

5 VOC scenario for the ρ meson

The Borel transformed OPE reads [Zsc04]

$$\tilde{\Pi}(M^2) = c_0 M^2 + \sum_{i=1}^{\infty} \frac{c_i}{(i-1)! \, M^{2(i-1)}} \qquad (5.1.3)$$

with coefficients up to mass dimension 6

$$c_0 = \frac{1}{8\pi^2}\left(1 + \frac{\alpha_s}{\pi}\right), \qquad (5.1.4a)$$

$$c_1 = -\frac{3}{8\pi^2}(m_u^2 + m_d^2), \qquad (5.1.4b)$$

$$c_2 = \frac{1}{2}(1 + \frac{\alpha_s}{4\pi}C_F)(m_u\langle \bar{u}u\rangle + m_d\langle \bar{d}d\rangle) + \frac{1}{24}\left\langle \frac{\alpha_s}{\pi}G^2\right\rangle + N_2, \qquad (5.1.4c)$$

$$c_3 = -\frac{112}{81}\pi\alpha_s\langle \mathcal{O}_4^V\rangle - 4N_4 \qquad (5.1.4d)$$

with $C_F = (N_c^2 - 1)/(2N_c) = 4/3$ for $N_c = 3$ colors. A mass dimension 2 condensate seems to be excluded in vacuum [Dom09]. In Eq. (5.1.4) we have introduced the vector channel (V) combination of four-quark condensates in compact notation

$$\langle \mathcal{O}_4^V\rangle = \frac{81}{224}\langle(\bar{\psi}\gamma_\mu\gamma_5\lambda^a\sigma^3\psi)(\bar{\psi}\gamma^\mu\gamma_5\lambda^a\sigma^3\psi)\rangle + \frac{9}{112}\langle \bar{\psi}\gamma_\mu\lambda^a\psi \sum_{f=u,d,s}\bar{f}\gamma^\mu\lambda^a f\rangle \quad (5.1.5)$$

with color matrices λ^a. (For a classification of four-quark condensates, cf. [Tho07].) It is also useful to introduce the averaged two-quark condensate $m_q\langle \bar{q}q\rangle = \frac{1}{2}\langle m_u\bar{u}u + m_d\bar{d}d\rangle$ and $m_q = (m_u + m_d)/2$. These terms constitute the contributions which already exist in vacuum (and might change in a medium) up to higher-order condensates encoded in c_i for $i > 3$ which are suppressed by higher powers in the expansion parameter M^{-2}. Additional non-scalar condensates come into play, in particular for in-medium situations. In Eqs. (5.1.4c) and (5.1.4d), only the twist-two non-scalar condensates [Hat93] are displayed, $N_i = -\frac{2}{3}i\langle \mathscr{ST}\bar{\psi}\gamma_{\mu_1}D_{\mu_2}\ldots D_{\mu_i}\psi\rangle g^{\mu_1 0}\ldots g^{\mu_i 0}$, where the operation $\mathscr{ST}$ is introduced to make the operators symmetric and traceless w. r. t. its Lorentz indices. Twist-four non-scalar condensates have been found to be numerically less important [Hat93, Hat95, Leu98a].

Using Eqs. (1.1.2a) and (1.1.2b) one can show that the only objects in the OPE (5.1.3) with coefficients (5.1.4) and (5.1.5) which are not chirally invariant[12] are

(i) the (numerically small) two-quark condensate

---

[12]Note that c_1 breaks the chiral symmetry explicitly. Its contribution is numerically completely negligible.

5.1 Chiral transformations and QCD condensates

and

(*ii*) a part of the (numerically important) four-quark condensate $\langle \mathcal{O}_4^V \rangle$ specified in Eq. (5.1.5).

The last term in Eq. (5.1.5) is a flavor singlet and, thus, invariant w. r. t. chiral transformations. Since the first term in Eq. (5.1.5) is the product of two axial-vector currents, the most intuitive way to split this four-quark condensate into a chirally symmetric and a chirally odd part is to add and to subtract the product of two vector currents. Owing to the definition of vector and axial-vector currents in terms of left- and right-handed currents (cf. App. A.1) one has

$$(\bar{\psi}\gamma_\mu\gamma_5 \lambda^a\sigma^3\psi)(\bar{\psi}\gamma^\mu\gamma_5\lambda^a\sigma^3\psi) - (\bar{\psi}\gamma_\mu\lambda^a\sigma^3\psi)(\bar{\psi}\gamma^\mu\lambda^a\sigma^3\psi)$$
$$= (\bar{\psi}\gamma_\mu\gamma_5\lambda^a\sigma^3\psi - \bar{\psi}\gamma^\mu\lambda^a\sigma^3\psi)(\bar{\psi}\gamma_\mu\gamma_5\lambda^a\sigma^3\psi + \bar{\psi}\gamma^\mu\lambda^a\sigma^3\psi)$$
$$= -4(\bar{\psi}_R\gamma_\mu\lambda^a\sigma^3\psi_R)(\bar{\psi}_L\gamma^\mu\lambda^a\sigma^3\psi_L), \qquad (5.1.6)$$

where we used the canonical ETC (cf. App. A.2). Therefore, we define the chirally odd condensate

$$\langle \mathcal{O}_4^{\text{odd}} \rangle \equiv -\frac{81}{112}\langle (\bar{\psi}_R\gamma_\mu\lambda^a\sigma^3\psi_R)(\bar{\psi}_L\gamma^\mu\lambda^a\sigma^3\psi_L) \rangle \qquad (5.1.7)$$

and the chirally even condensate

$$\langle \mathcal{O}_4^{\text{even}} \rangle \equiv \frac{81}{448}\langle (\bar{\psi}\gamma_\mu\gamma_5\lambda^a\sigma^3\psi)^2 + (\bar{\psi}\gamma_\mu\lambda^a\sigma^3\psi)^2 \rangle + \frac{9}{112}\langle \bar{\psi}\gamma_\mu\lambda^a\psi \sum_{f=u,d,s} \bar{f}\gamma^\mu\lambda^a f \rangle$$
$$(5.1.8)$$

with $\langle \mathcal{O}_4^V \rangle = \langle \mathcal{O}_4^{\text{even}} \rangle + \langle \mathcal{O}_4^{\text{odd}} \rangle$. The chirally odd part, which is the product of a left handed and a right handed chirality current, $\langle \mathcal{O}_4^{\text{br}} \rangle \propto g^{\mu\nu} j_\mu^{R,\sigma^3} j_\nu^{L,\sigma^3}$ (omitting color indices), can indeed be transformed into its negative by a proper chiral transformation and is therefore dubbed "chirally odd" condensate. In general it is clear that any transformation which transforms the vector current into the axial-vector current must generate a sign change for j_μ^R but not for j_μ^L (cf. Eq. (A.1.55)). Therefore, any finite chiral transformation which relates these particular parity partners and their spectral functions, results in a sign change of $\langle \mathcal{O}_4^{\text{odd}} \rangle$ but not of $\langle \mathcal{O}_4^{\text{even}} \rangle$. Indeed, the quantity $\langle \mathcal{O}_4^{\text{even}} \rangle - \langle \mathcal{O}_4^{\text{odd}} \rangle$ is the specific combination of four-quark condensates which enters the OPE for the a_1 meson – the chiral partner of the ρ.[13] It has been shown

---

[13]In App. A.1 we have shown, that any component of the chiral vector–isospin-vector current $\bar{j}_\mu^C$,

73

in [Tho08a] that $\langle \mathcal{O}_4^{\text{odd}} \rangle$ is an order parameter of the chiral symmetry (cf. App. A.2, Eq. (A.2.32)), but can also be concluded from chiral partner sum rules in the light meson sector. Indeed, the condensate (5.1.7) enters the Weinberg-Kapusta-Shuryak sum rules [Kap94], i.e. the OPE for the chiral difference of vector and axial-vector mesons with two light valence quarks, in order α_s (cf. Sec. 4).

The last term in Eq. (5.1.8) is an isospin-singlet. In an isospin invariant system (cf. App. A.2), the first two terms in Eq. (5.1.8) may be written as

$$\langle (\bar{\psi} \gamma_\mu \gamma_5 \sigma^3 \lambda^a \psi)^2 + (\bar{\psi} \gamma_\mu \sigma^3 \lambda^a \psi)^2 \rangle$$
$$= \frac{1}{3} \langle (\bar{\psi}_R \gamma_\mu \vec{\tau} \lambda^a \psi_R - \bar{\psi}_L \gamma_\mu \vec{\tau} \lambda^a \psi_L)^2 + (\bar{\psi}_R \gamma_\mu \vec{\tau} \lambda^a \psi_R + \bar{\psi}_L \gamma_\mu \vec{\tau} \lambda^a \psi_L)^2 \rangle$$
$$= \frac{2}{3} \langle (\bar{\psi}_R \gamma_\mu \vec{\tau} \lambda^a \psi_R)^2 + (\bar{\psi}_L \gamma_\mu \vec{\tau} \lambda^a \psi_L)^2 \rangle . \quad (5.1.9)$$

The latter two terms are separately invariant w.r.t. left-handed and right-handed isospin transformations as shown in App. A.1 (cf. Eq. (A.1.69)), i.e. they are chirally invariant.

It appears to be very natural that a four-quark condensate which breaks chiral symmetry can be related to the square of the two-quark condensate which also breaks chiral symmetry. Indeed, it has been shown in [Bor06] that the factorization of the four-quark condensate $\langle \mathcal{O}_4^{\text{odd}} \rangle$ given in Eq. (5.1.7) is completely compatible with the ALEPH data on the vector and axial-vector spectral distributions [Sch05]. Contrarily, it is not so obvious that a chirally symmetric four-quark condensate like $\langle \mathcal{O}_4^{\text{even}} \rangle$ in Eq. (5.1.8) is related directly to the chirally odd two-quark condensate $\langle \bar{q} q \rangle$.

5.2 VOC scenario for the ρ meson

Let us first describe briefly how we fix the numerical values for the QCD condensates. We note that the non-scalar condensates, which appear in Eqs. (5.1.4c) and (5.1.4d), are chirally symmetric and vanish in the vacuum, i.e. we can disregard them also for the VOC scenario. The chirally symmetric gluon condensate is determined from the QSR for the charmonium [Shi79b, Shi79a] as $\left\langle \frac{\alpha_s}{\pi} G^2 \right\rangle = (330\,\text{MeV})^4$. The running coupling has to be evaluated at the typical energy scale of 1 GeV. Following [Leu04]

---

with $C \in \{R, L\}$, can be transformed into its negative by an appropriate chiral transformation, cf. discussion after Eq. (A.1.62). Thus, a transformation may be applied which transforms the left-handed current into its negative but not the right handed, which results in a sign for $\langle \mathcal{O}_4^{\text{br}} \rangle$. Consequently, if chiral symmetry would be realized in the Wigner-Weyl phase, (5.1.7) must be zero.

5.2 VOC scenario for the ρ meson

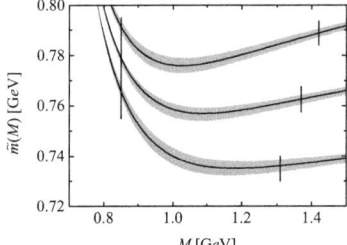
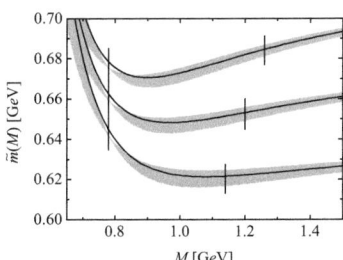

Figure 5.2.1: The mass parameter $\tilde{m}(M, s_+)$ as a function of the Borel mass M for various values of the continuum threshold s_+. The Borel windows are marked by vertical bars. The curves are for $c_4 = 0$, while the bands cover the range $c_4 = \pm \langle \frac{\alpha_s}{\pi} G^2 \rangle^2$. Left panel: $s_+ = 1.4$, 1.3 and $1.2\,\text{GeV}^2$ from top to bottom with condensates from Tab. 3.3.1 and the four-quark condensate as described in the text. Right panel: The VOC scenario $\langle \mathcal{O}_4^{\text{odd}} \rangle \to 0$ and $\langle \bar{q}q \rangle \to 0$ for continuum thresholds $s_+ = 1.1$, 1.0 and $0.9\,\text{GeV}^2$ from top to bottom.

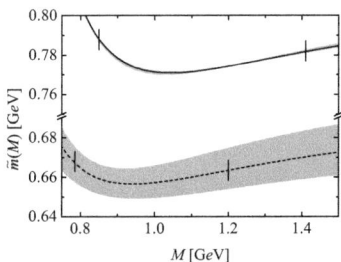

Figure 5.2.2: The mass parameter $\tilde{m}(M, s_+)$ for the optimized continuum threshold s_+ as a function of the Borel mass M for vacuum values of the condensates (upper curves) and the VOC scenario with $\langle \mathcal{O}_4^{\text{odd}} \rangle \to 0$ and $\langle \bar{q}q \rangle \to 0$ (lower curves). The Borel windows are marked by vertical bars. The curves are for $c_4 = 0$, while the bands cover the range $c_4 = \pm \langle \frac{\alpha_s}{\pi} G^2 \rangle^2$.

5 VOC scenario for the ρ meson

we use $\alpha_s = 0.38$. The chirally odd two-quark condensate is fixed by the Gell-Mann–Oakes–Renner relation [GM68] (cf. App. A.2) $m_q \langle \bar{q}q \rangle = -\frac{1}{2} f_\pi^2 m_\pi^2$ with the pion-decay constant $f_\pi \approx 92\,\text{MeV}$ and the pion mass $m_\pi = 139.6\,\text{MeV}$ [Ams08]. Using in addition $m_q \equiv (m_u + m_d)/2 = 5.5\,\text{MeV}$ [Nar99] one gets the vacuum value $\langle \bar{q}q \rangle = -(245\,\text{MeV})^3$. For vacuum, the chirally odd condensate $\langle \mathcal{O}_4^{\text{odd}} \rangle$ has been extracted from the experimental difference between vector and axial-vector spectral information [Sch05]. We use the result of [Bor06]: $\langle \mathcal{O}_4^{\text{odd}} \rangle_{\text{vac}} \approx \frac{9}{7} \langle \bar{q}q \rangle_{\text{vac}}^2$. To fix the vacuum value for the chirally symmetric $\langle \mathcal{O}_4^{\text{even}} \rangle$ defined in Eq. (5.1.8) one may resort to the vacuum ρ meson properties [Ams08].

For that purpose we rearrange Eq. (5.1.2) by splitting the integral $\int_0^\infty = \int_0^{s_+} + \int_{s_+}^\infty$ and putting the so-called continuum part to the OPE terms thus isolating the interesting hadronic resonance part below the continuum threshold s_+. This allows to define a normalized moment [Zsc02] of the hadronic spectral function

$$\tilde{m}^2(M, s_+) \equiv \frac{\int_0^{s_+} ds\, \Delta\Pi(s)\, e^{-s/M^2}}{\int_0^{s_+} ds\, \Delta\Pi(s)\, s^{-1} e^{-s/M^2}} \tag{5.2.1a}$$

$$= \frac{c_0 M^2 [1 - (1 + \frac{s_+}{M^2}) e^{-s_+/M^2}] - \frac{c_2}{M^2} - \frac{c_3}{M^4} - \frac{c_4}{2M^6}}{c_0 [1 - e^{-s_+/M^2}] + \frac{c_1}{M^2} + \frac{c_2}{M^4} + \frac{c_3}{2M^6} + \frac{c_4}{6M^8}}, \tag{5.2.1b}$$

where the semi-local duality hypothesis $\frac{1}{\pi} \int_{s_+}^\infty ds\, s^{-1} \Delta\Pi(s) e^{-s/M^2} = c_0 \int_{s_+}^\infty ds\, e^{-s/M^2}$, replacing the continuum integral by the integral over the perturbative term, is exploited in Eq. (5.2.1b). The second line emerges essentially from the OPE and "measures" how $\tilde{m}$ is determined by condensates. The meaning of the spectral moment (5.2.1a) becomes obvious for the pole ansatz [Shi79a]; generically, it is the s coordinate of the center of gravity of the distribution $\Delta\Pi(s) s^{-1} e^{-s/M^2}$ between 0 and s_+.

To determine $\langle \mathcal{O}_4^{\text{even}} \rangle$ for the vacuum case we identify the average of $\tilde{m}$ with the ρ vacuum mass. The averaged mass parameter is determined by $\overline{m}(s_+) = (M_{\max} - M_{\min})^{-1} \int_{M_{\min}}^{M_{\max}} \tilde{m}(M, s_+)\, dM$. According to [Zsc04] the Borel minimum $M_{\min}$ follows from the requirement that the mass dimension 6 contribution to the OPE is smaller than 10%:

$$\frac{\frac{56}{81 M_{\min}^4} \pi \alpha_s \langle \mathcal{O}_4^V \rangle}{\frac{1}{8\pi^2} (1 + \frac{\alpha_s}{\pi}) M_{\min}^2 + \frac{m_q}{M_{\min}^2} \langle \bar{q}q \rangle + \frac{1}{24 M_{\min}^2} \langle \frac{\alpha_s}{\pi} G^2 \rangle - \frac{56}{81 M_{\min}^4} \pi \alpha_s \langle \mathcal{O}_4^V \rangle} \leq 0.1. \tag{5.2.2a}$$

For the Borel maximum $M_{\max}$ we demand that the continuum contribution to the

5.2 VOC scenario for the ρ meson

spectral integral is smaller than 50%:

$$\frac{\frac{1}{8\pi^2}\left(1+\frac{\alpha_s}{\pi}\right)M_{\max}^2 e^{-s_+/M_{\max}^2}}{\frac{1}{\pi}\int_0^{s_+} ds\,\Delta\Pi(s)\,s^{-1}e^{-s/M^2} + \frac{1}{8\pi^2}\left(1+\frac{\alpha_s}{\pi}\right)M_{\max}^2 e^{-s_+/M_{\max}^2}} \leq 0.5. \quad (5.2.2b)$$

Both quantities do not depend on a certain parametrization of the spectral density, because the denominator of Eq. (5.2.2b) is given by Eq. (5.1.2) in terms of the OPE. Note that in Eq. (5.2.2a) the continuum contribution is not present. The threshold s_+ follows from the requirement of maximum flatness of $\tilde{m}(M,s_+)$ as a function of M within the Borel window. Employing this system of equations, $\langle \mathcal{O}_4^{\mathrm{even}}\rangle = (267\,\mathrm{MeV})^6$ is found from the requirement $\overline{m} = 775.5\,\mathrm{MeV}$, i. e. we identify $\overline{m}$ with the vacuum ρ mass. The Borel curves are exhibited in Figs. 5.2.1 (for different continuum thresholds) and 5.2.2 for the curves which are optimized for maximal flatness (upper solid curve embedded in the narrow band). The corresponding averaged mass parameters in Fig. 5.2.1 are $\overline{m} = 781.1, 761.0$ and $740.8\,\mathrm{MeV}$ from top to bottom for $c_4 = 0$.

Digression: Other moments

Formally, in order to obtain the OPE for spectral integrals of higher powers of s, i. e. $\Delta\Pi(s) \to s^n\Delta\Pi(s)$, the n-th derivative of Eq. (5.1.2) w. r. t. $-1/M^2$ can be taken and moments similar to (5.2.1a) weighting higher energy regimes of $\Delta\Pi(s)$ may be defined:

$$\frac{\int_0^{s_+} ds\, s^n \Delta\Pi e^{-s/M^2}}{\int_0^{s_+} ds\, s^{n-1}\Delta\Pi e^{-s/M^2}}$$

$$= \frac{(n+1)!}{n!}\frac{c_0 M^2\left[1 - e^{-s_+/M^2}\sum_{k=0}^{n+1}\frac{(s_+/M^2)^k}{k!}\right] + \frac{(-1)^{n+1}}{(n+1)!}\sum_{k=0}^{\infty}\frac{c_{k+n+2}}{k!(M^2)^{k+n+1}}}{c_0\left[1 - e^{-s_+/M^2}\sum_{k=0}^{n}\frac{(s_+/M^2)^k}{k!}\right] + \frac{(-1)^n}{n!}\sum_{k=0}^{\infty}\frac{c_{k+n+1}}{k!(M^2)^{k+n+1}}}. \quad (5.2.3)$$

The maximum number of terms of the asymptotic expansion of the spectral integral is limited in general by accuracy requirements [Win06, Hin95] and in practical calculations by the knowledge about condensates of higher mass dimension. Each derivative reduces the number of known condensates in the OPE beginning with the lowest mass dimensions, whereas each spectral integral basically has a different asymptotic expansion with different approximative properties. On the other hand the perturbative contribution, and thus the contribution of the continuum, remains in the sum and is enhanced compared to the non-perturbative corrections which is mirrored in the amplified weighting of the continuum in the moments. Therefore, using other moments of the spectral integral is expected to significantly reduce the numerical accuracy. Indeed, $\bar{m} = 453$ MeV and $s_+ = 0.92$ MeV^2 is obtained for $n = 1$ and $\langle \mathcal{O}_4^{\text{even}} \rangle = (267 \text{ MeV})^6$.

VOC analysis

Let us consider now the VOC scenario, where one sets the explicitly chiral symmetry breaking terms to zero. Thus, here one only needs the above quoted vacuum values for the chirally invariant terms. $\langle \mathcal{O}_4^{\text{odd}} \rangle \to 0$ and $\langle \bar{q}q \rangle \to 0$ but keeping the chirally symmetric condensate values causes a reduction of $\bar{m}$ to 659.8 MeV and the continuum threshold becomes $s_+ = 1.03$ GeV^2 (in vacuum it was $s_+ = 1.37$ GeV^2), see lower curves in Fig. 5.2.2. We emphasize the large impact on the averaged spectral moment $\bar{m}$ if $\langle \mathcal{O}_4^{\text{odd}} \rangle$ is set to zero.

We have convinced ourselves on the robustness of the quoted numbers, in particular the drop of $\bar{m}$, e.g. against variations of the criteria (5.2.2) for the Borel window. For instance, requiring for the r.h.s. of Eq. (5.2.2a) a value of 0.05 (0.20) instead of 0.1, one would get $\bar{m} = 753.8$ (801.3) MeV for the same condensates as before and a drop of $\bar{m}$ to 641.2 (685.0) MeV in the VOC scenario. (If the condensates are not frozen to the previous case but fixed to reproduce $\bar{m} = 775.5$ MeV, which gives $\langle \mathcal{O}_4^{\text{even}} \rangle = (276 \text{ MeV})^6$ ((256 MeV)^6), the mass would drop to $\bar{m} = 678.6$ (635.4) MeV in the VOC scenario.) Analogously, requiring 0.4 (0.6) instead of 0.5 in the r.h.s. of Eq. (5.2.2b) one would get $\bar{m} = 788.1$ (764.8) MeV and a drop of $\bar{m}$ to 672.0 (649.7) MeV in the VOC scenario. (A drop to $\bar{m} = 649.0$ (670.0) MeV and $\langle \mathcal{O}_4^{\text{even}} \rangle = (261 \text{ MeV})^6$ ((272 MeV)^6) if the condensates are fixed to reproduce $\bar{m} = 775.5$ MeV.)

To get an estimate of the possible importance of the poorly known next-order term c_4 of mass dimension 8 (cf. [Rei85]) we use, analogously to [Tho05], as an estimate the "natural scale" $\langle \frac{\alpha_s}{\pi} G^2 \rangle^2$ motivated by dimensional reasoning. The Borel curves for $|c_4| = \langle \frac{\alpha_s}{\pi} G^2 \rangle^2$ border the bands in Fig. 5.2.2. Of course, this estimate is quite rough as c_4 may contain also chirally odd condensates, whose change is not accounted for in the VOC scenario. Nevertheless, it supports the robustness of the VOC scenario which is characterized by a lowering of the spectral moment $\bar{m}$ by about 120 MeV.

Equation (5.2.1a) shows that the moment analysis is independent of the specific type of parametrization if the condensates are known. In contrast, setting the moment equal to the experimental mass is equivalent to a pole ansatz and, hence, fitting $\langle \mathcal{O}_4^{\text{even}} \rangle$ to a specific parametrization. However, if the experimental mass is a good approximation to the moment (5.2.1a), then $\langle \mathcal{O}_4^{\text{even}} \rangle$ is in good agreement to results that would have been obtained from experimental data. In this case the VOC scenario for the moment is a model independent result. Only the translation of the moment to the parameters of a certain spectral function is model dependent.

5.3 Mass shift vs. broadening

While for a narrow resonance in vacuum the often employed pole + continuum ansatz is reasonable, the spectral distribution may get a more complex structure in a medium [Zsc02, Ste06, Kwo08, Fri97, Rap97, Pet98c, Pos04, Hee06, Kli97, Lut02]. In particular, one cannot decide, within the employed framework of QSRs, whether the above observed reduction of $\tilde{m}$, and consequently $\overline{m}$, means a mass shift or a broadening or both. To make this fact explicit, we use a Breit-Wigner ansatz for the spectral function

$$\Delta \Pi(s) = \frac{F_0}{\pi} \left(\frac{\sqrt{s}}{m_0}\right)^a \frac{\sqrt{s}\Gamma(s)}{(s-m_0^2)^2 + s\Gamma^2(s)} \Theta(s_+ - s) + \pi s c_0 \Theta(s - s_+), \qquad (5.3.1a)$$

where the vacuum parametrization of the width is motivated by [Leu98b]

$$\Gamma(s) = \Theta(s - (km_\pi)^2)\Gamma_0 \left(\frac{\sqrt{s}}{m_0}\right)^b \left(1 - \frac{(km_\pi)^2}{s}\right)^c \left(1 - \frac{(km_\pi)^2}{m_0^2}\right)^{-c}. \qquad (5.3.1b)$$

The parameter k determines the threshold of the ρ decaying into k pions. In vacuum it is $k = 2$. F_0 determines the height of the spectral function. At the same time it is the coupling of the vector current to the ρ meson. As such it is also subject to in-medium changes and is given by inserting (5.3.1a) into (5.1.2) (with (5.1.3) for the r.h.s)

$$F_0 = \frac{c_0[1 - e^{-s_+/M^2}]M^2 + \frac{c_1}{M^0} + \frac{c_2}{M^2} + \frac{c_3}{2M^4} + \frac{c_4}{6M^6}}{\frac{1}{\pi}\int_0^{s_+} ds \frac{1}{\pi}\left(\frac{\sqrt{s}}{m_0}\right)^a \frac{\sqrt{s}\Gamma(s)}{(s-m_0^2)^2+s\Gamma^2(s)} s^{-1} e^{-s/M^2}}. \qquad (5.3.2)$$

Note that in a strict vector meson dominance scenario (cf. [Sak73]) the quantity F_0 would not be subject of medium-induced changes in contrast to what we find here. This signals an incompatibility between the sum rule approach and strict vector meson dominance as pointed out in [Leu06a] for the ω meson and generalized to the ω and ρ meson in [Ste06]. In an extended scenario of vector meson dominance, however, the coupling of the vector current to the ρ meson, encoded in F_0, can depend on

5 VOC scenario for the ρ meson

the density - and in addition also on the invariant mass squared s, as demonstrated, e. g. in [Fri97]. In-medium modifications of vector meson dominance have also been found in the framework of hidden local symmetry [Har03].

The Borel analysis is performed as described above, but with (5.3.1a) and (5.3.1b) in the r. h. s. of Eq. (5.2.1a) and w. r. t. m_0. Each trial value of $\langle \mathcal{O}_4^{\text{even}} \rangle$ delivers now a Borel window average $\overline{m_0}$ as a function of the width Γ_0.[14] A certain requirement for $\overline{m_0}(\Gamma_0)$ fixes $\langle \mathcal{O}_4^{\text{even}} \rangle$. Note that, in general, the peak position m_{peak} and the full width at half maximum Γ_{FWHM} of the spectral function do not coincide with the corresponding parameters m_0 (or $\overline{m_0}$) and Γ_0 of the ansatz (5.3.1a). For $\Gamma(s) = \Gamma_0 = \text{const.}$, e. g., m_0 is determined by $m_0^2 = \sqrt{4m_{\text{peak}}^4 + \Gamma_0^2 m_{\text{peak}}^2} - m_{\text{peak}}^2$. While for small Γ_0 the peak position m_{peak} and m_0 differ only by a few MeV (e. g. for the experimental values $\Gamma_0 = 149.4$ MeV and $\overline{m_0} = 775.5$ MeV one has $\Gamma_{\text{FWHM}} = 147.9$ MeV and $m_{\text{peak}} = 770.5$ MeV), they differ significantly for larger values of Γ_0. Especially, keeping the parameter $\overline{m_0}$ constant in the VOC scenario can cause a strong shift of the peak position. Note that Eq. (5.3.1a) represents a distribution with peak position at m_0^2 and a full width at half maximum of Γ_0, if (5.3.1b) is of the form $\Gamma(s) \propto \Gamma_0/\sqrt{s}$ and $a = 0$. Fitting $\langle \mathcal{O}_4^{\text{even}} \rangle$ may therefore be subject to a boundary condition for $m_{\text{peak}} = m_{\text{peak}}(\overline{m_0}, \Gamma_0)$ instead of $\overline{m_0}$. The strength $\overline{F_0}$ is determined by averaging F_0 in (5.3.2) within the Borel window.

Three parametrizations, which are typically used throughout the literature, are given by

(i) $a = b = c = k = 0$,

which corresponds to the special form $\Gamma = \Gamma_0 = \text{const.}$,

(ii) $a = b = 0$, $c = 3/2$ and $k = 2$,

which is a p-wave with two-pion threshold, and by

(iii) $a = b = 0$, $c = 1/2$ and $k = 1$,

which is an s-wave with a one-pion threshold. For all cases, a sum rule analysis can be performed with different assignments of the experimentally determined parameters of the ρ meson to the parameters of Eq. (5.3.1a) or the shape parameters m_{peak} and Γ_{FWHM}. The results are summarized in Tab. 5.3.1, columns 1-4, and displayed in Figs. 5.3.1-5.3.4. Note that the quoted values of the parameters, to be kept constant in the VOC scenario, are larger than their respective vacuum values. In this respect the results represent maximal broadening and lowering of the mass.

---

[14]More precisely, the moment defined in Eq. (5.2.1) with the distribution (5.3.1) has to be solved for $m_0 = m_0(M, s_+)$ for each Borel mass M and threshold s_+. Optimization w. r. t. s_+ for maximal flatness within the Borel window delivers the threshold and, thus, the Borel window average $\overline{m_0}$.

5.3 Mass shift vs. broadening

Table 5.3.1: Results of the finite width VOC scenario.

	$\langle \mathcal{O}_4^{\text{even}} \rangle^{\frac{1}{6}}$ [MeV]	267	172	242	244	261
vacuum	$\overline{m_0}$ [MeV]	775.5	775.5	775.5	780.5	774.3
	Γ_0 [MeV]	0	149.4	149.4	149.4	157.3
	m_{peak} [MeV]	775.5	771.9	770.5	775.5	767.0
	Γ_{FWHM} [MeV]	0	149.0	147.9	147.9	153.1
	$\overline{F_0}$ [GeV^4]	0.0362	0.0330	0.03985	0.0408	0.04626
	s_+ [GeV^2]	1.37	1.21	1.33	1.35	1.45
constant mass	$\overline{m_0}$ [MeV]	784.8	-	780.3	795.4	795.9
	Γ_0 [MeV]	130.0	-	250.0	250.0	280.0
	m_{peak} [MeV]	782.1	-	770.3	785.6	773.3
	Γ_{FWHM} [MeV]	130.7	-	248.3	248.3	277.1
	$\overline{F_0}$ [GeV^4]	0.03528	-	0.03179	0.0343	0.03931
	s_+ [GeV^2]	1.27	-	1.11	1.15	1.23
constant width	$\overline{m_0}$ [MeV]	659.8	174.4	669.0	682.0	715.0
	Γ_0 [MeV]	0	149.4	149.4	149.4	157.3
	m_{peak} [MeV]	659.8	158.4	664.8	677.9	710.6
	Γ_{FWHM} [MeV]	0	139.8	148.8	148.8	156.6
	$\overline{F_0}$ [GeV^4]	0.01976	0.00005	0.01893	0.02036	0.02766
	s_+ [GeV^2]	1.03	0.36	0.93	0.96	1.11
		1	2	3	4	5

5 VOC scenario for the ρ meson

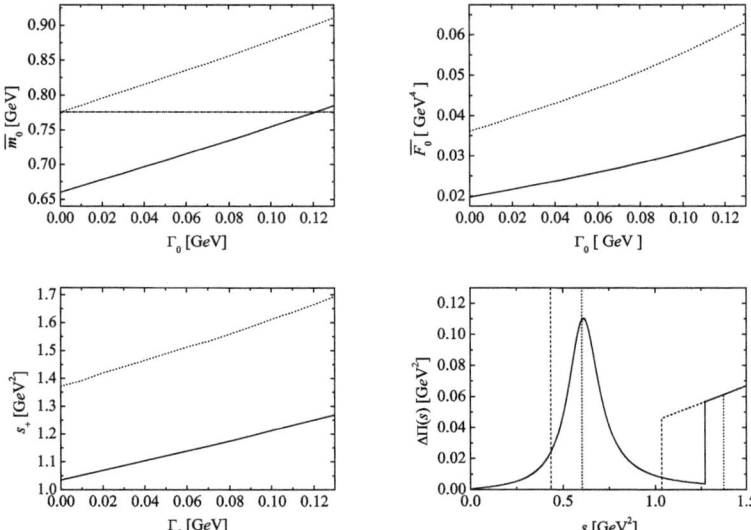

Figure 5.3.1: Averaged mass parameter $\overline{m_0}$ (upper left), strength $\overline{F_0}$ (upper right) and threshold s_+ (bottom left) as functions of the width parameter Γ_0. Solid: VOC scenario, dotted: vacuum. The spectral density $\Delta\Pi$ (bottom right) is plotted for the mass kept constant (solid), the width kept constant (dashed) and the vacuum case (dotted). $\langle \mathcal{O}_4^{\text{even}} \rangle = (267\,\text{MeV})^6$ as obtained from the moment analysis, the VOC scenario is for parametrization (i) and constant Breit–Wigner parameters. The horizontal dashed-dotted line in the mass–width panel marks the vacuum ρ mass. See column 1 of Tab. 5.3.1.

5.3 Mass shift vs. broadening

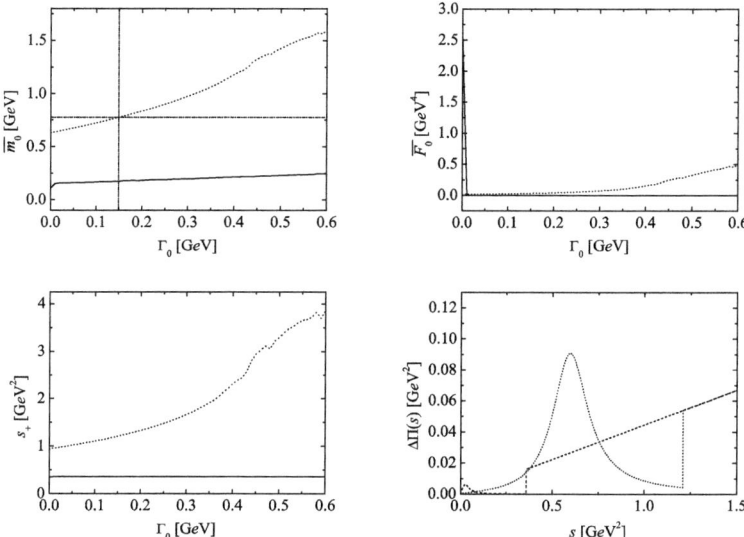

Figure 5.3.2: The same as in Fig. 5.3.1 but for $\langle \mathcal{O}_4^{\text{even}} \rangle = (172\,\text{MeV})^6$ being fitted to $\Gamma_0 = 149.4\,\text{MeV}$, $\overline{m_0} = 775.5\,\text{MeV}$ using parametrization (i). In the VOC scenario parametrization (i) has been used and the Breit-Wigner parameters have been kept constant. It is not possible to keep $\overline{m_0}$ or m_{peak} constant. See column 2 of Tab. 5.3.1.

5 VOC scenario for the ρ meson

In Fig. 5.3.1, with numbers given in column 1 of Tab. 5.3.1, we use $\langle \mathcal{O}_4^{\text{even}} \rangle = (267\,\text{MeV})^6$, obtained from the moment analysis of the previous section. The VOC scenario has been calculated using parametrization (i). Clearly, the anticipated lowering of the mass or the increasing of the width is found. However, applying a certain parametrization of the spectral density raises the question if the extraction of $\langle \mathcal{O}_4^{\text{even}} \rangle$ depends on the functional behavior of the assumed spectral density. In Fig. 5.3.2, with numbers given in column 2 of Tab. 5.3.1, $\langle \mathcal{O}_4^{\text{even}} \rangle$ has been adjusted such that the Breit-Wigner parameters $\overline{m_0}$ and Γ_0 of parametrization (i) equal the experimental values $m_\rho = 775.5\,\text{MeV}$ and $\Gamma_\rho = 149.4\,\text{MeV}$, respectively. The result for $\langle \mathcal{O}_4^{\text{even}} \rangle$ differs by a factor 14. As a consequence, it is impossible to obtain a VOC scenario, employing the same parametrization, with constant mass, neither for $\overline{m_0}$ nor for m_{peak}. For constant Γ_0 the ρ peak almost disappears and exploring the mass–width relation up to $\Gamma_0 = 0$, the sum rule cannot be fulfilled at all. The reason is the special parametrization. Lower values of $\langle \mathcal{O}_4^{\text{even}} \rangle$ correspond to more strength at lower energies and vice versa. The chosen parametrization with constant width is nonzero up to zero energy, which results in the very small value of $\langle \mathcal{O}_4^{\text{even}} \rangle$. In the VOC scenario, where $\langle \mathcal{O}_4^{\text{odd}} \rangle$ is set to zero, even more strength is required at lower energies. In this respect recall that the factorization of four-quark condensates into squares of the chiral condensate corresponds to $\langle \mathcal{O}_4^{\text{even}} \rangle = 0$ in the VOC scenario and, hence, to an even stronger enhancement of the spectral density at low energies.

Because the ρ meson spectral function must have a threshold dictated by its decay channels into pions, in [Leu98b] parametrization (ii) has been introduced. The spectral function has a two-pion threshold and an energy dependent width. The choice $\langle \mathcal{O}_4^{\text{even}} \rangle = (242\,\text{MeV})^6$ reproduces $\Gamma_0 = 149.4\,\text{MeV}$ and $\overline{m_0} = 775.5\,\text{MeV}$. The results are shown Fig. 5.3.3 with numbers given in column 3 of Tab. 5.3.1. For the VOC scenario, the parametrization (i) has been chosen. In Fig. 5.3.4 and column 4 of Tab. 5.3.1 the same parametrization has been used but instead of fitting $\langle \mathcal{O}_4^{\text{even}} \rangle$ such that the Breit-Wigner parameters $\overline{m_0}$ and Γ_0 are set to the experimental ρ values, we require that the peak position m_{peak} equals its experimental value.[15] The obtained value $\langle \mathcal{O}_4^{\text{even}} \rangle = (244\,\text{MeV})^6$ differs only slightly from the previous case. For the VOC scenario we used parametrization (i). Requiring that peak position and Γ_0 are kept constant in the VOC scenario gives slightly smaller mass shifts and broadenings as compared to fitting the Breit-Wigner parameters. All cases have in common that

---

[15] Of course, consistency demands also the fitting of the full width at half maximum Γ_{FWHM} to its experimental value, but the numerical effort to do so is excessive. On the other hand the width is not too large so that the difference between both quantities is expected to be rather small, cf. column 4 of Tab. 5.3.1, and the accuracy is sufficient for our purposes.

5.3 Mass shift vs. broadening

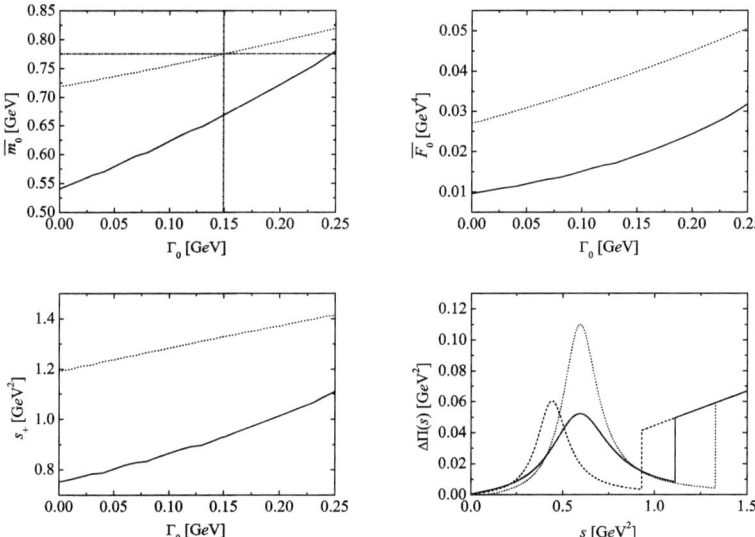

Figure 5.3.3: The same as in Fig. 5.3.1 but for $\langle \mathcal{O}_4^{\text{even}} \rangle = (242\,\text{MeV})^6$ being fitted to $\Gamma_0 = 149.4\,\text{MeV}$, $\overline{m_0} = 775.5\,\text{MeV}$ for parametrization (ii). In the VOC scenario parametrization (i) has been used and the Breit-Wigner parameters have been kept constant. See column 3 of Tab. 5.3.1.

5 VOC scenario for the ρ meson

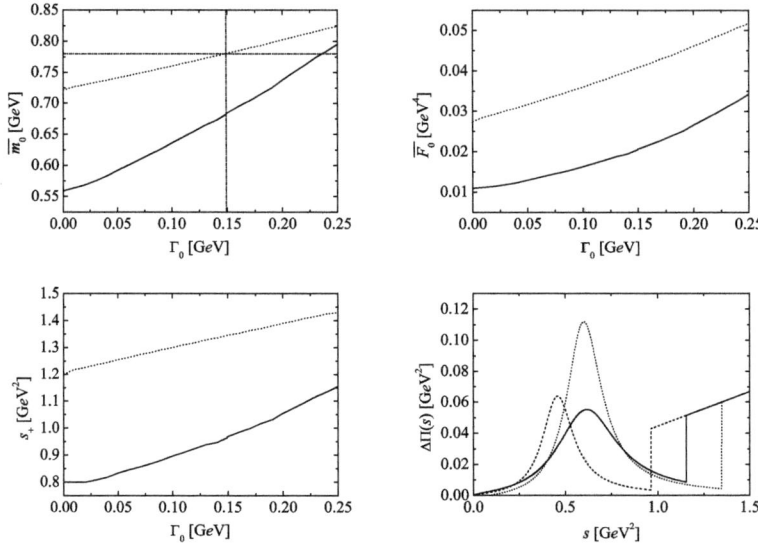

Figure 5.3.4: The same as in Fig. 5.3.1 but for $\langle \mathcal{O}_4^{\mathrm{even}} \rangle = (244\,\mathrm{MeV})^6$ being fitted to $\Gamma_0 = 149.4\,\mathrm{MeV}$, $m_{\mathrm{peak}} = 775.5\,\mathrm{MeV}$ for parametrization (ii). In the VOC scenario parametrization (i) has been used, whereas m_{peak} and Γ_0 have been kept constant. See column 4 of Tab. 5.3.1.

they require an enhancement of the spectral strength at lower energies which can be accomplished by a shift of the peak position or a broadening or a combination of both.

Extracting $\langle \mathcal{O}_4^{\mathrm{even}} \rangle$ from experimental data

As can be seen from the previous investigation, the extraction of $\langle \mathcal{O}_4^{\mathrm{even}} \rangle$ from the ρ meson sum rule depends on details of the parametrization. In particular the low energy behavior of the parametrization is of importance. Therefore a comparison to experimental data is necessary. In Fig. 5.3.5 the sum rule results for the vacuum cases are compared to the ρ spectral function obtained from the cross section $\tau \to \nu + n\pi$ for even n measured by the ALEPH detector [Sch05] and two fits which correspond to the above mentioned parametrizations (i) and (ii). These data do not suffer $\rho - \omega$ mixing effects because of charge conservation. The coincidence of the sum rule results with the experimental data is not very good. As can be seen, although the low energy

5.3 Mass shift vs. broadening

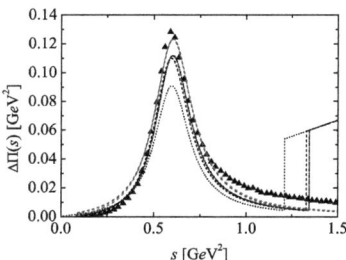

Figure 5.3.5: Comparison of experimental data for the ρ meson spectral function obtained from $\tau \to \nu + n\pi$ decays with the spectral functions obtained from a finite width QSR analysis and best fits for parametrizations (*i*) and (*ii*). Triangles: data from [Sch05]. Dashed blue: fit of parametrization (*i*); the obtained values are $m_0 = 780.7\,\text{MeV}$, $\Gamma_0 = 160.2\,\text{MeV}$ and $F_0 = 0.04834\,\text{GeV}^4$. Dashed red: fit of parametrization (*ii*); the obtained values are $m_0 = 780.9\,\text{MeV}$, $\Gamma_0 = 162.4\,\text{MeV}$ and $F_0 = 0.04884\,\text{GeV}^4$. Dashed black: QSR results for $\langle \mathcal{O}_4^{\text{even}} \rangle = (172\,\text{MeV})^6$ for parametrization (*i*) (see Fig. 5.3.2). Dotted black: QSR results for $\langle \mathcal{O}_4^{\text{even}} \rangle = (242\,\text{MeV})^6$ for parametrization (*ii*) and fitted $\overline{m_0}$ and Γ_0 (see Fig. 5.3.3). Dashed-dotted black: QSR results for $\langle \mathcal{O}_4^{\text{even}} \rangle = (244\,\text{MeV})^6$ for parametrization (*ii*) and fitted m_{peak} and Γ_0 (see Fig. 5.3.4).

tail is reproduced fairly well, the high energy tail is not. Furthermore, all three finite width vacuum scenarios have a too low $\overline{F_0}$ in common. Indeed, also earlier sum rule analyses of the ρ meson, e. g. [Leu98b], give a too low value for F_0 when the spectral density is compared to these data. In [Leu98b] an experimental value of $F_0 = 0.01\pi\,\text{GeV}^4 \approx 0.0314\,\text{GeV}^4$ is quoted. The vacuum sum rule results in [Leu98b] suggest $F_0 \leq 0.008\pi\,\text{GeV}$. A Breit-Wigner curve fitted to the experimental data for these cases does not coincide with the sum rule results and gives better agreement with the data. Nevertheless, the low and high energy tails still significantly differ from the data. In all fits a value of $F_0 \approx 0.048\,\text{GeV}^4$ is extracted, which is larger than the quoted value in [Leu98b]. Two conclusions may be drawn. First, the sum rule analysis using (*i*) or (*ii*) systematically give too low values for F_0. This is independent of whether standard values for the chirally symmetric four-quark condensate are used or if it is adjusted to reproduce the experimental mass and width of the ρ meson. Second, standard parametrizations of the spectral function, i. e. (*i*) or (*ii*), do not fit satisfactorily the low and high energy tails. Unfortunately, because of the exponential weighting of the integrand in Eq. (5.2.1a), Borel transformed sum rules are in particular sensitive to the low energy tail, which makes it necessary to model this part very carefully, especially when condensates are to be determined.

5 VOC scenario for the ρ meson

We adjust $\langle \mathcal{O}_4^{\text{even}} \rangle$ now in a two-step procedure. First, the experimental data for $\tau \to \nu + n\pi$ for even n [Sch05] given in Fig. 5.3.5 is fitted with Eq. (5.3.1) by $a = 0.84483$, $b = -0.33013$, $c = 6.72793$ and $k = 2$ (two-pion threshold). Consistency demands to use the charged pion mass $m_\pi = 139.6$ MeV. The thus obtained parameters are $m_0 = 774.6$ MeV, $\Gamma_0 = 157.3$ MeV and $F_0 = (467 \text{ MeV})^4 \approx 0.048 \text{ GeV}^4$. Note that with these parameter values the parametrization (5.3.1b) has no deeper physical motivation (in contrast to, e.g., an s-wave or p-wave parametrization). The purpose is solely to obtain an adequate description of the data. The quality of this description can be seen in Fig. 5.3.6. In a second step, the Borel analysis is performed as described above. The requirement $\overline{m_0}(\Gamma_0 = 157.3 \text{ MeV}) = m_0$, with a, b and c as obtained from the fit, fixes $\langle \mathcal{O}_4^{\text{even}} \rangle$. That is, the experimental spectral distribution is here used explicitly, while the above momentum analysis refers only to the spectral moment $\overline{m}$, i.e., the mean of the l.h.s. of Eq. (5.2.1a) is identified with the ρ peak mass. Any other parametrization, which fits the experimental data, may be used, e.g. a spline. We obtain $\langle \mathcal{O}_4^{\text{even}} \rangle = (261 \text{ MeV})^6$, which is somewhat smaller than the value obtained in the moment analysis but much closer to it than the above obtained values for the finite width scenarios. This points to the model dependence of extracting QCD condensates from hadron properties in such a manner. However, from the preceding investigations we conclude that equating the moment (5.2.1a) to the ρ vacuum mass gives the best approximation to $\langle \mathcal{O}_4^{\text{even}} \rangle$ compared to results obtained from using parametrizations which do not fit the experimental data. We obtain $\overline{F}_0 = (464 \text{ MeV})^4 = 0.046 \text{ GeV}^6$, which fits the experimental value. Note that this is a nontrivial consensus. The thus obtained spectral function, exhibited in Fig. 5.3.6, is in agreement with the experimental data and refers to a specific set of condensates and further parameters entering Eqs. (5.1.4) and (5.1.5) of the OPE from which Eq. (5.2.1b) is built up. This vacuum analysis evidences that one needs a prescribed parametrization of the spectral shape when attempting to quantify medium modifications. (The maximum entropy method, cf. [Asa01, Gub10], might overcome this restriction.)

We have chosen the option (iii) i.e. $a = b = 0$, $c = 0.5$ (s-wave) and $k = 1$ (one-pion threshold[16]) for studying implications of the VOC scenario. Setting $\langle \mathcal{O}_4^{\text{odd}} \rangle = \langle \bar{q}q \rangle = 0$ and repeating the Borel analysis w.r.t. m_0 for given values of Γ_0 one gets the correlation $\overline{m_0}(\Gamma_0)$. Instead of $\overline{m_0}(\Gamma_0)$ we show however the relation $m_{\text{peak}}(\Gamma_{\text{FWHM}})$ with the reasoning given below Eq. (5.3.2).

For the VOC scenario, the curve $m_{\text{peak}}(\Gamma_{\text{FWHM}})$ in Fig. 5.3.7 is significantly shifted

---

[16]This is the threshold for a pion and a nucleon going to a dilepton and nucleon [Leu98b].

5.3 Mass shift vs. broadening

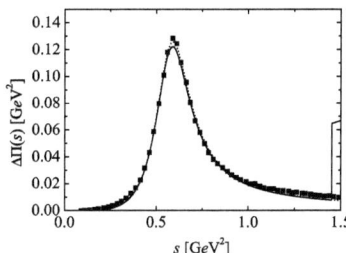

Figure 5.3.6: ρ meson spectral function from $\tau \to \nu + n\pi$ decays with even n (squares, data from [Sch05]) and a fit with Eqs. (5.3.1a) and (5.3.1b) (dotted). The parameters are $a = 0.84483$, $b = -0.33013$, $c = 6.72793$, $m_0 = 774.6\,\text{MeV}$, $\Gamma_0 = 157.3\,\text{MeV}$ and $F_0 = (467\,\text{MeV})^4$. The vacuum sum rule analysis results in the solid curve.

away from the vacuum physical point $(m_{\text{peak}}, \Gamma_{\text{FWHM}}) = (767.1\,\text{MeV}, 153.0\,\text{MeV})$. If one assumed that chiral restoration in the present spirit does not cause an additional broadening, one would recover the previously often anticipated "mass drop".

Figure 5.3.7 evidences, however, that an opposite interpretation is conceivable as well, namely pure broadening with keeping the vacuum value of m_{peak}. The NA60 [Arn06, Ada08] and CLAS [Dja08] data seem indeed to favor such a broadening effect. In fact, assuming that m_{peak} does not change by chiral restoration in the VOC scenario, the width is increased to $\Gamma_0 = 280\,\text{MeV}$ ($\Gamma_{\text{FWHM}} = 277\,\text{MeV}$), see Fig. 5.3.7. In this respect the broadening of a spectral function can be an indication for chiral restoration.

Figure 5.3.8 exhibits the spectral function $\Delta\Pi(s)$ as a function of s for the two extreme options above. The solid curve depicts the broadening when keeping the peak at the vacuum position. The dashed curve is for the option of keeping the full width at half maximum at its vacuum value. From the perspective of the employed QSR, both options are equivalent, as any other point on the curve $m_{\text{peak}}(\Gamma_{\text{FWHM}})$ in Fig. 5.3.7. The overall outcome seems to be that the Borel transformed QSR requires more strength of the spectral function at lower energies.

Relation to experimental data

As pointed out above, our investigation strictly separates between in-medium effects caused by the change of chirally odd condensates on the one hand and of chirally symmetric condensates on the other hand. Only the former are considered in our

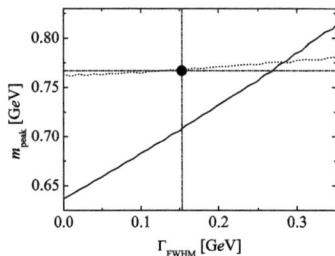
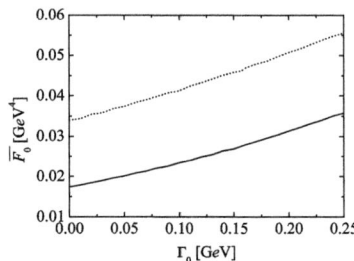

Figure 5.3.7: Peak position m_{peak} (left) as a function of the width parameter Γ_{FWHM} and the strength $\overline{F_0}$ (right) as a function of the Breit-Wigner parameter Γ_0 for $\langle \mathcal{O}_4^{\text{even}} \rangle = (261\,\text{MeV})^6$. The vacuum case is evaluated using the parametrization obtained from the fit, the VOC scenario is evaluated for parametrization (iii). Dotted lines mark the experimental values m_{peak} and Γ_{FWHM}, determining the physical vacuum point (heavy dot). Line code as in Fig. 5.3.1.

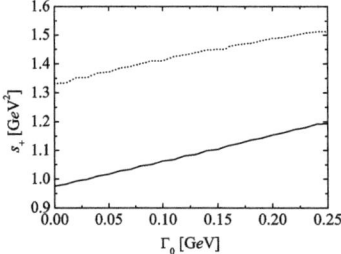
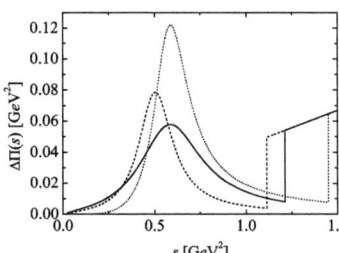

Figure 5.3.8: Continuum threshold $s_+(\Gamma_0)$ (left) and the spectral density $\text{Im}\Pi(s)$ (right) in the vacuum case (dotted curve) and in the VOC scenario for $\Gamma_{\text{FWHM}} = \text{const}$ (dashed curve) or $m_{\text{peak}} = \text{const}$ (solid curve). Line codes as in Fig. 5.3.1.

considerations, the VOC scenario. Of course, physically it is plausible that there are also in-medium effects which have no direct connection to chiral restoration. In turn this means that a comparison of our VOC scenario with experimental data can only be qualitative. In addition, experimental information about the current-current correlator emerges from the dilepton production in heavy-ion collisions. Here, in general we do not have a fully equilibrated, static and infinitely extended system but rather the time evolution of a cooling and expanding fireball of finite size leading to a time dependent radiation of dileptons from the bulk and from the surface. To simplify the anyway qualitative comparison we adopt the following strategy: In principle we would like to compare to the high-precision data of the NA60 collaboration [Dja08, Arn09]. In view of the mentioned complications we compare to a model based on equilibrium many-body theory which describes the NA60 data very well. For that purpose we use the results of van Hees and Rapp [Hee06, Hee08].

Since our VOC scenario describes the point of chiral restoration (all chirally odd condensates put to zero) we compare to the model for the assumed transition temperature to the quark-gluon plasma, namely 175 MeV. We compare the width obtained in the VOC scenario to the width obtained in the approach of [Hee06, Hee08]. The ρ meson propagator is related to the spectral function in the vector meson dominance model [Rap00] by

$$-\text{Im} D_\rho = \frac{g^2}{\left(m_\rho^{(0)}\right)^2} \text{Im} \Pi, \qquad (5.3.3)$$

with the bare ρ mass $m_\rho^{(0)} = 853$ MeV and the $\rho \to \pi\pi$ coupling $g_{\rho\pi\pi} = 5.9$. It is compared to the results obtained in the VOC scenario. As evidenced in Fig. 5.3.7 the sum rule does not allow to fix separately both the peak position and the width of the spectral information of the current-current correlator. It only correlates the two quantities. If we adjust the peak position to the one from [Hee06, Hee08], $m_{\text{peak}} = 830$ MeV, we find $\Gamma_{\text{FWHM}} = 380$ MeV in our VOC scenario which is smaller as compared to the width of the many-body model from [Hee06, Hee08] $\Gamma \approx 500$ MeV [Rap09]. We interpret this as additional chirally symmetric in-medium effects on top of the effects which we can pin down, i. e. the ones related to the chirally odd condensates and their in-medium change towards chiral restoration. As already pointed out we deem it quite natural to have these additional chirally symmetric in-medium effects. We note in passing that our simple one-peak parametrization (5.3.1a) cannot cover the subtle resonance-hole in-medium effects [Fri97, Hel95, Rap97, Pet98c, Pos04, Lut02]

5.4 Notes on ω and axial-vector mesons

The current $\bar{u}\gamma^\mu u + \bar{d}\gamma^\mu d$ has the quantum numbers of the ω meson and is a chiral singlet w. r. t. $SU_R(2) \times SU_L(2)$ chiral transformations. Consequently, the ω meson does not have a chiral partner in a world with two light flavors. Concerning three light flavors the current given above is a superposition of a member of the flavor octet and the singlet. The octet members do have chiral partners and one may assign a proper linear combination of the two f_1 [Ams08] mesons as the chiral partner of the respective linear combination of ω and ϕ.

All considerations made in the following concern two light flavors. The OPE side of the ω meson including terms up to mass dimension 6 contains only chirally symmetric terms [Shi79b, Shi79a, Hat93] – except for the two-quark condensate term $\propto m_q \langle \bar{q}q \rangle$. This term, however, breaks chiral symmetry explicitly by the quark mass and dynamically by the quark condensate. Considerations about symmetry transformations, on the other hand, concern the case where chiral symmetry is exact, i. e. without explicit breaking. Therefore the appearance of the term $\propto m_q \langle \bar{q}q \rangle$ is not in contradiction to the statement that the ω is a chiral singlet. Without explicit calculations it is clear that, in the VOC scenario, the ω meson does not change much of its mass since only the numerically very small term $\propto m_q \langle \bar{q}q \rangle$ is dropped while the chirally symmetric four-quark condensates do not change.

The current (1.1.3b) with the quantum numbers of the a_1 meson yields the same chirally symmetric OPE parts as the ρ meson. The chirally odd parts are of course different, they are the negative of the ones which appear in the OPE for the ρ meson [Shi79b, Shi79a, Hat93]. In addition, the hadronic side of the sum rule contains not only the a_1, but also the pion. The latter contribution is $\propto f_\pi^2$. Both effects, different chirally odd condensates and the appearance of the pion, lead to the fact that the sum rule method yields a mass for the a_1 which is significantly different from the ρ meson mass [Shi79b, Shi79a] – as it should be. In the VOC scenario the chirally odd condensates are put to zero. In addition, the pion-decay constant which is an order parameter of chiral symmetry breaking [Mei02] also vanishes. Then the sum rules for ρ and a_1 are the same. As expected the chiral partners become degenerate in the VOC scenario.

5.5 Interim summary

To summarize this chapter, we have successfully extracted the symmetric part of the four-quark condensate which enters the ρ meson QSR. In order to do so, we have directly used a spectral function measured in the high precision experiment ALEPH [Sch05] instead of relying on certain parametrizations and the corresponding parameters. In particular we found that equating the model independent moment $\tilde{m}$ to the ρ vacuum mass gives results which are in agreement to those obtained from a measured spectral function. In contrast, the results obtained by employing common Breit-Wigner parametrizations disagree to the result obtained from a measured spectral function. This is because these parametrizations fail to describe the low-energy domain, which is exponentially weighted within Borel transformed QSR. Along these lines we found that quite generally decreasing four-quark condensates correspond to an enhancement of the spectral density at low energies and vice versa. Moreover, within a QSR analysis common Breit-Wigner parametrizations fail to reproduce the strength as given by the measured spectral density. The non-trivial reproduction of the strength by the employed method in conjunction with the extracted value for the symmetric four-quark condensate may be considered as confirmation of both.

Two extreme and antagonistic statements concerning hadron masses and hadronic medium modifications could be raised:

(a) Basically all hadron masses are caused by chiral symmetry breaking. Consequently in a dense and/or hot strongly interacting medium the masses of hadrons vanish at the point of chiral restoration – apart from some small remainder which is due to the explicit breaking of chiral symmetry by the finite quark masses.

(b) The observed in-medium changes can be explained by standard hadronic many-body approaches and have no direct relation to chiral restoration.

Our findings do not support either of these extreme statements. If one sets the explicit chiral symmetry breaking condensates to zero in the QSR for the ρ meson one sees a significant change of the spectral mass moment. It neither vanishes (as statement (a) would suggest) nor does it stay unchanged (statement (b)).

In the VOC scenario we have kept the chirally invariant condensates at their vacuum values. Though, an adequate restoration mechanism might also change these condensates, the VOC scenario allows a discussion of chiral symmetry restoration which is not interfered by additional medium effects. This is clearly unrealistic for a

5 VOC scenario for the ρ meson

true in-medium situation. In particular, the contributions coming from the non-scalar twist-two operators are found to be sizable, e. g., for cold nuclear matter [Hat92]. Nonetheless, our study indicates that the connection between the vacuum spectral properties and chiral symmetry breaking or between in-medium altered spectral properties and chiral restoration is not direct. In principle, one could imagine conspiracies between chiral symmetry breaking and non-breaking condensates such that one of the extreme statements raised above becomes true. Such a conspiracy would be driven by the underlying microscopic mechanisms which cause spontaneous chiral symmetry breaking and/or its restoration. Clearly one needs a deeper understanding of these microscopic mechanisms.

We note that we talk about chiral restoration here as discarding chirally odd condensates from the QSR and not about any other in-medium effect. This is the strength of the VOC scenario. Of course, the statement that the broadening of a spectral function is an in-medium effect is a trivial statement. In contrast, a possible connection between broadening and chiral restoration is non-trivial. The present work demonstrates that such a connection exists.

6 Introduction to Dyson-Schwinger and Bethe-Salpeter equations

We now turn our attention to the DSEs and BSEs, which provide another non-perturbative tool to investigate quark-antiquark bound states.

6.1 Dyson-Schwinger equations

Despite of the fundamental nature of DSEs in quantum field theory, there are not many textbooks which deal with them. The derivation, which is outlined in the following, is based on the very extensive and didactic treatise of [Rom69], in which the DSE is derived for the pion-nucleon system. A tighter derivation which is closer to presentations found in applications of DSEs to the quark-gluon system can be found in [Itz80]. The major difference is that [Itz80] employs the path-integral formalism, whereas the derivation of [Rom69] directly relies on ground state expectation values of field operators in the Heisenberg picture and avails methods and techniques developed in the scope of canonical quantization. Both approaches are, however, equivalent and based on the same ideas. For the sake of a consistent presentation of the concepts and derivation of relations in the course of this thesis, the path integral formulation will not be used.

6.1.1 Derivation

We consider time-ordered ground state expectation values of unrenormalized bare field operators in the Heisenberg picture

$$S(x,y) \equiv -i \langle \text{T} \left[\psi(x) \bar{\psi}(y) \right] \rangle , \qquad (6.1.1\text{a})$$

$$D_{\mu\nu}^{AB}(x,y) \equiv -i \langle \text{T} \left[A_\mu^A(x) A_\nu^B(y) \right] \rangle , \qquad (6.1.1\text{b})$$

6 Introduction to Dyson-Schwinger and Bethe-Salpeter equations

which represent quark and gluon propagators. The derivation of DSEs is based on Eq. (2.0.8) (relating ground state expectation values of field operators in the Heisenberg picture to matrix elements in the interaction picture) and

$$\mathcal{O}'(x)\mathscr{S} = i \int d^4y \, T\left[\mathcal{O}'(x)\mathscr{L}'_{\text{int}}(y)\mathscr{S}\right], \tag{6.1.2}$$

where $\mathcal{O}'$ is an arbitrary interaction picture field operator, $\mathscr{L}_{\text{int}}$ is the interaction Lagrangian and

$$\mathscr{S} = \text{P}\exp\left\{i \int \mathscr{L}'_{\text{int}}(x) \, d^4x\right\} \tag{6.1.3}$$

is the S matrix operator in the interaction picture. Dyson's chronological P ordering is defined the same way as the T ordering but without a sign for fermion operators. In what follows a primed operator denotes an operator in the interaction picture.

Dyson's trick is to introduce external classical source terms which are to be set to zero at the end and which may interact with the corresponding fields. For that purpose we define the interaction Lagrangian as

$$\tilde{\mathscr{L}}_{\text{int}} = \mathscr{L}_{\text{int}} - J_A^\mu A_\mu^A, \tag{6.1.4}$$

where J denotes the non-quantized external gluon source. With the Lagrangian (6.1.4) the S-matrix (6.1.3) fulfills

$$\frac{\delta \mathscr{S}}{\delta J_A^\mu(z)} = -i\text{T}\left[A_\mu^{\prime A}(z)\mathscr{S}\right]. \tag{6.1.5}$$

These are the main ingredients which are needed to derive the coupled DSEs for quark and gluon propagators and the quark gluon vertex. Indeed, Eq. (6.1.5) in conjunction with the fundamental relation (6.1.2), can be used to show that the functional derivative of an n-point function w. r. t. to an external source in the limit of vanishing source is given by an $(n+1)$-point function. Consequently, quark and gluon propagators, as given in Eq. (6.1.1), may be obtained by the functional derivative of a one-point function w. r. t. an external quark or gluon source, respectively:

$$S(x,y) = i \left.\frac{\delta \Psi(x)}{\delta \eta(y)}\right|_{\eta \to 0}, \tag{6.1.6a}$$

6.1 Dyson-Schwinger equations

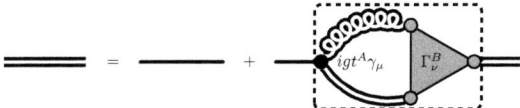

Figure 6.1.1: The quark DSE. Double lines denote exact propagators, single lines free propagators. The curly line is the gluon propagator and the gray triangle stands for the quark-gluon vertex function. The object within the dashed box is the quark self-energy.

$$D^{AB}_{\mu\nu}(x,y) = i \left. \frac{\delta \Phi^A_\mu(x)}{\delta J^\nu_B(y)} \right|_{J \to 0} . \tag{6.1.6b}$$

The external quark field is denoted by η, and

$$\Psi(x) \equiv \frac{1}{\langle 0|\mathscr{S}|0\rangle} \frac{\delta \langle 0|\mathscr{S}|0\rangle}{\delta \eta(x)} = -i\langle \psi(x) \rangle, \tag{6.1.7a}$$

$$\Phi^A_\mu(x) \equiv \frac{1}{\langle 0|\mathscr{S}|0\rangle} \frac{\delta \langle 0|\mathscr{S}|0\rangle}{\delta J^\mu_A(x)} = -i\langle A^A_\mu(x) \rangle \tag{6.1.7b}$$

are the respective one-point functions defined as ground state expectation values of Heisenberg field operators. Conversely, Eq. (6.1.6) may be considered as the definition of quark and gluon propagators. In the absence of external sources and if the field carries some symmetry property of the ground state the one-point function is zero. For nonzero external sources, in- and out-states are in general different and one-point functions are not necessarily zero. Furthermore, they may be nonzero for broken symmetries.

Applying Wick's theorem to the quark propagator (6.1.1a) (in order to do so, one has to apply Eq. (2.0.8) and, hence, to migrate to the interaction picture) and using the EoM for the free quark Green's function leads to a differential–functional-differential equation for the exact quark propagator:

$$\left[-i\hat{\partial} + ig\frac{\delta}{\delta J^\mu(x)} + ig\Phi_\mu(x) + m_0\right] S(x,y) = \delta^{(4)}(x-y). \tag{6.1.8}$$

Processing $\Phi^A_\mu(x)$ in a similar manner gives a differential equation which is coupled to the quark propagator and from which an integro-differential equation for the gluon propagator is obtained immediately by a functional derivative w.r.t. to the gluon source. The vertex function is defined as the three point function

$$\Gamma^A_\mu(x,y,z) \equiv -\frac{1}{g}\frac{\delta S^{-1}(x,y)}{\delta \Phi^\mu_A(z)}, \qquad (6.1.9)$$

which allows to derive a non-linear integral equation for the quark propagator

$$S(x,y) = S^{(0)}(x-y) + \int d^4x' \, S^{(0)}(x-x')\Sigma(x',x'')S(x'',y)\,d^4x'', \qquad (6.1.10a)$$

depicted in Fig. 6.1.1 and with the quark self-energy defined as

$$\Sigma(x',x'') = -ig^2 t^A \gamma^\mu \int d^4z\,d^4x'\, S(x,x')\Gamma^{B,\nu}(x',x'',z) D^{AB}_{\mu\nu}(z,x). \qquad (6.1.10b)$$

The thus obtained equation is a non-linear Fredholm integral equation of the second kind. Moreover, due to the infinite integration domain Eq. (6.1.10) is a singular integral equation. Unfortunately, the literature about the theory of non-linear integral equations is fairly scarce and the situation is even worse in case of singular equations. Essays about the solvability of non-singular non-linear integral equations can be found in e. g. [Pog66]. However, even for these cases only sufficient requirements for the solvability are given.

6.1.2 Renormalization

Introducing the Fourier transforms of $S^{(0)}(x-y)$, $S(x,y)$ and $\Sigma(x,y)$, Eq. (6.1.10) can be cast into

$$S(p) = S^{(0)}(p) + S^{(0)}(p)\Sigma(p)S(p) \qquad (6.1.11)$$

with the solution

$$S^{-1}(p) = m_0 - \hat{p} - \Sigma(p). \qquad (6.1.12)$$

Therefore, Σ is also called the mass shift operator and identifying $m_0 - \Sigma(\hat{p}_{\alpha\beta}) = m\delta_{\alpha\beta}) = m$ serves as a natural definition of the physical mass. Defining the mass shift of the renormalized Lagrangian as $\delta m \equiv m_0 - m$ and expanding the quark self-energy about the mass-shell, cancels δm and gives the dressed quark propagator in terms of

6.1 Dyson-Schwinger equations

Figure 6.1.2: The quark DSE in rainbow approximation.

the physical mass m as

$$S(p) = Z_2 \left[m - \hat{p} - Z_2 \Sigma_r \right]^{-1}, \qquad (6.1.13)$$

where $Z_2 \equiv [1 + \partial \Sigma / \partial \hat{p}|_{\hat{p}=m}]^{-1}$ and $\Sigma_r \equiv \Sigma - \Sigma(\hat{p} = m) - \partial \Sigma / \partial \hat{p}|_{\hat{p}=m} (\hat{p} - m)$ have been defined. Within multiplicative renormalization the renormalized propagator is defined as $S(p) \equiv Z_2 S_R(p)$, which can be shown to be finite and well defined in any order of perturbation theory. Consequently, the renormalized quark wave function is defined as $\psi \equiv Z_2^{1/2} \psi_R$.

Analog relations hold for the gluon propagator and quark-gluon vertex. Furthermore, also ghost terms have to be considered for the quantized Lagrangian of QCD. The gluon DSE couples also to the ghost propagator, which fulfills its own DSE (see App. A.3). However, neither ghost nor gluon propagator DSEs, nor the DSEs for the vertex functions will be considered within this thesis.

6.1.3 Rainbow approximation

Because $S(x, y)$ is defined as integral kernel, the kernel $S^{-1}(x, y)$ is implicitly defined such that it is the inverse operator w.r.t. to integration (and matrix multiplication). Thus, an explicit equation for the operator $S^{-1}(x, y)$ in the presence of an external source can be found by deriving Eq. (6.1.10) following the same steps but without the limit of a vanishing source and using the defining property of the inverse integral kernel for $S(x,y)$. From Eq. (6.1.9) it follows that the vertex function is directly related to the quark self energy

$$\Gamma_\mu^A(x, y, z) = \gamma_\mu t^A \delta^{(4)}(x - y) \delta^{(4)}(x - z) + \frac{1}{g} \frac{\delta \Sigma(x, u)}{\delta \Phi_A^\mu(z)}. \qquad (6.1.14)$$

The well-known rainbow approximation of Eq. (6.1.10) is to neglect the second term in Eq. (6.1.14), retaining only the lowest order contribution to the quark gluon vertex. The resulting equation is depicted in Fig. 6.1.2. Performing an iteration gives Fig. 6.1.3 and illustrates the origin of the notion. A range of diagrams which contribute to the quark propagator in rainbow approximation are shown in Fig. 6.1.4.

6 Introduction to Dyson-Schwinger and Bethe-Salpeter equations

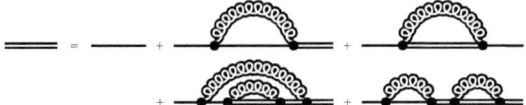

Figure 6.1.3: The first iteration of the quark DSE in rainbow approximation.

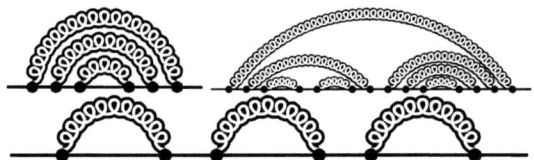

Figure 6.1.4: A selection of diagrams for radiative corrections of the quark propagator which contribute to the solution of the DSE in rainbow approximation.

6.2 Bethe-Salpeter equations

Apart from the treatment of the previous section, DSEs may also be derived by virtue of arguments which rely on the Feynman diagram technique and the notion of reducible and irreducible diagrams. To exemplify this, the two-particle BSE is now derived in this way.

6.2.1 Derivation

The BSE is a 16 dimensional integral equation for the BSA. Derivations can be found in many textbooks about quantum field theory, e. g. [Lur68, Rom69, Gre92, Sch61]. Let us start with an analysis of the two-particle propagator defined as the four-point function

$$K_{\alpha\beta\mu\nu}(x_1, x_2; y_1, y_2) \equiv -\langle\Omega|T\left[\psi^A_\alpha(x_1)\psi^B_\beta(x_2)\bar{\psi}^A_\mu(y_1)\bar{\psi}^B_\nu(y_2)\right]|\Omega\rangle, \qquad (6.2.1)$$

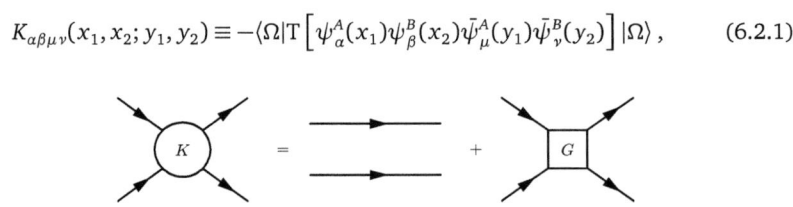

Figure 6.2.1: The exact four-point function is split into a part without interaction and a part which contains interaction between the two particles. All lines represent exact propagators.

6.2 Bethe-Salpeter equations

where the sign is just convention. Different particles are denoted by A and B respectively and Greek letters denote the spinor indices. Close to the QSR part of this thesis we are working with Heisenberg field operators. Following [GM51] a perturbative expansion for the two-particle propagator is given by Eq. (2.0.8). The diagrammatic representation is shown in Fig. 6.2.1. The diagram, which is labeled by G contains all diagrams where an interaction between the particles takes place. Some examples of this infinite sum are depicted in Fig. 6.2.2. We also note one diagram which describes a rather complicated process and includes self-interactions. The corresponding equation reads

$$K_{\alpha\beta\mu\nu}(x_1,x_2;y_1,y_2) = iS^A_{\alpha\mu}(x_1,y_1)iS^B_{\beta\nu}(x_2,y_2)$$
$$+ \int d^4y_3\,d^4y_4\,d^4y_5\,d^4y_6\,iS^A_{\alpha\alpha'}(x_1,y_3)iS^B_{\beta\beta'}(x_2,y_4)$$
$$\times G_{\alpha'\beta'\mu'\nu'}(y_3,y_4;y_5,y_6)iS^A_{\mu'\nu'}(y_5,y_6)iS^B_{\mu\nu}(y_1,y_2). \quad (6.2.2)$$

The important observation now is that the diagrams occurring at the r. h. s. of the equation given in Fig. 6.2.2, i. e. in G, can be categorized according to whether they are reducible or not. A reducible diagram is defined as a graph which can be separated into two diagrams by cutting two internal fermion lines, where each diagram has got two incoming and two outgoing fermion lines (see Fig. 6.2.3). Hence, each reducible diagram can be expressed by a combination of appropriate irreducible ones. If we define the sum of all irreducible diagrams as $\overline{G}$, we are able to express the exact four-point function K by the infinite sum of all irreducible diagrams only. One obtains the diagrammatical equation depicted in Fig. 6.2.4. The integral equation which corresponds to Fig. 6.2.4 is the DSE for K reading

$$K_{\alpha\beta\mu\nu}(x_1,x_2;y_1,y_2) = -S^A_{\alpha\mu}(x_1,y_1)S^B_{\beta\nu}(x_2,y_2)$$
$$- \int d^4y_3\,d^4y_4\,d^4y_5\,d^4y_6\,S^A_{\alpha\alpha'}(x_1,y_3)S^B_{\beta\beta'}(x_2,y_4)$$
$$\times \overline{G}_{\alpha'\beta'\mu'\nu'}(y_3,y_4;y_5,y_6)K_{\mu'\nu'\mu\nu}(y_5,y_6;y_1,y_2). \quad (6.2.3)$$

In coordinate space, it is an inhomogeneous 16 dimensional integral equation. In order to verify that this equation indeed reproduces the equation depicted in Figs. 6.2.1 and 6.2.2 one has to investigate the iterated solution of Eq. (6.2.3) or Fig. 6.2.4, respectively, which is depicted in Fig. 6.2.5. Clearly, every irreducible interaction is included in $\overline{G}$, while every reducible interaction is reproduced by an appropriate

6 Introduction to Dyson-Schwinger and Bethe-Salpeter equations

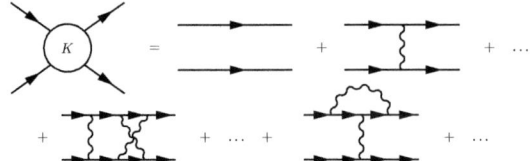

Figure 6.2.2: Perturbative expansion of the two-particle propagator (6.2.1). All the lines stand for free propagators.

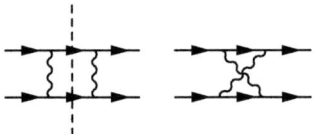

Figure 6.2.3: Two-particle irreducibility. Left panel: reducible diagram. Right panel: irreducible diagram.

combination of irreducible diagrams appearing somewhere in the second line of Fig. 6.2.5. The integration kernel $\overline{G}$ is also subject to the respective DSEs and apart from this rather artificially looking diagrammatic construction, the integral equation (6.2.3) can also be rigorously derived within the framework reviewed in Sec. 6.1.

The BSA and the adjoint BSA are defined as

$$\chi_{K,\alpha\beta}(x_1, x_2) \equiv \langle \Omega | T \left[\psi_\alpha^A(x_1) \psi_\beta^B(x_2) \right] | K \rangle \tag{6.2.4a}$$

$$\tilde{\chi}_{K,\mu\nu}(y_1, y_2) \equiv \langle K | T \left[\bar{\psi}_\mu^A(y_1) \bar{\psi}_\nu^B(y_2) \right] | \Omega \rangle, \tag{6.2.4b}$$

where $|K\rangle$ is an arbitrary physical eigenstate of the momentum operator carrying the momentum K. It can be a scattering stateas well as a bound state.Both quantities have 16 components in Dirac space. Often, they are defined as matrices by transposing the second spinor, but as long as indices are used there is no need to care about this. Unfortunately, the connection between them is non-trivial and not as simple as one might think at the first sight. Indeed, by adjoining the BSA one gets the anti time-ordered product of the adjoint spinors. Accordingly, the connection is not simply given by adjoining and multiplying Eq. (6.2.4) by γ_0 from both sides. Instead, an analytic relation between both quantities is given in [Man55].

With these definitions a complete set of physical eigenstates of the momentum operator can be inserted into Eq. (6.2.1) and the homogeneous BSE for the BSA and

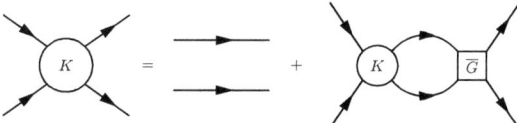

Figure 6.2.4: The DSE for the exact four-point function K in terms of irreducible interactions encoded in the infinite sum $\overline{G}$. All lines stand for the exact propagators.

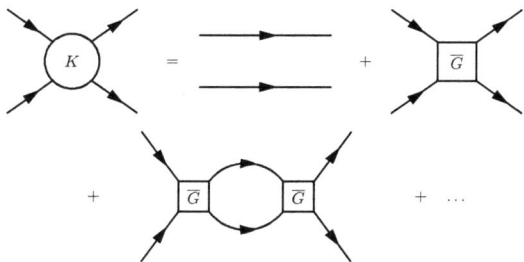

Figure 6.2.5: Iterated solution of Eq. (6.2.3) and Fig. 6.2.4.

the adjoint BSA can be derived from Eq. (6.2.3) as

$$\chi_{K,\alpha\beta}(x_1,x_2) = -\int d^4y_3\, d^4y_4\, d^4y_5\, d^4y_6\, S^A_{\alpha\alpha'}(x_1,y_1)S^B_{\beta\beta'}(x_2,y_2)$$
$$\times \overline{G}_{\alpha'\beta'\mu'\nu'}(y_1,y_2;y_3,y_4)\chi_{K,\mu'\nu'}(y_3,y_4)\,. \quad (6.2.5)$$

Since the inhomogeneity in Eq. (6.2.3) corresponds to cases where the two particles stop interacting at some time and the particles become free, i. e. to scattering states. To restrict ourselves to bound states only, we have to neglect the first term in Eq. (6.2.3). This restriction to the propagator function represents the claim that two particles which are bound never stop interacting. By this we also see that a finite number of diagrams in Fig. 6.2.2 and, hence, in Eq. (6.2.2) is not sufficient to describe bound states. Although the expansion of the four-point function (6.2.1) by means of perturbation theory and the integral equation (6.2.3) are equivalent, the latter one is more powerful, providing a broader range of physical applications.[17]

---

[17]In fact, comparing the power expansion of the four-point function (6.2.1) and its integral equation (6.2.3), the differences according to divergence or convergence of the one or the other are even more fundamental [GM51].

6 Introduction to Dyson-Schwinger and Bethe-Salpeter equations

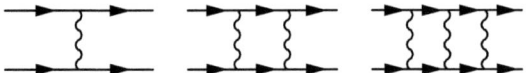

Figure 6.2.6: Typical contributions to the homogeneous BSE in ladder approximation.

6.2.2 Ladder approximation

In many applications of the BSE it is sufficient to include only a certain kind of interaction between the two particles. The so-called ladder approximation only includes single particle exchanges within the interaction kernel $\overline{G}$. The resulting iterated diagrammatical solution has the characteristic shape of a ladder, see Fig. 6.2.6. Strictly speaking, if the homogeneous BSE is considered, i. e. bound states, in each of these diagrams an infinite number of interactions takes place and the "ladder" is endless. In Fig. 6.2.7 an example for a diagram that is not included by the ladder approximation is shown. It has, however, to be taken into account, e. g., for positronium bound states [Gre92]. For the deuteron the lowest-order perturbative contribution to the fermion propagators $S^{A,B}$, i. e. free propagators, are inserted and give satisfactory results. However, due to the momentum dependence of the dressed quark mass, this approximation is not applicable in QCD. Indeed, in [Sou05, Sou10b] the constituent quark model for heavy quarks has been investigated in detail. Though undressed heavy quark masses yield very good results for the masses of the bound states, the decay constants strongly disagree with experimental data.

The free fermion propagators are given in (C.1.15). For a yet unspecified interaction, the lowest-order contribution to the irreducible interaction kernel in momentum space may be written as

$$\overline{G}^{(0)}(p,p',K)_{\alpha\beta\alpha'\beta'} = (2\pi)^4 \Gamma^A_{\alpha\alpha'}\Gamma^B_{\beta\beta'}\Delta^{(0)}(p-p'),\tag{6.2.6}$$

where $\Gamma^{(A,B)}_{\alpha\beta}$ are the vertex functions and $\Delta^{(0)}$ is the lowest order interaction particle propagator. The vertex functions as matrices in Dirac space may be decomposed over $\{\mathbb{1},\gamma_\mu,\sigma_{\mu<\nu},i\gamma_5\gamma_\mu,\gamma_5\}$ and represent the coupling of the interaction particle to the fermions. In Fig. 6.2.8, we give the diagrammatic representation of the integral equation for the four-point function in ladder approximation.

The homogeneous BSE for the BSA in ladder approximation and momentum space thus reads

6.2 Bethe-Salpeter equations

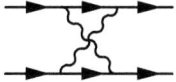

Figure 6.2.7: A diagram contributing to the BSE beyond the ladder approximation.

$$\left[S^A\left(\eta_+ K + p\right)\right]^{-1}_{\alpha\alpha'}\left[S^B\left(\eta_- K - p\right)\right]^{-1}_{\beta\beta'}\chi_{K,\alpha'\beta'}(p)$$
$$= -\int\frac{d^4p'}{(2\pi)^4}\Delta^{(0)}(p-p')\Gamma^A_{\alpha\alpha'}\Gamma^B_{\beta\beta'}\chi_{K,\alpha'\beta'}(p'). \quad (6.2.7)$$

The momentum partitioning parameter is denoted by $\eta \equiv \eta_+ = 1 - \eta_-$, the bound state momentum is K, and p is the relative momentum of the two constituents. If the mass dressing is negligible, e.g. for the deuteron, a natural choice is given by the reduced mass $\eta = \mu/m_A$, with $\mu = m_A m_B/(m_A + m_B)$, which represents the bound states rest frame. For obvious reasons, in case of diquark bound states, the mass dressing renders this choice useless. However, due to the Poincaré invariance of the BSE any value $\eta \in (0, 1)$ may be chosen.

Equation (6.2.7) is a four dimensional integral equation for the 16 components of the BSA. Whether a solution exists or not depends, beside other requirements, on the four-momentum K of the bound state. Considered as an eigenvalue problem, Eq. (6.2.7) has solutions only for discrete values of the bound state four-momentum. Unfortunately, the BSA does not have a direct physical meaning and calculating observables from it requires knowledge of the adjoint BSA. Hence, in general, if no further restrictions to the time ordering are made and a simple algebraic relation between the BSA and its adjoint counterpart is provided, one also has to solve the BSE for the adjoint BSA.

In any case, as Eq. (6.2.7) is solvable only for discrete values of K, solving the BSE for the BSA is a problem of finding eigenvalues K and eigenstates χ_K to the operator defined in Eq. (6.2.7). Thus, a solution at least provides the determination of the bound state mass K_0, if one is working in the bound states rest frame. Because the BSE is Lorentz invariant and the BSA transforms covariantly, one is free to choose any frame that is convenient to simplify calculations. Choosing the center of mass system does not mean any loss of generality, but enables a direct determination of the bound state mass by solving the eigenvalue problem defined in Eq. (6.2.7). Any other frame complicates the solution but may be useful in the one or the other context [Mar06]. If it is desired to calculate observables, e.g. the electromagnetic form factor, both functions are essential, the BSA in the rest frame of the bound state and in other

6 Introduction to Dyson-Schwinger and Bethe-Salpeter equations

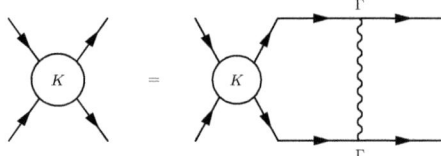

Figure 6.2.8: Diagrammatic representation for the BSE (6.2.3) and Fig. 6.2.4 in ladder approximation.

frames. The latter one can be obtained as the Lorentz transform of the first one.

6.2.3 Solving the Bethe-Salpeter equation in ladder approximation

In order to simplify notations, in what follows, we will use the matrix notation and introduce a redefined BSA as

$$\Psi_K(p) \equiv -\chi_K(p)C, \qquad (6.2.8)$$

with $C = i\gamma_0\gamma_2$ being the charge conjugation matrix. Furthermore, we introduce the Bethe-Salpeter vertex function (BSV) $\Gamma(K,p)$ as

$$\Psi_K(p) = S^A(\eta_+ K + p)\Gamma(K,p)\widetilde{S}^B(\eta_- K - p), \qquad (6.2.9)$$

where we have implicitly introduced

$$\widetilde{S}(p) = CS^T C, \qquad (6.2.10)$$

which means for the free propagator $[\widetilde{S}^{(0)}]^{-1} = \hat{p} + m$. For the BSA these definitions translate to

$$\chi_K(p) = S^A(\eta_+ K + p)\widetilde{\Gamma}(K,p)\left[S^B(\eta_- K - p)\right]^T \qquad (6.2.11)$$

and

$$\Gamma(K,p) \equiv \widetilde{\Gamma}(K,p)C. \qquad (6.2.12)$$

106

6.2 Bethe-Salpeter equations

Transforming into the bound states rest frame, the BSE for the vertex-function of a bound state in ladder approximation reads

$$\Gamma(K,p) = i \int \frac{d^4p'}{(2\pi)^4} \Delta(p-p') \Gamma^A S^A \left(\eta_+ K + p'\right) \Gamma(K,p') \widetilde{S}^B \left(\eta_- K - p'\right) \widetilde{\Gamma}^B. \quad (6.2.13)$$

Within a numerical approach, linear integral equations, such as Eq. (6.2.13), are matrix equations and the solution of a homogeneous linear integral equation may be obtained by solving the eigenvalue problem. In the following, the matrix which corresponds to Eq. (6.2.13) is called Bethe-Salpeter matrix (BSM). Similarly, the solution of an inhomogeneous linear integral equation may be obtained by matrix inversion. Once the matrices are known, using modern computers and programs both tasks can be accomplished fast and reliable in most cases without big efforts. However, the high dimensionality of Eq. (6.2.13) (16 components of the vertex function and 4 momentum integrations), cause a large amount of work. In order to solve the BSE, the Dirac structure of the vertex function Γ is projected onto a set which respects the quantum numbers and symmetries of the respective bound state. The resulting coefficient functions are then expanded over hyperspherical harmonics. In case of the deuteron, one is left with a set of coupled one-dimensional integral equations if the appropriate bases are employed for the expansions [Dor08, Hil08]. The other integrations can be done analytically, which results in analytical expressions for the entries of the BSM. However, the quark dressing of QCD again prevents such a simplification. Thus, for hadrons one is left with a set of coupled integral equations where the entries must be obtained numerically.

7 Dyson-Schwinger and Bethe-Salpeter approach

In this chapter we apply the DSEs and BSEs introduced in Secs. 6.1 and 6.2 to quark-antiquark bound states. Possibilities of an application to heavy-light mesons are investigated. For this purpose, the analytic structure of the quark propagator in the complex plane is analyzed numerically and Poincaré invariance of the BSE is used to avoid poles in the integration domain. As an extension of the investigation in Sec. 5, the BSE is solved for the Wigner-Weyl solutions of the DSE at nonzero quark masses.

7.1 Dyson-Schwinger equation for the quark propagator

The renormalized Euclidean DSE in momentum space for the quark propagator in vacuum reads [Fis05] (cf. Sec. 6.1, Eq. (6.1.10))

$$S^{-1}(p) = Z_2\left(i\gamma \cdot p + Z_m m_0\right) - Z_1 \int^{\Lambda} \frac{d^4q}{(2\pi)^4} \Gamma_\mu^{0,a}(p,q) S(q) \Gamma_\nu^b(q,p) G_{\mu\nu}^{ab}(p-q), \quad (7.1.1)$$

where Z_1, Z_2 and Z_m are gluon, quark and mass renormalization functions, respectively. $\Gamma_\nu^b(p,q)$ is the quark-gluon vertex function, $G_{\mu\nu}^{ab}(p-q)$ the gluon propagator. Bare quantities have the superscript 0. The regularization mass scale is denoted by Λ and the renormalized quark mass in the QCD Lagrangian is m_0.

The gluon propagator may be decomposed by virtue of Eq. (B.3.3) and Tab. B.3.1 as [Rob94]

$$G_{\mu\nu}^{ab}(k) = \delta^{ab}\left[\left(g_{\mu\nu} - \frac{q_\mu q_\nu}{q^2}\right)G(q^2) + \alpha\frac{q_\mu q_\nu}{q^2}G_L(q^2)\right] \equiv \delta^{ab} G_{\mu\nu}(k). \quad (7.1.2)$$

The choice $\alpha = 0$ is called Landau gauge, the choice $\alpha = 1$ refers to the Feynman gauge. As has been argued in [Fra96], the rainbow-ladder truncation, which will be

7 Dyson-Schwinger and Bethe-Salpeter approach

employed in the following, is numerically reliable in Landau gauge. In particular, the rainbow truncation, i. e. the replacement of the quark-gluon vertex by its tree-level counterpart $\Gamma_\nu^b \to igt^b\gamma_\nu$, is not reliable for $\alpha \neq 0$. In [Cha07, Zho08], where the DSE approach has been used to study quark and gluon condensates, however, the Feynman gauge has been used.

Recall that all n-point functions, e. g. quark and gluon propagators and quark-gluon vertex, satisfy their own DSEs. They form an infinite set of coupled integral equations, relating n-point functions to $(n+1)$-point functions. A suitable truncation scheme must respect global symmetries of QCD, in particular chiral symmetry and its dynamical breaking. One such scheme is the rainbow-ladder truncation which amounts to the replacement of gluon propagator and quark-gluon vertex by their tree-level counterparts.

The exact quark propagator in vacuum and Euclidean space may be decomposed by virtue of two quark dressing functions as

$$S(p) = \left[i\gamma \cdot pA(p) + B(p)\right]^{-1} = \frac{-i\gamma \cdot pA(p) + B(p)}{p^2 A^2(p) + B^2(p)} = -i\gamma \cdot p\sigma_\nu(p) + \sigma_s(p), \quad (7.1.3)$$

where $M(p) = B(p)/A(p)$ is the quark mass function and $1/A$ is called the quark wave function. The two sets $\{\sigma_s, \sigma_\nu\}$ and $\{A, B\}$ are related via

$$A = \frac{\sigma_\nu}{p^2 \sigma_\nu^2 + \sigma_s^2}, \qquad B = \frac{\sigma_s}{p^2 \sigma_\nu^2 + \sigma_s^2}. \quad (7.1.4)$$

Any phenomenological ansatz for the gluon propagator must be capable of dynamically generating sufficiently large dressed quark masses in order to reproduce the observed hadronic mass spectra. A simple, but surprisingly successful ansatz, is a two-parameter function which models the infrared part of the interaction, which is the most important part for diquark bound states [Alk02]

$$g^2 G^0(q) = 4\pi^2 D \frac{q^2}{\omega^2} e^{-\frac{q^2}{\omega^2}}. \quad (7.1.5)$$

Gluon propagator $G_{\mu\nu}^{ab}$ and the function G^0 are related via Eq. (7.1.2). The momentum dependence of the coupling strength g is thereby effectively encoded in the gluon-propagator. In the chosen model, the strength of the interaction is described by D, while ω is a measure of the range of the interaction. Since the effective interaction (7.1.5) is exponentially damped in the ultraviolet, the momentum integration in

7.1 DSE for the quark propagator

Eq. (7.1.1) is well-defined and the renormalization constants may be set to one [Fis05]. Equation (7.1.5) is a simplified version of the Maris-Tandy model [Mar99] (see also [Mar97])

$$g^2 G^0(q) = 4\pi^2 D \frac{q^2}{\omega^2} e^{-\frac{q^2}{\omega^2}} + \frac{8\pi^2 \gamma_m F(k^2)}{\ln\left[\tau + \left(1 + \frac{k^2}{\Lambda_{QCD}^2}\right)\right]}, \tag{7.1.6}$$

where the second term ensures the correct ultraviolet asymptotics, which is dictated by perturbation theory [Mar97]. We will not continue to investigate this model, since the effective quark-gluon interaction is qualitatively well described by (7.1.5). With these conventions, the DSE for the quark propagator in Euclidean space, Landau gauge and rainbow-ladder approximation for the effective interaction (7.1.5) (or (7.1.6)) reads

$$S^{-1}(p) = i\gamma \cdot p + m_0 + \frac{4}{3} \int \frac{d^4 l}{(2\pi)^4} \left[g^2 G^0_{\mu\nu}(p-l)\right] \gamma_\mu S(l) \gamma_\nu. \tag{7.1.7}$$

Since the employed interaction is rotational invariant w.r.t. spatial coordinates, Eq. (7.1.7) reduces to a two-dimensional integral equation.

For $q = p - l$ one obtains

$$\left(g_{\mu\nu} - \frac{q_\mu q_\nu}{q^2}\right) \gamma^\mu S(l) \gamma^\nu = \frac{i\gamma \cdot l A(l) + 3B(l)}{l^2 A^2(l) + B^2(l)} + \frac{\gamma \cdot q}{q^2} \frac{2iq \cdot l A(l)}{l^2 A^2(l) + B^2(l)}. \tag{7.1.8}$$

Inserting (7.1.8) into (7.1.7) results in

$$S^{-1}(p) = i\gamma \cdot p + m_0$$
$$+ \frac{4}{3} \int \frac{d^4 l}{(2\pi)^4} g^2 G^0(p-l) \left[\frac{i\gamma \cdot l A(l) + 3B(l)}{l^2 A^2(l) + B^2(l)} + \frac{\gamma \cdot q}{q^2} \frac{2iq \cdot l A(l)}{l^2 A^2(l) + B^2(l)}\right]. \tag{7.1.9}$$

Noting that p_μ is the only four-vector which the integral depends on, which allows to project $\gamma \cdot l = (\gamma \cdot p)(l \cdot p)/p^2$ and, hence, to use

$$i(q \cdot l)(\gamma \cdot q) = i(\gamma \cdot p)(p \cdot l - l^2) \frac{p \cdot l}{p^2}, \tag{7.1.10}$$

one can read off and collect the components which belong to different elements of the Clifford base. Equation (7.1.7) thus defines a coupled system of non-linear,

7 Dyson-Schwinger and Bethe-Salpeter approach

two-dimensional integral equations for the propagator functions:

$$A(p) = 1 + \frac{1}{3} \int_0^\infty \int_{-1}^{+1} \frac{dl\, l^3}{\pi^3} dt \sqrt{1-t^2} g^2 G^0(p-l)$$
$$\times \frac{A(l)}{l^2 A^2(l) + B^2(l)} \frac{l}{p}\left(t + 2\frac{(p-lt)(pt-l)}{p^2+l^2-2plt}\right), \quad (7.1.11a)$$

$$B(p) = m_0 + \int_0^\infty \int_{-1}^{+1} \frac{dl\, l^3}{\pi^3} dt \sqrt{1-t^2} g^2 G^0(p-l) \frac{B(l)}{l^2 A^2(l) + B^2(l)}, \quad (7.1.11b)$$

where $p \cdot l = plt$. From Eqs. (7.1.3) and (7.1.4), an analog system of coupled integral equations can be read off for $\{\sigma_v, \sigma_s\}$ instead of $\{A, B\}$. The integrand is then linear in the functions to be solved for, i.e. σ_v and σ_s, but the system itself is still non-linear. For gauges which differ from the Landau gauge it is easy to obtain the additional structures in Eq. (7.1.11) by noting

$$\gamma_\mu S(l) \gamma_\mu = \frac{2i\gamma \cdot lA(l) + 4B(l)}{l^2 A^2(l) + B^2(l)}. \quad (7.1.12)$$

However, it has to be checked whether the employed gauge and truncation scheme respect the global symmetries of the interaction and features DCSB. In fact, as has been argued, e.g., in [Fis09a, Rob94, Rob07, Rob00] and references therein, the ladder-rainbow truncation is such a scheme.

7.1.1 Analytical angle integration of the infrared part of the potential

As has been noted in [Alk02] the main technical benefit of the choice (7.1.5) is that the angle integration in (7.1.11) can be done analytically if the arguments of (7.1.5) are real. To see this, note that the Chebyshev polynomials of the second kind C_n^1 form a complete orthogonal set in the space of functions over the interval $[-1,1]$ [Abr72, Boy00]. Hence, the angle dependence of the potential $gG^0(p-l) = gG^0(p,l,t)$ may be expanded by virtue of

$$gG^0(p,l,t) \equiv \sum_n D_n(p,l) C_n^1(t). \quad (7.1.13)$$

7.1 DSE for the quark propagator

The Chebyshev polynomials of the second kind are a special case of the Gegenbauer polynomials C_n^α for $\alpha = 1$, which emerge when investigating four-dimensional spherical harmonics, sometimes related to as hyperspherical harmonics, and which are a solution to the Gegenbauer differential equation [Erd55a]. The coefficient functions can then be evaluated by virtue of the appropriate inner product which has to be chosen according to the polynomials. In view of the generic structure of the angle integration in (7.1.11), it is convenient to chose the Chebyshev polynomials of the second kind rather than the first kind owing to their orthogonality relation

$$\int_{-1}^{+1} dt \sqrt{1-t^2} C_n^1(t) C_m^1(t) = \frac{\pi}{2} \delta_{nm}. \tag{7.1.14}$$

For the replacement $\sqrt{1-t^2} \to \left(\sqrt{1-t^2}\right)^{-1}$ the Chebyshev polynomials of the first kind C_n^0, which satisfy the according orthogonality relation, are the most convenient choice. Equation (7.1.14) defines the coefficient functions as

$$D_n(p,l) \equiv \frac{2}{\pi} \int_{-1}^{+1} dt \sqrt{1-t^2} C_n^1(t) g G^0(p,l,t). \tag{7.1.15}$$

Similarly, it is useful to define the expansion coefficients for $gG^0(q)/q^2$ as

$$Z_n(p,l) \equiv \frac{2}{\pi} \int_{-1}^{+1} dt \sqrt{1-t^2} C_n^1(t) \frac{gG^0(p,l,t)}{(p-l)^2}. \tag{7.1.16}$$

In the following the coefficient functions will be called diagonal (D_n) and non-diagonal (Z_n) partial potentials, respectively. For the special choice of the potential (7.1.5) it follows that

$$Z_n(p,l) = 8\pi \frac{D}{\omega^2} e^{-x} H_n(z), \tag{7.1.17}$$

where we have defined

$$H_n(z) \equiv \int_{-1}^{+1} dt \sqrt{1-t^2} C_n^1(t) e^{zt} \tag{7.1.18}$$

7 Dyson-Schwinger and Bethe-Salpeter approach

with $x \equiv (p^2 + l^2)/\omega^2$ and $z \equiv 2pl/\omega^2$. By mathematical induction one can show that

$$H_n(z) = (n+1)\pi \frac{I_{n+1}(z)}{z}, \qquad (7.1.19)$$

where $I_n(z)$ are the modified Bessel functions of the first kind.

Equation (7.1.19) can be confirmed for $n = 0$ and $n = 1$. From Eq. (7.1.18) and the recurrence relation of the Chebyshev polynomials $C_{n+1}^1(t) = 2tC_n^1(t) - C_{n-1}^1(t)$, the following recurrence relation can be given for H_n:

$$H_{n+1}(z) = 2\frac{\partial}{\partial z}H_n(z) - H_{n-1}(z). \qquad (7.1.20)$$

Using $\frac{\partial}{\partial z}I_{n+1}(z) = \frac{1}{2}I_n(z) + \frac{1}{2}I_{n+2}(z)$ the derivative of $H_n(z)$ by assumption (7.1.19) reads

$$\begin{aligned}\frac{\partial}{\partial z}H_n(z) &= (n+1)\pi \left(\frac{1}{z}\frac{\partial}{\partial z}I_{n+1}(z) - \frac{I_{n+1}(z)}{z^2}\right) \\ &= (n+1)\pi \frac{1}{2z}\left(I_n(z) + I_{n+2}(z) - \frac{2}{z}I_{n+1}(z)\right).\end{aligned} \qquad (7.1.21)$$

Inserting Eqs. (7.1.21) and (7.1.19) into (7.1.20) one obtains

$$H_{n+1}(z) = \frac{\pi}{z}\left(I_n(z) - \frac{2(n+1)}{z}I_{n+1}(z) + (n+1)I_{n+2}(z)\right). \qquad (7.1.22)$$

Finally, using the recurrence relation $I_{n+2}(z) = I_n(z) - 2(n+1)I_{n+1}(z)/z$ proves Eq. (7.1.19) by mathematical induction. Hence, the integration in Eq. (7.1.16) can be performed analytically yielding the final result for the non-diagonal partial potentials

$$Z_n(p,l) = (n+1)8\pi^2\frac{D}{\omega^2}e^{-x}\frac{I_{n+1}(z)}{z}. \qquad (7.1.23)$$

The diagonal partial potentials can now be easily evaluated by noting

$$D_n(p,l) = 8\pi De^{-x}\left(xH_n(z) - z\frac{\partial}{\partial z}H_n(z)\right). \qquad (7.1.24)$$

Using Eq. (7.1.20) and the recurrence relation for the modified Bessel functions of

7.1 DSE for the quark propagator

the first kind, the diagonal partial potentials are given by

$$D_n(p,l) = (n+1)8\pi^2 D e^{-x}\left[\frac{x+n+2}{z}I_{n+1}(z)-I_n(z)\right]. \quad (7.1.25)$$

Having the analytical expressions for the partial potentials at our disposal, we note that the modified Bessel functions of the first kind are strongly growing functions. Therefore, caution should be exercised when the numerical implementation is addressed. It is clear that, if the implementation is limited to double-float numbers, the numerical difference of two very large numbers might differ significantly from the true result. Despite of the exponential damping a careless treatment would lead to severe numerical discrepancies. Furthermore, for complex arguments the strongly growing modified Bessel functions of the first kind become rapidly oscillating Bessel functions of the first kind. This leads to difficulties in solving the DSE in the complex plane.

Due to the success of Eq. (7.1.5) in evaluating many hadronic bound state masses, cf. e. g. [Alk02], one might argue that most of the infrared part of the interaction is indeed contained there or must at least follow a similar functional behavior. Thus, any refinement might be considered as a small disturbance to the ansatz (7.1.5) featuring similar properties, e. g. rapid oscillations for large complex momenta.

As a result of Eqs. (7.1.23) and (7.1.24) the angle integration in Eq. (7.1.11) in case of the potential (7.1.5) can be performed analytical, yielding [Alk02] coupled, non-linear, one-dimensional integral equations:

$$A(p) = 1 + \frac{1}{3}\int_0^\infty \frac{dl\, l^4}{4\pi^2}\frac{A(l)}{l^2 A^2(l)+B^2(l)}$$
$$\times\left[\frac{1}{p}D_1(p,l)+2p\left(1+\frac{l^2}{p^2}\right)Z_1(p,l)-l\left(5Z_0(p,l)+Z_2(p,l)\right)\right], \quad (7.1.26a)$$

$$B(p) = m_0 + 4\int_0^\infty \frac{dl\, l^3}{8\pi^2}\frac{B(l)}{l^2 A^2(l)+B^2(l)}D_0(p,l), \quad (7.1.26b)$$

In principle, any potential (7.1.2) may be split into a part which is of the functional type of Eq. (7.1.5) and may be treated by virtue of Eq. (7.1.26) and a part which has to be treated numerically according to Eq. (7.1.11).

7 Dyson-Schwinger and Bethe-Salpeter approach

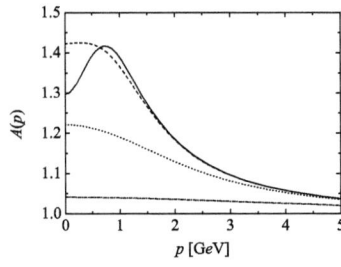 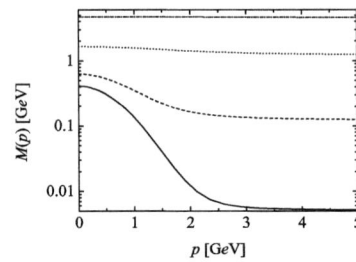

Figure 7.1.1: Solutions of the DSE (7.1.11) with the phenomenological potential given in Eq. (7.1.5) for propagator functions $A(p)$ (left panel) and $M(p)$ (right panel) along the real axis for $\omega = 0.5\,\text{GeV}$ and $D = 16.0\,\text{GeV}$. Solid: $m_0 = 5\,\text{MeV}$, dashed: $m_0 = 120\,\text{MeV}$, dotted: $m_0 = 1.2\,\text{GeV}$, dashed-dotted: $m_0 = 4.5\,\text{GeV}$.

7.1.2 Solution along the real axis

Equation (7.1.26), or Eq. (7.1.11) for more sophisticated potentials, can easily be solved along the real axis by fixed-point iteration, see Fig. 7.1.1. The convergence of the iteration depends on the quark mass and the initial function. If the initial function is constant, then the iteration converges faster for heavy quarks.

For $p \to \infty$ the quark mass function approaches its perturbative limit, that is $\lim_{p \to \infty} M(p) = m_0$. Furthermore, the solutions exhibit the anticipated dynamical mass generation for small momenta which cannot be obtained within a perturbative treatment. As the most dominant energy domain for diquark bound states is below 1 GeV, we find constituent quark masses of $\approx 400\,\text{MeV}$ for current quark masses of 5 MeV. However, as can be seen from Fig. 7.1.2 at least for the assumed potential (7.1.5), the systematic of dynamically generated mass is non-trivial, because the lightest quark does not exhibit the largest dynamically generated mass. For the chosen potential and parameters, we find a maximal dynamical generated mass increase of the quark mass function $M(p)$ at zero momentum for $m_0 \approx 350\,\text{MeV}$. On the other hand, the right panel of Fig. 7.1.2 shows that the quark mass function $M(p)$ decreases faster for light quarks than for heavy quarks. Although for meson masses the low energy domain is more important than the high energy domain, Fig. 7.1.2 shows that in view of a simple constituent quark model with constant quark masses, the constituent masses will be generated by a subtle balance between maximal mass gain at low momenta and the decreasing of the mass function at large momenta.

Considering the functions A and B as functions of p^2, Schwartz's reflection principle tells us that the analytic continuation of $A(p^2)$ and $B(p^2)$ in the complex plane is

7.1 DSE for the quark propagator

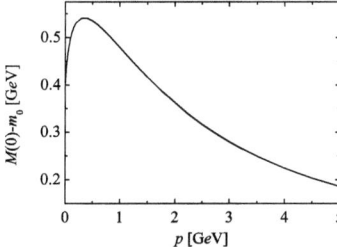
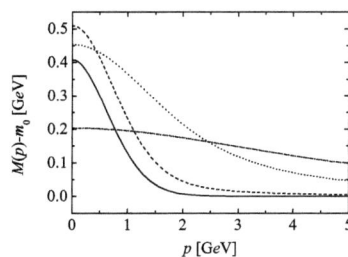

Figure 7.1.2: Comparison of dynamically generated masses obtained from (7.1.11) with phenomenological potential given in Eq. (7.1.5) for $\omega = 0.5\,\text{GeV}$ and $D = 16.0\,\text{GeV}$. Left panel: $\lim_{p\to 0} M(p) - m_0$ as a function of m_0. Riht panel: $M(p) - m_0$ as a function of p. Line code as in Fig. 7.1.1.

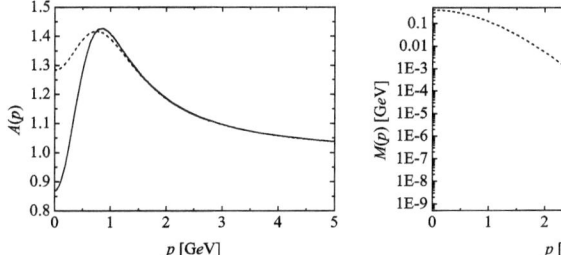

Figure 7.1.3: Solutions of the DSE (7.1.11) for propagator functions $A(p)$ (left panel) and $M(p)$ (right panel) along the real axis obtained for different initial functions $B_0(p)$ in the chiral limit $m_0 = 0$ for $\omega = 0.5\,\text{GeV}$ and $D = 16.0\,\text{GeV}$. Solid: $B_0(p) = 0$, dashed: $B_0(p) \neq 0$.

subject to the condition $A^*(p^2) = A((p^2)^*)$ (and analogously for B), i.e. their imaginary parts are antisymmetric w.r.t. to $p^2 \to (p^2)^*$.

Apart from the case $m_0 = 0$ the result does not depend on the trial function in case of a fixed point iteration.[18] For $m_0 = 0$ the initial choice $B_0 = 0$ reproduces itself and converges to a different solution A and B of Eq. (7.1.11) than for any other choice. In Fig. 7.1.3 both solutions are depicted. As $M(p) = 0$ (and, as will become clear later on, $\langle \bar{q}q \rangle = 0$) this represents a chirally symmetric solution to Eq. (7.1.11). From Fig. 7.1.3 we conclude, that DCSB affects the quark wave function only below 1.5 GeV.

---

[18]Note, however, that there are, in general, multiple solutions to the DSE [LE07].

117

7 Dyson-Schwinger and Bethe-Salpeter approach

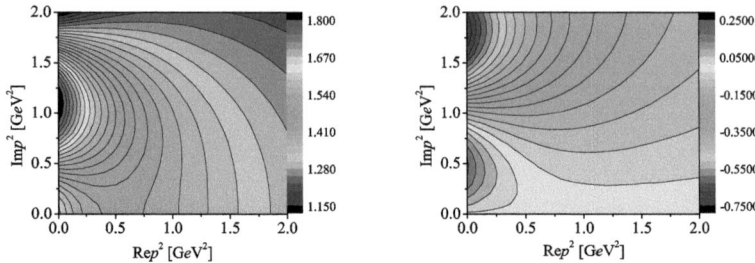

Figure 7.1.4: Re$A(p^2)$ (left) and Im$A(p^2)$ (right) for $m_0 = 5\,\text{MeV}$, $\omega = 0.5\,\text{GeV}$ and $D = 16.0\,\text{GeV}$ in the complex p^2 plane.

7.1.3 Solution in the complex plane

As will be shown in Sec. 7.2 the quark propagator must be evaluated along and inside a parabola in the complex plane in order to solve the BSE. Various approaches can be found in the literature and three of them have been used by us to study the quark propagator in the complex plane.

Complex gluon momenta and solution along the real axis

The first method, cf. e. g. [Alk02], is to solve Eq. (7.1.11) along the real axis to very high precision by fixed-point iteration. Having obtained a solution along the real axis, the propagator functions in the complex plane can be evaluated by virtue of Eq. (7.1.11) for complex gluon momenta, see Figs. 7.1.4 and 7.1.5. This method is reliable if the imaginary part of the gluon momentum is not too large. According to the discussion given previous to Eq. (7.1.26) the integrand is a rapidly oscillating function in l for arguments with large imaginary part. Hence, numerical stability of the integration routine has to be controlled carefully. Finally note that the assumption of the same functional form of the gluon propagator for complex arguments as for real arguments is an ad hoc ansatz [Alk02]. However, the results are in agreement to the other two approaches.

7.1 DSE for the quark propagator

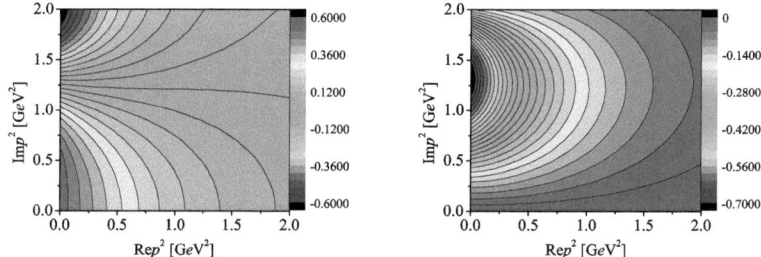

Figure 7.1.5: $\text{Re}B(p^2)$ (left) and $\text{Im}B(p^2)$ (right) in the complex p^2 plane for parameters as in Fig. 7.1.4.

Complex gluon momenta and solution in the complex plane

The second method, cf. [Sta92], relies on analyticity of the integrand in Eq. (7.1.11), which allows to perform a Wick rotation by an angle Θ:

$$p \to p_0 e^{i\Theta}. \qquad (7.1.27)$$

Obviously, the integrand has to tend to zero rapidly enough for $|p| \to \infty$. The angle defines the area of analyticity of the integrand. Denoting the integrands of the DSE by $K_{A,B} = K_{A,B}(p,l,t;A,B)$, Eq. (7.1.11) reads[19]

$$A(p) = 1 + \int_0^\infty \int_{-1}^{+1} dl\, dt\, K_A(p,l,t;A,B), \qquad (7.1.28a)$$

$$B(p) = m_0 + \int_0^\infty \int_{-1}^{+1} dl\, dt\, K_B(p,l,t;A,B), \qquad (7.1.28b)$$

which, by virtue of the Wick rotation (7.1.27), becomes

$$A(p = p_0 e^{i\Theta}) = 1 + \int_0^\infty \int_{-1}^{+1} dl_0\, e^{i\Theta} dt\, K_A(p_0 e^{i\Theta}, l_0 e^{i\Theta}, t; A, B), \qquad (7.1.29a)$$

$$B(p = p_0 e^{i\Theta}) = m_0 + \int_0^\infty \int_{-1}^{+1} dl_0\, e^{i\Theta} dt\, K_B(p_0 e^{i\Theta}, l_0 e^{i\Theta}, t; A, B). \qquad (7.1.29b)$$

---

[19] Indeed, the integrands can even be written as $K_{A,B} = \tilde{K}_{A,B}(p,l,t) F_{A,B}(l, A(l), B(l))$ and Eq. (7.1.11) is thus a Hammerstein equation [Ham30, Tri85].

7 Dyson-Schwinger and Bethe-Salpeter approach

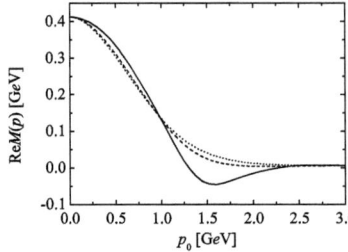
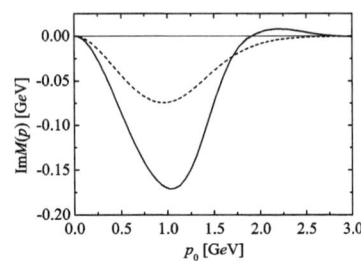

Figure 7.1.6: Solutions of the DSE (7.1.29) for real (left panel) and imaginary (right panel) part of the quark mass function $M(p)$ in the complex plane at angles $\Theta = 0$ (dotted), $\Theta = 0.25$ (dashed) and $\Theta = 0.5$ (solid). The parameters are $m_0 = 5\,\mathrm{MeV}$, $\omega = 0.5\,\mathrm{GeV}$ and $D = 16.0\,\mathrm{GeV}$.

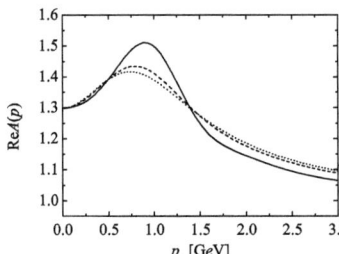
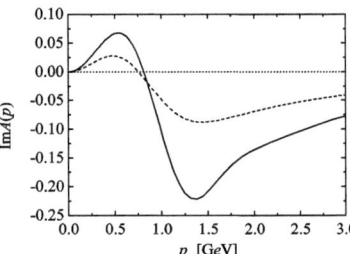

Figure 7.1.7: Solutions of the DSE (7.1.29) for real (left panel) and imaginary (right panel) part of the quark wave function $A(p)$ in the complex plane. Line code as in Fig. 7.1.6.

Similarly, to the solution along the real axis, Eq. (7.1.29) can be solved by fixed-point iteration for given trial functions A and B. Note that, in contrast to Eq. (7.1.11), the DSE is solved directly in the complex plane and the solutions for A and B along the real axis are not needed. The results are shown in Figs. 7.1.6 and 7.1.7. Performing the Wick rotation (7.1.27) requires analyticity in the enclosed domain. As argued in [Sta92] one should therefore gradually increase the rotation angle in order to detect an emerging singular behavior before the singularity enters the contour. However, no indications for an irregular behavior of the functions A and B could be found within our investigation. Though, this does not mean that the integration kernel $K_{A,B}$ is regular.

Despite of the advantage that Eq. (7.1.29) allows to solve the DSE directly in the

7.1 DSE for the quark propagator

complex plane, the gluon momenta are still imaginary. Hence, the integrand of the momentum integration is again an oscillating function for large complex momenta which leads to severe numerical complications. Because of this, the method cannot be applied to arbitrary large angles.

Real gluon momenta and solution in the complex plane

In [Fis05] an approach has been proposed that allows to solve the DSE directly in the complex plane, whereas only real gluon momenta are involved. Shifting the integration variable in Eq. (7.1.9) by $l \to p - l$, using (7.1.10) and expanding over the Clifford base one obtains

$$A(p) = 1 + \frac{4}{3}\int_0^\infty \int_{-1}^{+1} \frac{dl\, l^3}{\pi^3}\, dt\, \sqrt{1-t^2}g^2 G^0(l)$$
$$\times \frac{A(p-l)}{(p-l)^2 A^2(p-l) + B^2(p-l)}\left(1 - 3\frac{l}{p}t + 2t^2\right), \quad (7.1.30a)$$

$$B(p) = m_0 + 4\int_0^\infty \int_{-1}^{+1}\frac{dl\, l^3}{\pi^3}\, dt\, \sqrt{1-t^2} g^2 G^0(l) \frac{B(p-l)}{(p-l)^2 A^2(p-l) + B^2(p-l)}.$$
$$(7.1.30b)$$

The integration now runs over the gluon momentum instead of the quark momentum. Note that the Wick rotation (7.1.27) can also be applied to Eq. (7.1.30), but would lead to imaginary gluon momenta again.

Consider a contour given by

$$z(\lambda) = \frac{\lambda^2}{4\eta^2 M^2} - \eta^2 M^2 + i\lambda, \quad (7.1.31)$$

which is plotted in Fig. 7.1.8. Its relevance for diquark bound states is revealed in Sec. 7.2. The crucial point which allows the solution of the DSE by virtue of Eq. (7.1.30) is the insight that the arguments of the quark propagator functions $A(p-l)$ and $B(p-l)$ which enter the integrand lie within the contour (7.1.31) for $p^2 = z(\lambda)$.

In order for a fixed momentum q^2 to lie within or on the parabola (7.1.31) the following condition has to be satisfied:

$$|\mathrm{Im}\, q^2| \leq |\mathrm{Im}\, z(\lambda_q)| \quad \text{with} \quad \mathrm{Re}\, z(\lambda_q) = \mathrm{Re}\, q^2, \quad (7.1.32)$$

7 Dyson-Schwinger and Bethe-Salpeter approach

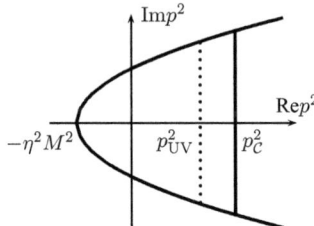

Figure 7.1.8: The contour (7.1.31) in the complex p^2-plane with vertex at $-\eta^2 M^2$.

where equality implies q^2 on the parabola. With

$$\lambda_q = 2\eta M \sqrt{\text{Re}q^2 + \eta^2 M^2} \tag{7.1.33}$$

Eq. (7.1.32) translates to

$$\frac{(\text{Im}q^2)^2}{4\eta^2 M^2} - \eta^2 M^2 \leq \text{Re}q^2 . \tag{7.1.34}$$

Solving along the contour requires the propagator functions at arguments $q^2 = (p-l)^2 = z + l^2 - 2\sqrt{z}lt$. Hence,

$$\text{Im}(p-l)^2 = \lambda - 2lt\eta M , \tag{7.1.35a}$$

$$\text{Re}(p-l)^2 = \frac{\lambda^2}{4\eta^2 M^2} - \eta^2 M^2 + l^2 - lt\frac{\lambda}{\eta M} , \tag{7.1.35b}$$

which fulfills Eq. (7.1.34) for all $t \in [-1,+1]$ and, therefore, proves the statement. Applying a Wick rotation (7.1.27) to Eq. (7.1.30) for the contour given in Eq. (7.1.31) results in a rotated parabola.

The fixed-point iteration along the parabola (7.1.31) can now be done in the following way. Given the functions A_n and B_n along the contour $\mathscr{C}$, where n refers to the iteration step, the functions inside of the contour can be evaluated by virtue of Cauchy's integral formula for an analytic function f

$$f(z_0) = \frac{1}{2\pi i} \oint_{\mathscr{C}} dz \frac{f(z)}{z - z_0} \tag{7.1.36}$$

provided A and B are analytic inside of the contour $\mathscr{C}$. Thereby, the contour is closed at some $p^2_{\mathscr{C}} > p^2_{\text{UV}}$. An improvement of the numerical implementation of Cauchy's

7.1 DSE for the quark propagator

integral formula has been proposed in [Ioa91] and first applied in the context of DSEs and BSEs in [Kra08]. The idea is to suppress the numerical error, which occurs when z_0 approaches the contour $\mathscr{C}$, by introducing the Cauchy formula representation of $1 = \frac{1}{2\pi i}\oint_{\mathscr{C}} dz \frac{1}{z-z_0}$ in the r.h.s. of Eq. (7.1.36). For arguments with $\text{Re} p^2 > \text{Re} p_{UV}^2$ the functions may be fitted to their asymptotic form [Fra96, Fis05]

$$A(p^2) \stackrel{\text{Re}p^2 \to \infty}{=} 1 + \sum_{n=1}^{N} \frac{a_n}{p^{2n}}, \tag{7.1.37a}$$

$$B(p^2) \stackrel{\text{Re}p^2 \to \infty}{=} m_0 + \sum_{n=1}^{N} \frac{b_n}{p^{2n}}. \tag{7.1.37b}$$

Having evaluated the functions inside of the parabola, integration in Eq. (7.1.30) can be performed, giving A_{n+1} and B_{n+1} along the parabola.

The procedure works for all quark masses m_0 up to a specific value of the parameter

$$\chi \equiv \eta M, \tag{7.1.38}$$

which completely determines the parabola (7.1.31). The maximum $\chi_{\max}$ depends on the quark mass. The reason for the failure of this procedure can be traced back to poles of the integrands in Eq. (7.1.30), namely for σ_v and σ_s in the complex plane. Note that so far no poles have been found for A and B, which we therefore consider as regular.

Analytic structure of the quark propagator

One way to determine the position of these poles is to search for the roots of the denominator of $\sigma_{v,s}$. The denominator can be written as

$$p^2 A^2(p) + B^2(p) = \big(pA(p) + iB(p)\big)\big(pA(p) - iB(p)\big) \equiv g^*(p^*)g(p), \tag{7.1.39}$$

where we use the reflection property of the propagator functions (see Sec. 7.1.2), which therefore allows us to restrict our studies to g. From Eq. (7.1.39) we see that a root of $g(p)$ corresponds to a pair of complex conjugated poles of the quark propagator. Furthermore, due to $g(-p^*) = -g^*(p)$, the roots of $g(p)$ are symmetric w.r.t. to reflection at the imaginary axis. Hence, the analysis may be restricted to the first quadrant of the momentum plane p. This is not the case in the squared

7 Dyson-Schwinger and Bethe-Salpeter approach

momentum plane p^2.

To locate the roots of g we partly follow the method used in [Mar92]. We enclose the domain where poles are expected by a rectangular contour and apply Rouchés theorem, which states that if a function f is meromorphic (has only isolated poles) along and inside of the contour $\mathscr{C}$, then

$$N - P = \frac{1}{2\pi i} \oint_{\mathscr{C}} \frac{f'(z)}{f(z)} dz , \qquad (7.1.40)$$

where N is the number of roots and P is the number of poles within $\mathscr{C}$. Thereby it is assumed that g does not have poles within the considered contour, which can be seen from Figs. 7.1.4 and 7.1.5. The derivative of g w. r. t. p can easily be evaluated by means of Eq. (7.1.11) and is given by

$$\frac{\partial}{\partial p}\left[pA(p)\right] = 1 + \frac{1}{3}\int_0^\infty \int_{-1}^{+1} \frac{dl\, l^3}{\pi^3} dt\, \sqrt{1-t^2} \frac{lA(l)}{l^2 A^2(l) + B^2(l)}$$
$$\times \frac{\partial}{\partial p} g^2 G^0(p-l)\left(t + 2\frac{(p-lt)(pt-l)}{p^2 + l^2 - 2plt}\right), \quad (7.1.41a)$$

$$\frac{\partial}{\partial p}\left[B(p)\right] = \int_0^\infty \int_{-1}^{+1} \frac{dl\, l^3}{\pi^3} dt\, \sqrt{1-t^2} \frac{B(l)}{l^2 A^2(l) + B^2(l)} \frac{\partial}{\partial p} g^2 G^0(p-l).$$
$$(7.1.41b)$$

The first equation can be transformed by virtue of

$$\frac{\partial}{\partial p} g^2 G^0(p-l)\left(t + 2\frac{(p-lt)(pt-l)}{p^2 + l^2 - 2plt}\right)$$
$$= 2(p-lt)\left(\frac{\partial}{\partial q^2} D(q^2)\right)_{q=p-l}\left(t + 2\frac{(p-lt)(pt-l)}{p^2 + l^2 - 2plt}\right)$$
$$+ D(p-l)2\left(\frac{pt-l}{p^2+l^2-2plt} + \frac{(p-lt)t}{p^2+l^2-2plt} - 2\frac{(p-lt)^2(pt-l)}{p^2+l^2-2plt}\right),$$
$$(7.1.42)$$

For the potential given in Eq. (7.1.5) one has

$$\frac{\partial}{\partial q^2} D(q^2) = \left(\frac{1}{q^2} - \frac{1}{\omega^2}\right) D(q^2). \qquad (7.1.43)$$

7.1 DSE for the quark propagator

Having found one or more roots inside of the contour, the rectangle is divided into four smaller rectangles and Eq. (7.1.40) is applied to each of them. The results are shown in Fig. 7.1.9 and Tab. 7.1.1 for up, strange and charm quark masses. We conclude from the left panel, that the imaginary part of the lowest pole is larger than $m_0 + 500$ MeV. Furthermore, as the Wick rotation (7.1.27) is performed within the p-plane rather than the p^2-plane, the lowest poles enter the corresponding contour, which makes an account of the poles necessary. If the rotation would be performed in the complex p^2-plane according to $p^2 \to \exp\{i\Theta\}$ the poles would not enter the contour, because $\text{Re}p_0^2 < 0$ for any pole. Apart from the numerical information contained in Fig. 7.1.9 nothing is known about the analytic structure. It is worth noting that a line in the p-plane with constant imaginary part corresponds to a parabola in the p^2-plane with direction as in Fig. 7.1.8, vertex at $-(\text{Im}p)^2$ and curvature $(2(\text{Im}p)^2)^{-1}$. Contrarily, a line with constant real part corresponds to a parabola with vertex at $(\text{Re}p)^2$ and curvature $-(2(\text{Re}p)^2)^{-1}$, i.e. opposite direction as compared to Fig. 7.1.8.

By Eq. (7.1.31) the upper limit for χ, such that a pole at momentum p_0 does not enter the parabola, is given by

$$\chi^2 = -\frac{\text{Re}p_0^2}{2} + \sqrt{\left(\frac{\text{Re}p_0^2}{2}\right)^2 + \left(\frac{\text{Im}p_0^2}{2}\right)^2}, \qquad (7.1.44)$$

which holds true for positive χ^2. Using $p^2 = (\text{Re}p)^2 - (\text{Im}p)^2 + 2i\,\text{Re}p\,\text{Im}p$ one can show that

$$\chi = \text{Im}p_0. \qquad (7.1.45)$$

Therefore, it is preferable to study the analytic properties within the p-plane rather than the p^2-plane, although the parabola is given in the latter one. Having found the pole with the lowest imaginary part in the p-plane determines the maximum value χ. Consequently, having found one pole it is sufficient to look for poles with lower imaginary part if the maximum χ has to be determined. If the position of a pole is given in the p^2-plane such a simple relation does not exist. Furthermore, in the p^2-plane both, real and imaginary parts, have to be known to decide whether a pole enters the parabola and, thus, is problematic or not. Contrarily, knowledge of the imaginary part of the pole is sufficient in the momentum plane.

The minimum value of χ for all poles determines the largest parabola where all poles are outside of the parabola. Note that, for all $\chi < \chi_0$, all parabolas are in the

7 Dyson-Schwinger and Bethe-Salpeter approach

Table 7.1.1: Position of the poles and maximum values of χ for various values of m_0 corresponding to up, strange and charm quarks.

m_0[GeV]	0.005	0.12	1.2
p_0[GeV]	$0.181 + i0.539$	$0.349 + i0.753$	$0.750 + i1.940$
p_0^2[GeV^2]	$-0.258 + i0.196$	$-0.444 + i0.526$	$-3.203 + i2.910$
χ_{max}[GeV]	0.540	0.753	1.940

interior of the parabola which corresponds to χ_0. For our choice of quark masses, these parabolas are shown in the right panel of Fig. 7.1.9. Note that these parabolas only exclude the poles which we found focusing on a limited domain. In particular, the charm quark is expected to have additional poles entering the integration domain of the depicted parabola. In our analysis these poles correspond to larger Rep. As will become clear in Sec. 7.2, the parameter χ determines the maximum bound state mass which can be evaluated without introducing additional pole handling.

The outcome of this study therefore is the following. Although the propagator functions A and B are analytic in the complex plane and the Cauchy formula may be used to evaluate A and B inside of the parabola provided they are known along the parabola, the integrands of the DSE with real gluon momenta (7.1.30) possess singularities within the integration domain by virtue of σ_v and σ_s, see Fig. 7.1.10, which leads to a failure of the iteration along the parabola. As will be demonstrated in Sec. 7.2 these singularities also infer the solution of the bound state BSE. Therefore, an extensive study is in order. Apart from a direct localization, the position of the poles may also be obtained by fitting the quark propagator along the real axis to a complex conjugate mass pole parametrization in the complex plane, as was done in [Sou10a]. This could be advantageous as the propagators can be determined along the real axis to very high precision.

7.2 Bethe-Salpeter equation for mesons

Within a Poincaré invariant quantum field theory, two-particle interactions may be described by virtue of the BSE. As discussed in Sec. 6.2, the BSE is a linear inhomogeneous Fredholm integral equation of the second kind for the BSV $\Gamma(P,p)$. As distinguished from potential approaches to two-particle interactions, recoil effects are naturally taken into account. For pure bound states the BSE becomes a homogeneous one and may be considered as an eigenvalue problem for the integration kernel,

7.2 Bethe-Salpeter equation for mesons

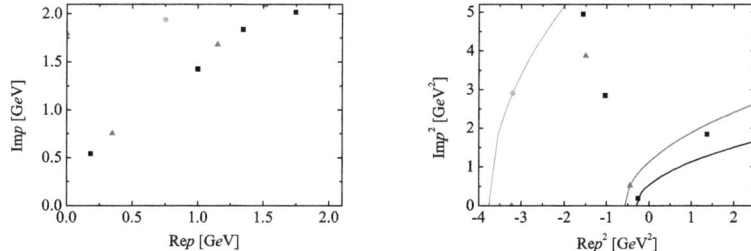

Figure 7.1.9: The roots of $g(p^2)$ in the complex plane for $\omega = 0.5\,\text{GeV}$ and $D = 16\,\text{GeV}$. $m_0 = 5\,\text{MeV}$ (black squares), $m_0 = 120\,\text{MeV}$ (red upwards triangle) and $m_0 = 1.2\,\text{GeV}$ (green downwards triangle). Left: p-plane; right: p^2-plane. The curves correspond to the parabolas which meet the lowest poles of the respective quark flavor.

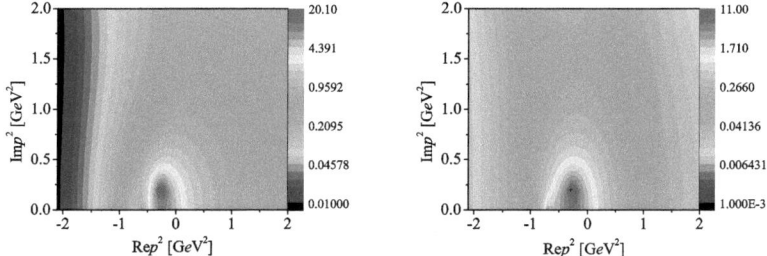

Figure 7.1.10: The quantities $|\sigma_v|$ (left) and $|\sigma_s|$ (right) in the complex p^2 plane. Parameters are $m_0 = \text{MeV}$, $\omega = 0.5\,\text{GeV}$ and $D = 16\,\text{GeV}$.

127

having solutions only for discrete values of the total momentum of the two-particle bound state. For two-quark bound states, the quark and gluon propagators and the quark-gluon vertex function constitute the integration kernel. Employing the rainbow-ladder truncation, the Euclidean BSE for a two-quark bound state in momentum space reads

$$\Gamma(P,p) = -\frac{4}{3}\int\frac{d^4k}{(2\pi)^4}\gamma_\mu S_1(k_+)\Gamma(P,k)\tilde{S}_2(k_-)\gamma_\nu\left[g^2 G^0_{\mu\nu}(p-k)\right], \qquad (7.2.1)$$

where $k_\pm = \eta_\pm P \pm k$, with k being the relative and P the total momentum of the two quarks; $S(p)$ is the quark propagator and $\left[g^2 D(p-l)\right]_{\mu\nu}$ denotes the gluon propagator (and QCD coupling strength). Using the charge conjugation matrix $C = i\gamma_0\gamma_2$ we have defined

$$\tilde{S}(k) = CS^T(p)C = \frac{1}{A^2(k)}\frac{-i\gamma\cdot k - M^2(k)}{k^2 + M^2(k)} \qquad (7.2.2)$$

and implicitly used

$$\tilde{\gamma}_\mu = C\gamma_\mu^T C = C^2\gamma_\mu C^2 = \gamma_\mu. \qquad (7.2.3)$$

The momentum partitioning parameter η, with $\eta \equiv \eta_+ = 1 - \eta_-$, is a free parameter which serves to chose the frame we are working in. Note that, for constant constituent masses, the momentum partitioning parameter $\eta_{1/2} = m_{1/2}/(m_1 + m_2)$ is uniquely fixed by the constituent masses. Due to the momentum dependence of the quark mass functions, η would depend on the momenta $k_\pm$ of both particles. One therefore abandons the definition by virtue of the particles momenta and/or masses and defines it w. r. t. to the relative momentum of the quarks and the bound state momentum. It therefore allows to choose the frame in which the BSE is solved. As the equation is Poincaré invariant, the results must not depend on η. But as approximations and truncations are employed which may break the invariance, the independence of the results has to be checked. In Eq. (7.2.1), the exact gluon propagator has been replaced by a phenomenological one, which describes a one gluon interaction, and the exact quark-gluon vertex has been replaced by the free one $\propto \gamma_\mu$.

The BSV is a 4×4 matrix in Dirac space which can be expanded in a base which is complete in the space of physical states for the system under consideration. In particular, the expansion has to reflect the transformation properties of the bound state. The general structure of BSVs for bound states of spinor particles has been

7.2 Bethe-Salpeter equation for mesons

Table 7.2.1: Commutators and anticommutators of the elements of the base defined in (7.2.4) with $\hat{q}$.

	$T_1^\dagger(\vec{p})$	$T_2^\dagger(\vec{p})$	$T_3^\dagger(\vec{p})$	$T_4^\dagger(\vec{p})$
$[\cdot,\hat{q}]_+$	0	$-2|\vec{q}|T_3^\dagger(\vec{q})$	$-2\frac{\vec{p}\vec{q}}{|\vec{p}|}T_2^\dagger(\vec{p}) - 2q_0 T_4^\dagger(\vec{p})$	$[\hat{q},\vec{p}\vec{\gamma}]\frac{1}{|\vec{p}|}T_1^\dagger(\vec{p})$
$[\cdot,\hat{q}]_-$	$2q_0 T_2^\dagger(\vec{p}) + 2|\vec{q}|T_4^\dagger(\vec{p})$	$2q_0 T_1^\dagger(\vec{q})$	$[\vec{p}\vec{\gamma},\vec{q}\vec{\gamma}]\frac{1}{|\vec{p}|}T_2^\dagger(\vec{p})$	$2\frac{\vec{p}\vec{q}}{|\vec{p}|}T_1^\dagger(\vec{p})$

investigated in, e.g., [Kub72]. Specifying parity and angular momentum, for a pseudoscalar $^1S_0$ state such a base is given by [Dor08]

$$\begin{aligned} T_1(\vec{p}) &= \tfrac{\gamma_5}{2} &=\ T_1^\dagger(\vec{p}) &= -\tfrac{1}{4}\gamma_\mu T_1^\dagger(\vec{p})\gamma_\mu, \\ T_2(\vec{p}) &= \tfrac{\gamma_0\gamma_5}{2} &=\ -T_2^\dagger(\vec{p}) &= -\tfrac{1}{2}\gamma_\mu T_2^\dagger(\vec{p})\gamma_\mu, \\ T_3(\vec{p}) &= \tfrac{\vec{p}\vec{\gamma}}{2|\vec{p}|}\gamma_0\gamma_5 &=\ T_3^\dagger(\vec{p}) &, \\ T_4(\vec{p}) &= \tfrac{\vec{p}\vec{\gamma}}{2|\vec{p}|}\gamma_5 &=\ T_4^\dagger(\vec{p}) &= \tfrac{1}{2}\gamma_\mu T_4^\dagger(\vec{p})\gamma_\mu \end{aligned} \qquad (7.2.4)$$

and fulfills the orthogonality relation

$$\int \frac{d\Omega_{\vec{p}}}{4\pi} \mathrm{Tr}_D\left[T_i T_j^\dagger\right] = \delta_{ij} \qquad (7.2.5)$$

with the infinitesimal solid angle $d\Omega_{\vec{p}} = \sin\theta\, d\theta\, d\phi$. Their commutators and anticommutators with $\hat{q}$ are given in Tab. 7.2.1. This base originates from BSE studies of deuterons, where it allows to reduce analytically the four-dimensional integral equation (7.2.1) to a one-dimensional. Due to the momentum dependence of the quark mass this is not possible in the case of mesons. However, it differs from the standard basis used in Bethe-Salpeter studies of mesons, cf. e.g. [Rob07], where the basis is given by

$$\tilde{T} \in \left\{\gamma_5, -i\gamma_5\hat{P}, -i\gamma_5\hat{p}, -\gamma_5\left[\hat{P},\hat{p}\right]_-\right\}. \qquad (7.2.6)$$

The two sets are related via

$$\tilde{T}_1 = \frac{1}{2}T_1, \qquad (7.2.7a)$$

$$\tilde{T}_2 = 2iP_0 T_2, \qquad (7.2.7b)$$

$$\tilde{T}_3 = 2i\left(p_0 T_2 - |\vec{p}|T_4\right), \qquad (7.2.7c)$$

7 Dyson-Schwinger and Bethe-Salpeter approach

$$\tilde{T}_4 = 4|\vec{p}|P_0 T_4. \tag{7.2.7d}$$

In the scalar channel, the BSV may be decomposed over the basis given by

$$\begin{array}{llllll}
T_1^S(\vec{p}) = & \frac{1}{2} & = & T_1^{S\dagger}(\vec{p}) & = T_1\gamma_5 & = \gamma_5 T_1 & = T_1^\dagger\gamma_5, \\
T_2^S(\vec{p}) = & \frac{\gamma_0}{2} & = & T_2^{S\dagger}(\vec{p}) & = T_2\gamma_5 & = -\gamma_5 T_2 & = -T_2^\dagger\gamma_5, \\
T_3^S(\vec{p}) = & \frac{\vec{p}\vec{\gamma}}{2|\vec{p}|}\gamma_0 & = & T_3^{S\dagger}(\vec{p}) & = T_3\gamma_5 & = \gamma_5 T_3 & = T_3^\dagger\gamma_5, \\
T_4^S(\vec{p}) = & \frac{\vec{p}\vec{\gamma}}{2|\vec{p}|} & = & -T_4^{S\dagger}(\vec{p}) & = T_4\gamma_5 & = -\gamma_5 T_4 & = T_4^\dagger\gamma_5
\end{array} \tag{7.2.8}$$

which fulfills the same orthogonality relation (7.2.5).

By virtue of Eq. (7.2.5) the BSE in Euclidean space can be decomposed into a set of four coupled linear integral equations

$$g_i(P; p_4, |\vec{p}|) = -\frac{4}{3} \int \frac{d\Omega_{\hat{p}}}{4\pi} \int \frac{d^4k}{(2\pi)^4} \frac{g^2 G^0(p-k) \sum_j M_{ij} g_j(P; k_4, |\vec{k}|)}{A_1(k_1)C_1(k_1)A_2(k_2)C_2(k_2)} \tag{7.2.9}$$

with the traces

$$M_{ij}(p,k) = \text{Tr}_D\left[\gamma_\mu T_i^\dagger(\vec{p})\gamma_\mu F_1(k_+)T_j(\vec{k})\tilde{F}_2(k_-) \right.$$
$$\left. -\frac{1}{q^2}\hat{q}\,T_i^\dagger(\vec{p})\hat{q}F_1(k_+)T_j(\vec{k})\tilde{F}_2(k_-)\right]. \tag{7.2.10}$$

With the sets given in Eqs. (7.2.4) and (7.2.8), the traces for the scalar and pseudo-scalar BSE (7.2.9) are related via the following prescription

$$M_{ij}^S(M_1(p), M_2(k)) = (-1)^{j+1-\delta_{i2}} M_{ij}^P(-M_1(p), M_2(k)), \tag{7.2.11}$$

which can be shown by virtue of Tab. 7.2.1. The elaborate and lengthy explicit evaluation of these traces has been done using the algebra package HIP for Maple [Hsi92, Yeh92]. For the sake of brevity, nominator and denominator of the quark propagator have been denoted by

$$C(k) = k^2 + M^2(k^2), \quad F(k) = -i\gamma \cdot k + M(k), \quad \tilde{F}(k) = -i\gamma \cdot k - M(k).$$

$$\tag{7.2.12}$$

7.2 Bethe-Salpeter equation for mesons

Expanding the potential over hyperspherical harmonics

$$g^2 G^0(p-k) = \sum_n \frac{2\pi^2}{n+1} D_n(p,k) \sum_{lm} Z_{nlm}(\chi_p, \Omega_p) Z_{nlm}(\chi_k, \Omega_k) \tag{7.2.13}$$

and

$$\frac{g^2 G^0(p-k)}{q^2} = \frac{1}{2} \sum_n \frac{2\pi^2}{n+1} Z_n(p,k) \sum_{lm} Z_{nlm}(\chi_p, \Omega_p) Z_{nlm}(\chi_k, \Omega_k), \tag{7.2.14}$$

where $\cos \chi_p = p_4/p$ defines the four-dimensional polar angle and

$$C_n^1(t) = \frac{2\pi^2}{n+1} \sum_{lm} Z_{nlm}(\chi_p, \theta_p, \phi_p) Z_{nlm}(\chi_k, \theta_k, \phi_k), \tag{7.2.15}$$

with $t = \cos \chi_{p-k}$, has been used, allows to expand the g_i's into partial amplitudes

$$g_i(P; p, \chi_p) = \sum_{m=0}^{\infty} g_{im}(P; p) X_{mm'}(\chi_p). \tag{7.2.16}$$

The equation to be solved, thus, reads

$$g_{im}(P,p) = -\frac{4}{3} \int_0^\infty \frac{k^3 \, dk}{(2\pi)^4} \sum_j \sum_n A_{ij,mn}(P,p,k) g_{jn}(P,k) \tag{7.2.17}$$

with

$$A_{ij,mn}(P,p,k) = \int \sqrt{1-t_p^2}\, dt_p \sqrt{1-t_k^2}\, dt_k \, \frac{d\Omega_{\tilde{p}}}{4\pi}\, d\Omega_{\tilde{k}}$$
$$\times \frac{g^2 G^0(p-k) M_{ij}(P,p,\chi_p,k,\chi_k) X_{mm'}(\chi_p) X_{nn'}(\chi_k)}{A_1(k_+) C_1(k_+) A_2(k_-) C_2(k_-)}. \tag{7.2.18}$$

Equation (7.2.17) can be solved as an eigenvalue problem for the Bethe-Salpeter integral operator, which has solutions only for discrete values of the bound state momentum P. Numerically, the integral operator in Eq. (7.2.17) becomes a matrix, which we call BSM. Once the BSM is obtained, eigenvalues, eigenstates and determinants are easily calculated using modern standard libraries and are less time consuming than the evaluation of the BSM itself. The propagator functions have to be known in the complex plane within the region defined by $k_\pm^2 = k^2 - \eta_\pm^2 M^2 \pm 2i\eta_\pm Mkt$ for $k \in [0, \infty)$ and $t \in [-1, +1]$, which is the contour analyzed in Sec. 7.1.3 for

7 Dyson-Schwinger and Bethe-Salpeter approach

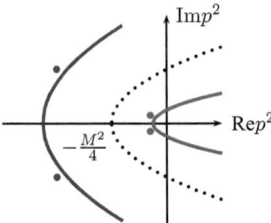

Figure 7.2.1: Parabolas and poles for asymmetric momentum partitioning (red and blue) in case of different quark masses. The dotted line is for $\eta = 0.5$ and includes the pair of poles of the light quark in the integration domain of DSE and BSE.

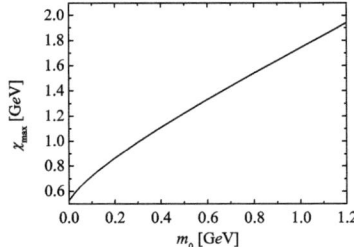

Figure 7.2.2: Maximal χ as a function of the bare quark mass.

$\lambda = 2\eta M k$. The analytic structure of the propagator functions σ_v and σ_s therefore restricts the analysis of the BSE if no additional schemes of handling the poles are introduced. Considering light quarks, e.g. $m_0 = 5\,\text{MeV}$ or $m_0 = 120\,\text{MeV}$, bound states of equal quarks can thus be evaluated up to an energy of $P_0 \approx 1.05\,\text{GeV}$ or $P_0 \approx 1.5\,\text{GeV}$, respectively. The gap between the lightest bound state and the pole of the corresponding quark ensures that the BSE can be solved.

As the position of the poles and, hence, the parameter χ depends on the quark mass, the maximum bound state mass to be evaluated depends on both quark masses, i.e. χ_1 and χ_2, and the frame, i.e. the momentum partitioning parameter η. The quark mass determines the maximal parabolic domain for the argument of the quark propagator in the complex q^2-plane as obtained from the DSE. The momentum partitioning parameter determines the required parabolic domains for the arguments of the quark propagators within the BSE. Because the BSE is Poincaré invariant, we may choose η such that Eq. (7.2.17) may be solved for the largest possible bound state mass without the pole entering the integration domain of neither of the two quarks. For equal

7.2 Bethe-Salpeter equation for mesons

Table 7.2.2: Maximal bound state masses and corresponding momentum partitioning parameter given as tuple $(M_{\max}, \eta(M_{\max}))$ for different current quark masses. The momenta indicate the integration contour for the respective quark. All masses are in GeV.

$k_\pm$		$(1-\eta)P-k$		
	m_0	0.005	0.115	1.05
$\eta P + k$	0.005	(1.07922, 0.5)	(1.28479, 0.42)	(2.33122, 0.23147)
	0.115	(1.28479, 0.58)	(1.49036, 0.5)	(2.53679, 0.29375)
	1.05	(2.33122, 0.76853)	(2.53679, 0.70625)	(3.58322, 0.5)

quark masses it is clear that $\eta = 0.5$ provides the frame in which the largest bound state mass can be evaluated. However, from Tab. 7.1.1 we see that, for heavy-light quark mesons, η may be changed in order to increase the maximal bound state mass, see Fig. 7.2.1. In fact, demanding $\chi_1/\eta_+ \stackrel{!}{=} \chi_2/\eta_-$ gives

$$\eta = \eta_+ = \frac{\chi_1}{\chi_1 + \chi_2} = 1 - \eta_-, \qquad (7.2.19)$$

and the maximum bound state mass of two quarks for which the BSE can be solved reads

$$M_{\max} = \chi_1 + \chi_2. \qquad (7.2.20)$$

For given valence quark mass the location of the poles directly translates by virtue of Eq. (7.1.45) to the maximum accessible parameter by determining $\chi_{\max}(m_0)$, see Fig. 7.2.2. For various combinations of bare quark masses, chosen in view of the investigation presented at the end of this chapter, the largest accessible bound state mass and the corresponding momentum partition are given in Tab. 7.2.2. However, as in practical calculations the BSE is only approximately Poincaré invariant due to the truncation of the expansion into Chebyshev polynomials, cf. Eq. (7.2.16), the results are not independent of the momentum partitioning. The thus obtained results can therefore only be considered as a proof of concept and may be improved by virtue of extra numerical effort. Indeed, increasing the number of Chebyshev polynomials improves the stability w. r. t. variations of η [Alk02]. The dependence on η also depends on the considered channel [Alk02]. Moreover, shifting the momentum partitioning results in increasing bound state masses obtained as solutions of the BSE. This numerical η-dependence is particularly strong for large differences in the bare quark masses of the valence quarks. Therefore, the bound state mass, which would have been obtained as a solution of the numerical BSM, might be larger than the bound state mass which would have been obtained from the complete BSE and might even exceed the maximal bound state mass which is calculable by shifting the momentum partition. However, the situation can in principle be improved by according numerical efforts.

The bound state masses of scalar and pseudo-scalar mesons are evaluated for the potential (7.1.5) with $\omega = 0.5\,\text{GeV}$ and $D = 16\,\text{GeV}$ as follows. First, we evaluate the pseudo-scalar bound state mass curve M_{xx} for equal quarks up to quark masses of $m_x = 1.125\,\text{GeV}$. The curve is used to determine the bare up quark mass such that the pion mass $m_\pi = 135\,\text{MeV}$ is reproduced. We obtain $m_u = 5\,\text{MeV}$. This mass is then used to fit the strange quark mass at the Kaon mass $m_K = 498\,\text{MeV}$ and the charm quark mass at the D meson mass $m_D = 1.865\,\text{GeV}$ by evaluating the bound state mass curve M_{ux} for one quark being fixed to the up quark. We obtain $m_s = 115\,\text{MeV}$ and $m_c = 1.05\,\text{GeV}$. The thus obtained quark masses may now be used to evaluate the bound state mass curves M_{sx}, with one quark being fixed to the strange quark, and M_{cx}, with one quark being fixed to the charm quark. In this way the masses of $s\bar{s}$ (hypothetical), $s\bar{c}$ (D_s) and $c\bar{c}$ (η_c) states are predicted. The quark masses obtained in the pseudo-scalar channel serve as input for the scalar channel.

In Fig. 7.2.3 the results for pseudo-scalar and scalar bound states are given together with the maximal accessible bound state mass of the respective quark combination. The hypothetical pseudo-scalar $s\bar{s}$ state is found at a mass of 693 MeV. The D_s mass is $m_{D_s} = 1.934\,\text{GeV}$, which fits the experimental value 1.968 GeV [Nak10]. The η_c mass is $m_{\eta_c} = 2.825\,\text{GeV}$. Experimentally, a mass of $m_{\eta_c} = 2.980\,\text{GeV}$ [Nak10] is found. As can be seen, the method is in principle capable of determining the pseudo-scalar bound state masses including states with charm quarks. This allows to predict three meson masses in the pseudo-scalar channel for D_s, J/ψ and a hypothetical pseudo-scalar $s\bar{s}$ state. Note that the pseudo-scalar meson mass becomes zero if both quark masses approach zero. Hence, the character of the pion as the (massless) Goldstone boson of the strong interaction which only acquires a nonzero mass due to the explicit symmetry breaking is fulfilled in the employed model. Concerning the parameters of the interaction, ω and D, this can be used as a constraint such that only one free parameter is left; i.e. $\omega = \omega(D)$. Indeed, for the chosen values $\omega = 0.5\,\text{GeV}$, $D = 16\,\text{GeV}$ we have $M = 0.1\,\text{MeV}$ for both bare quark masses being zero. Once the bound state mass has been evaluated, the BSV is known and further observables are calculable. Hence, also the leptonic decay constant or other quantities can be used

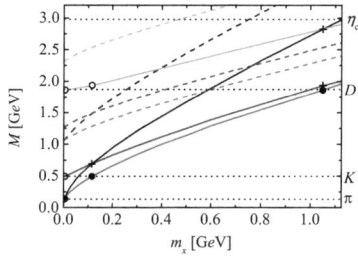 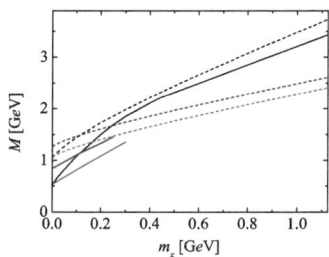

Figure 7.2.3: Bound state masses in the pseudo-scalar (left) and scalar (right) channel for equal (black solid) and unequal bare quark masses: up (red solid, $m_u = 5\,\text{MeV}$), strange (blue solid, $m_s = 115\,\text{MeV}$) and charm quark (green solid, $m_c = 1.05\,\text{GeV}$). The dashed curves depict the upper limits of bound state masses for the respective quark combination. The horizontal dotted lines are the experimental meson masses [Nak10]: $m_\pi = 135\,\text{MeV}$, $m_K = 498\,\text{MeV}$, $m_D = 1.865\,\text{GeV}$ and $m_\eta = 2.980\,\text{GeV}$. Dots mark the meson masses which are used to fit the quark masses, circles mark internal consistency checks and crosses mark predicted meson masses.

to specify parameters of the model such as quark masses, range and strength of the phenomenological interaction or any other more sophisticated model. Unfortunately, the mesons of the scalar channel are too heavy to reach the charm quark sector for unequal quarks. Bound states with a strange or an up quark can only be handled up to $m_x \approx 250\,\text{MeV}$. Furthermore, numerical instabilities trouble the solution when the (expected) bound state mass is close to the maximal bound state mass. This can be traced back to the numerical dependence of the bound state mass on the momentum partitioning parameter due to the truncation of the Chebyshev expansion (7.2.16). The thus expected increased masses exceed the maximal accessible bound state mass. Of course, the precise position of the poles depends on the interaction and is altered for more sophisticated gluon propagators [Sou10a].

7.3 Wigner-Weyl solution

Experimentally, the pion dynamics and the pion mass in the Wigner-Weyl phase is not known. However, calculations within the Nambu-Jona-Lasinio model at nonzero temperature and baryon density [Rat04, Ber87] point to an (monotonically) increasing pion mass. Inspired by the investigation of Sec. 5 one may employ the coupled Dyson-Schwinger–Bethe-Salpeter approach to ask for meson properties within a chirally symmetric solution without incorporating medium effects and, hence, probing the effect of chiral symmetry restoration on hadronic properties as such and disentangled from many-body effects. Similar investigations have been done in [Bic06] and references therein for simplified confining models.

As pointed out in Sec. 7.1.2, the system (7.1.11) has multiple solutions. In particular, it obeys a chirally symmetric solution if $m_0 = 0$, which can easily be obtained by choosing $B(p) = 0$ as initial function for the fixed-point iteration. It is called the Wigner-Weyl solution. It features $M(p) = 0$ with $A(p) \neq 0$. It is indeed a chirally symmetric solution, i. e. Eq. (7.2.9) gives the same BSM for scalar and pseudo-scalar mesons and both mesons are degenerate in their mass, as can be shown in the case of zero bare quark masses in a two-fold way. First, the chiral condensate, which is given within the employed model (7.1.5) for the gluon propagator by [Zon03b, Lan03]

$$\langle :\bar{q}q: \rangle = -\frac{3}{2\pi^2} \int \mathrm{d}l \, l^3 \sigma_s(l), \qquad (7.3.1)$$

is zero, because $B(p) = 0$. In the Nambu-Goldstone phase we obtain $\langle :\bar{q}q: \rangle = (-251\,\mathrm{MeV})^3$ for the employed set of parameters, which is in agreement with the Gell-Mann–Oakes–Renner relation (A.2.28). As it is an order parameter of DCSB, vanishing of the chiral condensate points to the restoration of the symmetry. However, strictly speaking the vanishing of an order parameter, which is qualified as such by means of Eq. (A.2.19), is merely a necessary but not sufficient requirement for the realization of a symmetry. The realization of other symmetries may lead to a vanishing condensate as well. Second, it can be seen as degeneracy of the solutions to the BSE for chiral partners. With the relation between scalar and pseudo-scalar BSE given by Eq. (7.2.11), the only difference of the BSMs is due to sign changes of certain traces and an additional sign according to $M_1(p) \rightarrow -M_1(p)$ in all traces. As $M_1(p) = M_2(p) = 0$ the latter prescription is without effect for the BSM in the Wigner-Weyl phase, and one is left with the following changes

$$M^\mathrm{P}_{ij} = -M^\mathrm{S}_{ij} \quad : \left(i \in \{1,3,4\} \wedge j \in \{2,4\}\right) \vee \left(i = 2 \wedge j \in \{1,3\}\right). \qquad (7.3.2)$$

Apart from the cases $(i,j) = (4,2)$ and $(i,j) = (4,4)$ all of the affected terms are proportional to $M_i(p)$ and are, thus, zero. For the other two cases it turns out, that the difference $M^\mathrm{P}_{ij} - M^\mathrm{S}_{ij}$ is proportional to $M_i(p)$ and therefore must be zero for $M_i(p) = 0$, too, which proves the stated degeneracy. Since the BSEs for scalar and pseudo-scalar mesons are identical in the Wigner-Weyl phase, the BSVs and, hence, observables are

7.3 Wigner-Weyl solution

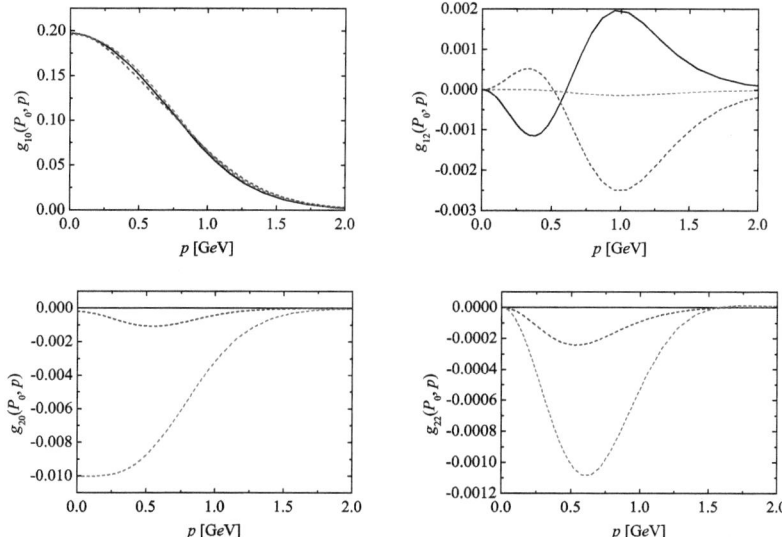

Figure 7.3.1: The scalar and pseudo-scalar partial amplitudes g_{10} (upper left panel), g_{12} (upper right panel), g_{20} (lower left panel) and g_{22} (lower right panel) in the chirally broken phase (blue dashed: scalar, red dashed: pseudo-scalar) and the Wigner-Weyl phase (black). For g_{2j} (lower panels), the scalar partial amplitudes in the Nambu-Goldstone phase have been scaled (multiplied) by 10^{15}.

too.

Having the propagator functions in the Wigner-Weyl phase for zero bare quark mass at our disposal, see Fig. 7.1.3, one can solve the BSE in the Wigner-Weyl phase and obtain an explicit value of the bound state mass. As in [Bic06], the lowest bound state mass where the BSE (7.2.9) can be solved is found at negative squared masses $M^2 = -0.1172\,\text{GeV}^2 = -(342.1\,\text{MeV})^2$ and are therefore called tachyonic solutions. As argued in [Jai07] the Wigner-Weyl solution corresponds to a maximum of the effective action. Therefore, the squared mass must be negative and signals the instability of the chiral symmetric ground state. An arbitrary small disturbance drives the system from the Wigner-Weyl realization to the Nambu-Goldstone realization.

In the upper panels of Fig. 7.3.1 the two largest partial amplitudes of the Wigner-Weyl solution are compared to the corresponding functions of the solution in the chirally broken phase. Note that the largest contribution g_{10} remains almost unchanged, whereas g_{12} completely changes. The partial amplitudes for g_2 are shown in

7 Dyson-Schwinger and Bethe-Salpeter approach

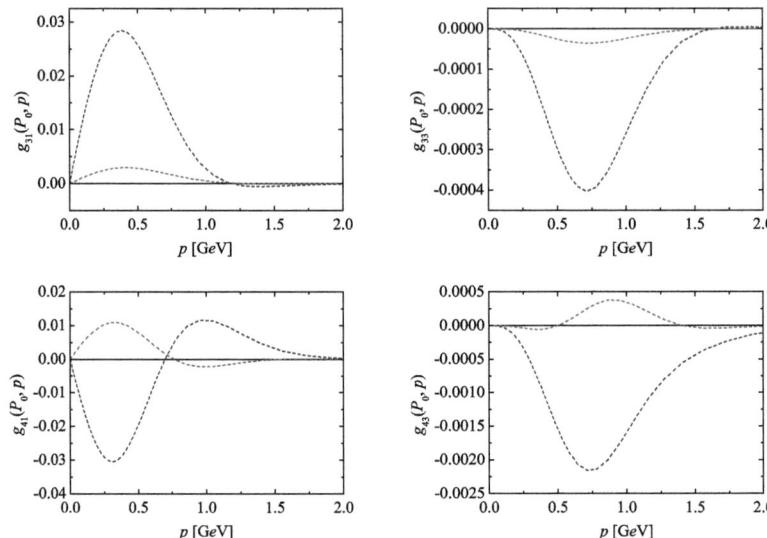

Figure 7.3.2: The scalar and pseudo-scalar partial amplitudes g_{31} (upper left panel), g_{33} (upper right panel), g_{41} (lower left panel) and g_{43} (lower right panel). Line codes as in Fig. 7.3.1. For g_{33} (upper right panel), the pseudo-scalar partial amplitude for the Nambu-Goldstone phase in the right panel has been scaled (multiplied) by 10. For g_{4j} (lower panels), the pseudo-scalar partial amplitudes in the Nambu-Goldstone phase have been scaled (multiplied) by 10^{16}.

the lower panels of Fig. 7.3.1 and g_3 and g_4 are depicted in Fig. 7.3.2. Note that, in the Wigner-Weyl phase, g_2 and g_4 are identically zero to any order in the expansion over Chebyshev polynomials, whereas g_3 turns out to be numerically zero. Nevertheless, it is very illustrative to compare the partial amplitudes of scalar and pseudo-scalar mesons in the Nambu-Goldstone phase as the pseudo-scalar amplitude is by magnitudes larger than its scalar counterparts for g_2. The contrary is the case for g_3 and g_4.

In [Wil07b, Wil07d, Wil07c, Wil07a] the Wigner-Weyl solution and so-called noded solutions[20] to the DSE have been used to discuss the chiral condensate beyond the chiral limit. Thereby, linear combinations of the quark propagators have been

---

[20]Solutions of the DSE in the Nambu-Goldstone phase do not have roots along the positive real axis. Solutions in the Wigner-Weyl phase have one root (node). Other solutions have more than one node and are, therefore, dubbed noded solutions. In analogy to vibrating strings, they are sometimes referred to as excited solutions [LE07].

7.3 Wigner-Weyl solution

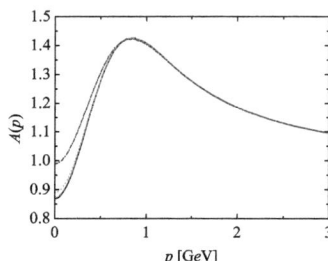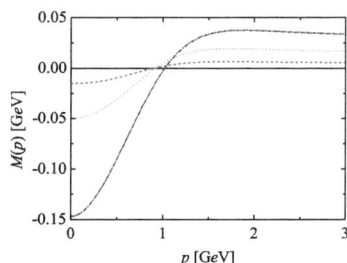

Figure 7.3.3: The propagator functions in the Wigner-Weyl phase. Solid black line: $m_0 = 0$ MeV, dashed red line: $m_0 = 5$ MeV, dotted green line: $m_0 = 15$ MeV, dashed-dotted blue line: $m_0 = 30$ MeV.

introduced, which all generate identical condensates in the chiral limit. However, due to the non-linearity of the DSE, a linear combination of solutions cannot fulfill the corresponding DSE and, therefore, is questionable in its physical relevance. It has been pointed out, that Wigner-Weyl solutions exist up to a critical quark mass, but cannot be obtained by a fixed-point iteration method as they correspond to maxima in the effective action. Instead, Newton methods or improvements thereof need to be employed. This allows to study a scenario without DCSB, but with explicit symmetry breaking by small quark masses. Employing a Newton-Krylov iteration method to find Wigner-Weyl solutions, a critical bare quark mass of $m_0^{cr} = 30$ MeV has been found. Its value is model dependent and may be interpreted as the maximum quark mass for which chiral symmetry may be restored. The obtained propagator functions are shown in Fig. 7.3.3.

Having the Wigner-Weyl solutions for the quark DSE at our disposal, we are now able to study a scenario only with explicit symmetry breaking in the scope of a coupled Dyson-Schwinger–Bethe-Salpeter approach. Note that the thus obtained bound state mass may only be considered as a formal solution, in the sense that it is the mass parameter at which the BSE has a solution rather than a state which is realized in nature. Such a scenario may be useful to investigate the amount of mass splitting in the parity doublet caused by finite quark masses. For the scalar and pseudo-scalar channel the masses are shown in Fig. 7.3.4. They are complex over the whole mass region. Moreover, the pseudo-scalar bound state mass M^2 is even decreasing. Finally, evaluating the formal splitting of scalar and pseudo-scalar mesons at a quark mass of $m_0 = 5$ MeV yields $|\Delta M| = 1.6$ MeV, which is tiny as compared to the splitting due to DCSB ($|\Delta M| \approx 350$ MeV). However, it is of the order of the mass splitting in the

7 Dyson-Schwinger and Bethe-Salpeter approach

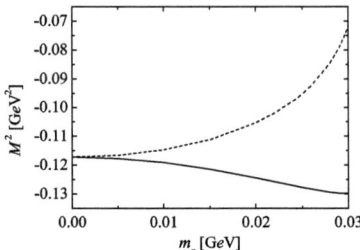

Figure 7.3.4: The bound state mass in the Wigner-Weyl phase for equal quarks in the pseudo-scalar (black) and scalar (dashed) channel up to the critical mass $m_0^{\text{cr}} = 30\,\text{MeV}$.

isospin multiplet.

In view of the investigation of Sec. 5 the extension of the above presented analysis to the spin-1 channel is in order. Furthermore, based on a phenomenological interaction which successfully describes the hadronic spectrum, the solutions of the DSE in the Wigner-Weyl phase may be used to determine the condensates, in particular the symmetric four-quark condensate $\langle \mathcal{O}_4^{\text{odd}} \rangle$, in the chiral symmetric ground state. This provides us with a reliable relation between changes of chirally symmetric and chirally odd condensates.

8 Summary and outlook

This thesis is devoted to the determination of medium induced modifications of hadron properties. In particular, we concentrate on changes of the mass and the spectral function of heavy-light mesons, such as the pseudo-scalar D meson, and the ρ meson. The motivation is a quantitative prediction of possible mass shifts and splittings which may be observed in upcoming experiments on the one hand and on the other hand on determining the relation of medium modifications of hadrons to the anticipated restoration of the chiral symmetry of quantum chromodynamics and its order parameters. Medium modifications of mesons might be signals for the restoration or partial restoration of the chiral symmetry. Summarizing the main findings we mention the successful quantification of the in-medium mass splittings of heavy-light pseudo-scalar and scalar D mesons. Furthermore, chiral partner sum rules for heavy-light spin-1 and spin-0 mesons in the medium have been derived. It has been demonstrated that chiral symmetry restoration indeed affects the ρ meson spectral function and that the observed in-medium effects may be caused by a partial restoration of the chiral symmetry. However, it has also been found that in general the relation of medium modifications of heavy-light mesons as well as the ρ meson is not as direct as often anticipated. In particular the correlation of parameters which characterize the spectral function of a meson to condensates which serve as order parameters of the chiral symmetry is rather intricate.

The in-medium sum rule analysis up to and including mass dimension 5, has been performed for the heavy-light pseudo-scalar D, $\bar{D}$ and $\bar{B}$, B and the scalar D^*, $\bar{D}^*$ mesons. In all cases a fairly robust mass splitting (for the employed set of condensates) has been found, while the mass shift of the centroids depends on the in-medium modeling of the continuum threshold. The chiral condensate $\langle \bar{q}q \rangle$ enters the mass splitting in next-to-leading order of the density. Nevertheless, at nuclear saturation density its impact is comparable to $\langle q^\dagger g \sigma \mathcal{G} q \rangle$ due to the heavy charm quark mass. The leading contribution to the mass splitting in case of up or down valence quarks in nuclear matter at saturation density, however, is $\langle q^\dagger q \rangle$, which measures the net quark density. For D_s mesons it measures the net quark density of strange quarks,

8 Summary and outlook

which is zero in nuclear matter at saturation density, and results in mass splittings of opposite sign as compared to D mesons. As the sign of the condensate $\langle q^\dagger g\sigma\mathcal{G}q\rangle$ is not known, the size of the $D-\bar{D}$ mass splitting may serve as an indicator for it.

We extended Weinberg-Kapusta-Shuryak sum rules in the medium for light quark spin-1 mesons to heavy-light spin-1 and spin-0 mesons in the medium. While the only non-trivial contributions to the former stems from power corrections to the $\mathcal{O}(\alpha_s)$ perturbative expansion and is of mass dimension 6, i.e. a four-quark condensate, the chiral condensate enters the latter one as the lowest-order power correction to the $\mathcal{O}(\alpha_s^0)$ perturbative term, i.e. it is of mass dimension 4, and is amplified by the heavy charm quark mass. Indeed, the chiral condensate only enters chiral partner sum rules in $\mathcal{O}(\alpha_s^0)$ for heavy-light mesons. Addressing chiral symmetry restoration, $m_c\langle\bar{q}q\rangle$, $m_c\langle\bar{q}g\sigma\mathcal{G}q\rangle$, and $m_c(\langle\bar{q}g\sigma\mathcal{G}q\rangle - 8\langle\bar{q}D_0^2q\rangle)$ may be considered as "order parameters". Conversely, vanishing of these condensates in the medium would mean the degeneracy of chiral partners. Evaluation of the sum rules at zero temperature and nuclear saturation density exemplifies a significant change of the operator product expansion side, which, consequently requires an according modification of the integrated spectral difference of chiral partners in the heavy-light meson sector.

We have extracted the symmetric part of the four-quark condensate which enters the ρ meson QCD sum rule in multiple ways. It turns out that equating the first moment of the spectral function to the ρ mass, often abbreviated as pole-ansatz, gives values which are closer to the values obtained by directly invoking experimental data than a common Breit-Wigner ansatz. The weakness of the latter one is the modeling of the low-energy region, which is a crucial domain due to the exponential weighting of the spectral function within Borel transformed sum rules. The vanishing of chirally odd condensates scenario for the ρ meson exemplifies the impact of chirally odd condensates on the spectral function. We have shown, that setting all chirally odd condensates to zero, requires an enhancement of the spectral function at lower energies. However, there has been neither evidence for a vanishing of the meson masses at the point of chiral restoration, nor that the observed in-medium changes are not linked to chiral restoration.

We applied the coupled Dyson-Schwinger–Bethe-Salpeter approach to diquark bound states for equal and unequal quarks. In order to do so, an exhaustive investigation concerning the analytic structure of the quark propagator in the complex plane, i.e. the position of its poles, was necessary. It has been shown that in the complex energy plane the imaginary part of the lowest pole determines the maximal parabola in the complex plane along which the Dyson-Schwinger equation can be solved. As this parabola is chosen such that it corresponds to the integration domain of the respective quark within the Bethe-Salpeter equation, the same quantity also limits the maximal accessible bound state mass. Concerning the lowest bound state masses for equal quarks, the analytic structure is such that the singularities are always outside of the integration domain. For unequal quarks, such as the D meson, it has been shown that the momentum partitioning parameter provides a restricted possibility to increase the largest accessible bound state mass, which in turn depends on the imaginary parts of the lowest poles of both quarks. This allows to evaluate pseudo-scalar mesons for heavy-light quarks up to the D and D_s meson. However, it also turned out that the applied truncation of the expansion over Chebyshev polynomials and the associated breakdown of Lorentz covariance of the Bethe-Salpeter equation, leads to limitations in the scalar channel, where the meson masses are naturally close to the maximal bound state mass.

We used the possibility to solve the Dyson-Schwinger equation in the Wigner-Weyl mode for nonzero quark masses up to a critical mass to study a scenario of explicit but without dynamical chiral symmetry breaking within the BS approach. The determined bound state masses are complex and may therefore only be regarded as formal solutions to the Bethe-Salpeter equation. For equal quarks the results gave an impression of the tiny mass splitting of scalar and pseudo-scalar mesons due to finite quark masses.

Outlook

On the basis of the operator product expansion techniques developed within this thesis, the evaluation of the mass dimension 6 power corrections to heavy-light meson sum rules beyond the factorization of four-quark condensates is seizable. An interesting aspect will be the cancellation of infrared divergences introduced by gluon condensates in this mass dimension. Also the QCD sum rule analysis in the spin-1 channel for heavy-light mesons can be performed within the presented formalism.

The coupled Dyson-Schwinger–Bethe-Salpeter approach may be extended to spin-1 mesons. However, these studies are yet restricted to vacuum properties of mesons as bound states of quarks and antiquarks. Once the singularity structure of the approach is under control, the formalism can be extended to finite temperature and density by introducing Matsubara frequency summations [Kap06]. Furthermore, observables beyond the bound state mass, such as decay constants and form factors, should be considered. Additionally, the gluon and ghost Dyson-Schwinger equations

8 Summary and outlook

and the corresponding vertex equations may be included. Asymmetric momentum routings for the quark Dyson-Schwinger equation and a complex conjugate mass pole parametrization of the quark propagator may be applied to analyze and understand its analytic structure in the complex plane. The Bethe-Salpeter equation may also be solved in the Wigner-Weyl phase of heavy-light mesons and provides the possibility to investigate consequences of the chiral symmetry restoration for these mesons.

As the Dyson-Schwinger equation approach allows to evaluate condensates, a possibility to check the consistency of both methods and to systematically determine condensates and their medium dependence is at our disposal. The Wigner-Weyl solution of the Bethe-Salpeter equation and the vanishing of chirally odd condensates scenario for the ρ meson may be compared in the spin-1 channel. An interesting combination of both approaches would be to determine four-quark condensates, in particular the chirally symmetric condensate $\langle \mathcal{O}_4^{\text{even}} \rangle$, in the Wigner-Weyl phase. This allows to check if the mechanism which causes the restoration of chiral symmetry also affects these condensates.

Appendices

A Chiral symmetry, currents and order parameters

As symmetries play a significant role in this thesis we will review basic ingredients of Lagrangian field theory, which is a common framework to study symmetries within quantum field theory. Exhaustive introductions to the topic can be found in e. g. [Wei95, Koc95, Koc97, Tre85, Sak73, Mos99, Adl68, Che82a, Pok00, Gre96, Rom69, Gol50, Ryd85]. The subsequent outline not only distillates aspects which are of importance for what is dealt with in this thesis but also gives a self contained derivation of the fundamental relations and thereby follows a path which, to the best of our knowledge, has not been presented in the literature so far. In particular we will emphasize the classical field theory part to shed some light on algebraical equalities which both theories, classical and quantum field theory, have in common. This will allow to identify more clearly such features of quantum field theory which do not exist in a classical field theory. In particular, it allows to derive a classical algebra of currents. The existence of such an algebra is not surprising. On the way we will generalize the constant time concept of Poisson brackets to arbitrary space-like surfaces. It is clear, however, that only the constant time brackets determine the canonical commutators or anticommutators in canonical quantization. Nevertheless, the generalization clearly exposes the nature behind the specific role played by the time coordinate in the context of Legendre transformations. Having a classical algebra of currents at hand, one may ask for the difference between the classical and the quantum algebra of currents. The difference appears in form of Schwinger terms and is intimately connected to the subject of anomalies, divergent operator products and OPEs. As this is a fairly wide subject, we will merely review the main ideas and collect the rather scarce literature on this complicated but nevertheless fundamental topic.

Later on, emphases is put on internal chiral transformations, their intimate connection to currents and symmetries of the theory. Thereby we point out in detail the role chiral transformations and dynamical chiral symmetry breaking play in the sector of heavy-light quark mesons. Furthermore, transformation properties of and

A Chiral symmetry, currents and order parameters

commutation relations between meson currents are derived for arbitrary flavor content. In contrast to the well-known transformation properties and current-current commutation relations of light-quark meson currents, relations for arbitrary flavor content are needed to investigate, in particular, mesons consisting of a heavy and a light quark.

Apart from these developments, this appendix is not designed to give a comprehensive overview about chiral symmetry breaking or even a complete treatise thereof, but rather tends to collect the main ideas and relations which are necessary to guide the reader from a common base of knowledge to the theoretical background of this work. This also includes an introduction to order parameters and QCD.

A.1 The formalism for classical fields

Many algebraic properties of currents and charges already exist on the level of classical field theory. This is clear, as the canonical quantization relies on the replacement of Poisson brackets of classical fields by commutators of quantum field operators. In particular, symmetries are often considered by virtue of the operator algebra whereas the classical Lagrangian is used. Quantization of classical gauge field Lagrangians, such as the classical Lagrangian of chromodynamics, necessitates the introduction of gauge-fixing and ghost terms. Although throughout this work emphasis is put on infinitesimal internal transformations of the quark fields, which do not act on gauge fields, it is didactically advantageous to see how the fundamental relations arise from a classical field theory by virtue of classical algebraic relations rather than by virtue of quantum field theory commutators. Moreover, as we will see later on in the context of anomalies and Schwinger terms, the operator algebra cannot be complete.

Equations of motion and infinitesimal transformations

Let the action be given as $S = \int_R d^4x\, \mathscr{L}(\phi, \partial_\mu \phi)$, with R a four-volume and $\mathscr{L} = \mathscr{L}(\phi, \partial_\mu \phi)$ a Lagrangian density of the c-number fields ϕ and their first derivatives $\partial_\mu \phi$ (possible additional indices, e. g. flavor or color or Dirac indices, are suppressed).[21] A set of infinitesimal transformations may be specified by [Rom69]

$$\delta\phi(x) = \phi'(x') - \phi(x), \qquad (A.1.1a)$$

---

[21] The discussion is restricted to the case of second order EoM for the fields. Otherwise the Lagrangian could additionally depend on higher derivatives of the fields and a more extended investigation would be in order.

A.1 The formalism for classical fields

$$\delta x^\mu = x'^\mu - x^\mu, \tag{A.1.1b}$$

$$\delta^\Lambda \phi(x) = \phi'(x) - \phi(x), \tag{A.1.1c}$$

where (A.1.1a) are the local and (A.1.1c) the total field variations. Clearly, two types of transformations have to be distinguished. Pure space-time transformations ($\delta\phi = 0$), such as e.g. shifts or rotations, and pure internal transformations ($\delta x^\mu = 0$) which only act on the field degrees of freedom leaving the space-time coordinates invariant, such as e.g. phase transformations or rotations in flavor space. Then, the integral measure transforms by virtue of the Jacobian as $\delta d^4x = d^4x' - d^4x = \left(\partial_\mu \delta x^\mu\right) d^4x$, whereas partial derivative and total field variation commute $\partial_\mu \left(\delta^\Lambda \phi(x)\right) = \delta^\Lambda \partial_\mu \phi(x)$. It is also useful to relate local and total field variation: i.e. $\delta\phi(x) \approx \delta^\Lambda \phi(x) + \left(\partial_\mu \phi(x)\right) \delta x^\mu$ for infinitesimal transformations. The induced variation of the Lagrangian density is

$$\delta\mathscr{L} = \left(\frac{\partial\mathscr{L}}{\partial\phi} - \partial_\mu \frac{\partial\mathscr{L}}{\partial\left(\partial_\mu\phi\right)}\right)\delta^\Lambda\phi + \partial_\mu\left(\frac{\partial\mathscr{L}}{\partial\left(\partial_\mu\phi\right)}\delta^\Lambda\phi\right) + \left(\partial_\mu\mathscr{L}\right)\delta x^\mu \tag{A.1.2}$$

and the action varies as

$$\delta S = \int_R d^4x \left[\left(\frac{\partial\mathscr{L}}{\partial\phi} - \partial_\mu \frac{\partial\mathscr{L}}{\partial\left(\partial_\mu\phi\right)}\right)\delta^\Lambda\phi + \partial_\mu j^\mu\right]$$

$$= \int d^4x \left(\frac{\partial\mathscr{L}}{\partial\phi} - \partial_\mu \frac{\partial\mathscr{L}}{\partial\left(\partial_\mu\phi\right)}\right)\delta^\Lambda\phi + \int_\sigma d\sigma_\mu j^\mu, \tag{A.1.3}$$

where we have defined the associated Noether current j_μ, the energy-momentum tensor $T^{\mu\nu}$ and the canonical conjugate momentum π^μ as

$$j^\mu = -T^{\mu\nu}\delta x_\nu + \pi^\mu \delta\phi, \tag{A.1.4a}$$

$$T^{\mu\nu} = \pi^\mu \left(\partial^\nu \phi\right) - g^{\mu\nu}\mathscr{L}, \tag{A.1.4b}$$

$$\pi^\mu = \frac{\partial\mathscr{L}}{\partial\left(\partial_\mu\phi\right)}. \tag{A.1.4c}$$

Schwinger's action principle then requires that the variation of the action is merely a functional of the surface

$$\delta S = F[\sigma_1] - F[\sigma_2]. \tag{A.1.5}$$

149

As the four-volume R and the local variation $\delta^\wedge \phi$ are arbitrary this requirement leads to the well-known EoM, called Euler-Lagrange equations

$$\partial_\mu \pi^\mu = \frac{\partial \mathscr{L}}{\partial \phi}. \tag{A.1.6}$$

Note that no boundary conditions for the variations have been introduced. The quantity F is the generator of the infinitesimal transformation defined as

$$F[\sigma] = \int_\sigma d\sigma^\mu \, j_\mu, \tag{A.1.7}$$

and we assumed in Eq. (A.1.5) for simplicity that the four-volume R may be bounded by two space-like surfaces $\sigma_{1,2}$. For a constant-time surface the generator will be referred to as

$$Q(t) = \int_t d^3x \, j_0(x). \tag{A.1.8}$$

A non-space-like surface, i.e. time-like or light-like, leads to causality constraints of the boundary conditions and one would have to take care for choosing them non-contradictory. On a space-like surface, i.e. space-like separation of all points on the surface, instead each point lies in the exterior of the light-cone of all the other points on the surface and causality guarantees a free choice of boundary conditions. The variation of a functional as given in Eq. (A.1.7) with respect to the surface σ is [Rom69]

$$\frac{\delta F[\sigma]}{\delta \sigma(x)} = \partial_\mu j^\mu. \tag{A.1.9}$$

An infinitesimal transformation (A.1.1) is called a symmetry transformation if it leaves the action invariant, i.e. $\delta S = 0$. According to Eqs. (A.1.5), (A.1.7) and (A.1.9) this means that the generator F does not depend on the surface σ. Hence, its variation w.r.t. the surface is zero, which is equivalent to the conservation of the associated Noether current $\partial^\mu j_\mu = 0$. The generator F is then a constant of motion, e.g. in case of a constant time surface it is time independent, $\partial_t F = 0$. In [Rom69] the most common symmetry transformations such as shifts, rotations and phase transformations and the corresponding conservation laws (energy-momentum, angular momentum and charge conservation) can be found.

Poisson brackets

In the following Poisson brackets are introduced. They will allow to obtain a classical current algebra. Consider the transformation given by constant shifts of the coordinates, i.e. $\delta x^\mu = \epsilon^\mu = \text{const}$. It follows that $\delta^\wedge \phi(x) = -\epsilon^\mu \partial_\mu \phi(x)$ and $\delta \phi(x) = 0$. Poisson brackets are then defined such that they reproduce the EoM. The generator becomes

$$F[\sigma] = -\epsilon^\nu P_\nu \qquad (A.1.10)$$

with the energy-momentum four-vector

$$P_\nu[\sigma] = \int_\sigma d\sigma_\mu \, T^\mu{}_\nu . \qquad (A.1.11)$$

Furthermore, consider the functional

$$G(t) = \int d^3x \, \mathscr{G}^0(\phi, \pi_0, \phi_{,i}) , \qquad (A.1.12)$$

i.e. a functional of a constant-time surface with density $\mathscr{G}^0$. Note that $\mathscr{G}^0$ depends on the canonical conjugate momentum which, in terms of Legendre transformations, is the slope of the density w.r.t. to the time-derivative of the field. The time derivative of such a functional is

$$\partial_t G(t) = \int d^3x \left[\left(\frac{\partial \mathscr{G}^0}{\partial \phi} - \partial_i \frac{\partial \mathscr{G}^0}{\partial \phi_{,i}} \right) \partial_t \phi + \left(\frac{\partial \mathscr{G}^0}{\partial \pi_0} - \partial_i \frac{\partial \mathscr{G}^0}{\partial \pi_{0,i}} \right) \partial_t \pi_0 \right] \qquad (A.1.13)$$

which can be expressed in terms of the functional derivative

$$\frac{\delta}{\delta \eta(x)} G(t) = \frac{\partial \mathscr{G}^0}{\partial \eta} - \partial_i \frac{\partial \mathscr{G}^0}{\partial \eta_{,i}} \qquad (A.1.14)$$

with η being a field or a canonical conjugate momentum. On the other hand,

$$P_0(t) = \int d^3x \, T_{00} = \int d^3x \, (\pi_0 \partial_t \phi - \mathscr{L}) \qquad (A.1.15)$$

is the Legendre transform of the Lagrangian density w. r. t. the time derivative of ϕ and fulfills

$$\frac{\delta}{\delta\phi}P_0 = -\partial_t\pi_0, \qquad (A.1.16a)$$

$$\frac{\delta}{\delta\pi_0}P_0 = \partial_t\phi. \qquad (A.1.16b)$$

Insertion into Eq. (A.1.13) gives the time derivative of the functional $G(t)$

$$\partial_t G(t) = \int d^3x \left[\frac{\delta G}{\delta\phi}\frac{\delta P_0}{\delta\pi_0} - \frac{\delta G}{\delta\pi_0}\frac{\delta P_0}{\delta\phi} \right]. \qquad (A.1.17)$$

Finally, the change of the functional G under infinitesimal shifts of the coordinates (A.1.10) is $\delta G = -\epsilon^0 \partial_t G$. Thus, Eq. (A.1.17) may serve as a definition of the constant-time surface Poisson bracket. The quantity δG may, therefore, be written as

$$\delta G(t) = \{G(t), F(t)\}_t. \qquad (A.1.18)$$

So far the presented treatment can be found in many textbooks, such as e. g. [Rom69, Rom65, Gol50]. We will now generalize it to arbitrary flat space-like surfaces σ. Consider the functional

$$G[\sigma] = \int_\sigma d\sigma_\alpha \, \mathscr{G}^\alpha(\phi, n\cdot\pi, O^{\kappa\lambda}\phi_{,\lambda}) \qquad (A.1.19)$$

with the orthogonal projector

$$O^{\kappa\lambda} = g^{\kappa\lambda} - n^\kappa n^\lambda, \qquad (A.1.20)$$

satisfying $O_{\mu\alpha}O^{\alpha\nu} = O_\mu{}^\nu$ and n^κ being the surface orthogonal unit vector $d\sigma_\alpha(x) \equiv n_\alpha(x)d\sigma$, $n^2 = 1$, defining the surface, e. g. for a flat surface $n_\alpha x^\alpha - \tau = 0$. The surface element may be defined by $d\sigma_\alpha \equiv dV/dx_\alpha = d^4x/dx_\alpha$. The notation in Eq. (A.1.19) means that $\mathscr{G}^\alpha$ depends on the component of π^α perpendicular to the surface element $d\sigma^\alpha$, i. e. parallel to n^α. The remaining components of $\phi_{,\alpha}$ lie in the surface element $d\sigma^\alpha$. Orthogonality implies $(nO)^\nu = 0$ and any four-vector V_μ may be decomposed via $V_\mu = n_\mu(n\cdot V) + O_\mu{}^\nu V_\nu$. Hence, one obtains

$$\partial_\mu G[\sigma] = n_\mu(n\cdot\partial)\int_\sigma d\sigma \, (n\cdot\mathscr{G}) + O_{\mu\nu}\partial^\nu \int_\sigma d\sigma \, (n\cdot\mathscr{G}). \qquad (A.1.21)$$

A.1 The formalism for classical fields

Because of the orthogonal projection in the last term, the derivative only acts on coordinates which lie in the surface σ and have been integrated out. Thus, the integral does not depend on these coordinates anymore. This also exemplifies the special role of the time-derivative for constant-time surfaces. In particular, no information is lost when applying the projected derivative to the functional $G[\sigma]$

$$(n \cdot \partial) G[\sigma] = \int_\sigma d\sigma \ (n \cdot \partial)(n \cdot \mathcal{G}) \tag{A.1.22}$$

which is merely the formal notation of the trivial statement that the functional only depends on coordinates perpendicular to the surface. Of course, one is free to derive the functional w. r. t. any coordinate, but only the projection perpendicular to the surface gives a nonzero result. Furthermore, the functional derivative of G is

$$\frac{\delta G[\sigma]}{\delta \eta} = \frac{\partial (n \cdot \mathcal{G})}{\partial \eta} - O^{\kappa \lambda} \partial_\kappa \frac{\partial (n \cdot \mathcal{G})}{\partial \eta_{,\lambda}} . \tag{A.1.23}$$

In particular, using $O^{\kappa \lambda} \partial (n \cdot \pi)/\partial \phi_{,\lambda} = 0$, because the components of the canonical momentum perpendicular to the surface do not depend on field derivatives within the surface, the functional derivative of the energy-momentum four-vector P_ν reads

$$\frac{\delta P_\nu[\sigma]}{\delta \phi} = -\partial_\nu (n \cdot \pi), \tag{A.1.24a}$$

$$n_\alpha \frac{\delta P_\nu[\sigma]}{\delta \pi_\alpha} = \partial_\nu \phi . \tag{A.1.24b}$$

The change of the functional G under infinitesimal transformations (A.1.10) is $\delta G = -(\epsilon \partial) G$, which, by virtue of Eq. (A.1.21) and the subsequent discussion, can be written as $\delta G = -(n \cdot \epsilon)(n \cdot \partial) G$ and evaluates to

$$(n \cdot \partial) G[\sigma]$$
$$= \int_\sigma d\sigma \left(\frac{(n \cdot \mathcal{G})}{\partial \phi}(n \cdot \partial)\phi + \frac{(n \cdot \mathcal{G})}{\partial \phi_{,\nu}} O_\nu{}^\lambda (n \cdot \partial)\phi_{,\lambda} \right.$$
$$\left. + n_\alpha \frac{(n \cdot \mathcal{G})}{\partial \pi_\alpha}(n \cdot \partial)(n \cdot \pi) \right)$$
$$= \int_\sigma d\sigma \left[\left(\frac{\partial (n \cdot \mathcal{G})}{\partial \phi} - O^{\kappa \lambda} \partial_\kappa \frac{\partial (n \cdot \mathcal{G})}{\partial \phi_{,\lambda}} \right)(n \cdot \partial)\phi + n_\alpha \frac{(n \cdot \mathcal{G})}{\partial \pi_\alpha}(n \cdot \partial)(n \cdot \pi) \right]$$

A Chiral symmetry, currents and order parameters

$$= \int_\sigma d\sigma \left[\frac{\delta G[\sigma]}{\delta \phi} n_\alpha \frac{\delta(n \cdot P)}{\delta \pi_\alpha} - \frac{\delta G[\sigma]}{\delta \pi_\alpha} n_\alpha \frac{\delta(n \cdot P)}{\delta \phi} \right]$$
$$\equiv \{G[\sigma], n[\sigma] \cdot P\}_\sigma \,. \tag{A.1.25}$$

In the second line an integration by parts has been performed. In the third line, Eqs. (A.1.23) and (A.1.24b) have been used. The derivation clearly exposes the necessity for both functionals being evaluated at the same surface σ, e.g. equal times for a constant-time surface. Equation (A.1.25) also defines the Poisson brackets for two functionals $A[\sigma] = \int_\sigma d\sigma_\mu \, \mathscr{A}^\mu$ and $B[\sigma] = \int_\sigma d\sigma_\mu \, \mathscr{B}^\mu$

$$\{A,B\}_\sigma = \int_\sigma d\sigma_\mu \left[\frac{\delta A}{\delta \phi} \frac{B}{\delta \pi_\mu} - \frac{B}{\delta \phi} \frac{\delta A}{\delta \pi_\mu} \right]. \tag{A.1.26}$$

It is clear that one has to take care of the correct order and matrix structure of the functional derivatives if there are internal degrees of freedom such as color or flavor. For an arbitrary transformation the generator fulfills

$$n_\mu \frac{\delta F[\sigma]}{\delta \pi_\mu} = \delta^\Lambda \phi \tag{A.1.27}$$

and the variation of the fields can be written as

$$\{\phi(x), F[\sigma]\}_\sigma = \delta^\Lambda \phi(x). \tag{A.1.28}$$

Field and canonical conjugate momentum may be considered as functionals of themselves with parametric dependence on space and time and the thus defined Poisson brackets fulfill per construction

$$\{\phi(y), n \cdot \pi(x)\}_\sigma = \delta^{(3)}\left(O^{\mu\nu}(x-y)_\nu\right), \tag{A.1.29a}$$
$$\{n \cdot \pi(x), n \cdot \pi(y)\}_\sigma = \{\phi(x), \phi(y)\}_\sigma = 0, \tag{A.1.29b}$$

where evaluation at the same surface means $n \cdot y = n \cdot x$, and the three-dimensional Dirac-distribution is meant to act on the coordinates within the surface. This expression also directly follows from Eq. (A.1.28) using $\{AB,C\} = A\{B,C\} + \{A,C\}B$, a fundamental algebraic property of the Poisson brackets. In analogy to Eqs. (A.1.25) and (A.1.28) a transformation given by its generator $F[\sigma]$ is called canonical transformation if

$$\{G,F\}_\sigma = \delta G. \tag{A.1.30}$$

Note the sign, as for a constant shift of the coordinates the generator is given by Eq. (A.1.10).

Infinitesimal internal transformations

An infinitesimal internal transformation is given by Eq. (A.1.1) with $\delta x^\mu = 0$ and $\delta\phi = \delta^\Lambda \phi$. If the transformation describes a symmetry of the system, using the Euler-Lagrange equations, a conserved current $\partial^\mu j_\mu^\Lambda = 0$ is given by

$$j_\mu^\Lambda = \pi_\mu \delta^\Lambda \phi, \tag{A.1.31}$$

where Λ labels the transformation.

However, if the Lagrangian density is not invariant w. r. t. the considered infinitesimal internal transformation one may, of course, nevertheless define a current as in Eq. (A.1.31). Consequently, such a current is not conserved and, using the Euler-Lagrange equations, its divergence is given by Eq. (A.1.2)

$$\partial^\mu j_\mu^\Lambda = \delta^\Lambda \mathscr{L}. \tag{A.1.32}$$

Note, as discussed above, Eq. (A.1.32) does neither assume nor imply that Λ is a symmetry transformation. As no boundary constraints for the variation of the fields on the surface have been introduced, the point that the variation of the Lagrangian can be written as a total divergence does not imply invariance of the action. Indeed, from Eq. (A.1.2) and the Euler-Lagrange equations it can be seen that for internal transformations the variation of the Lagrangian can always be cast into a total divergence. However, as has already pointed out before the current's divergence must be zero in order to generate a symmetry transformation. Also note that (A.1.2) can only be cast into (A.1.32) for internal transformations.

Finally, the classical algebra of currents and charges can be derived. Consider an infinitesimal transformation of the type

$$\delta^\Lambda \psi = T_\epsilon \phi, \tag{A.1.33}$$

where T_ϵ is an infinitesimal matrix acting on internal degrees of freedom, e. g. flavor or color. For a system with, e. g., N_f flavors, such a matrix may be given by $T_\epsilon = \epsilon^a t^a$, where $t^a \in U(N_f) = U(1) \times SU(N_f)$. Then the Poisson bracket of two currents j_μ^Λ and $j_\nu^{\Lambda'}$ reads

$$\{n \cdot j^\Lambda(x), n \cdot j^{\Lambda'}(y)\}_\sigma = \delta^{(3)}(O^{\mu\nu}(x-y)_\nu) n \cdot \pi \left[T_\epsilon, T'_\epsilon\right]_- \phi$$
$$\equiv \delta^{(3)}(O^{\mu\nu}(x-y)_\nu) n \cdot j^{[\Lambda,\Lambda']}_-(x). \qquad (A.1.34)$$

As the currents are linear in the infinitesimal parameters ϵ^a, Eq. (A.1.34) also holds for currents constructed of non-infinitesimal matrices T. Integrating Eq. (A.1.34) w. r. t. $O^{\mu\nu} y_\nu$ gives the change of the current j_μ^Λ, induced by the transformation Λ, under the transformation Λ'

$$\{n \cdot j^\Lambda(x), F^{\Lambda'}(n \cdot x)\}_\sigma = n \cdot j^{[\Lambda,\Lambda']}(x) = n^\alpha \delta^{\Lambda'} j_\alpha^\Lambda(x). \qquad (A.1.35)$$

Though completely derived within a classical framework, these relations are algebraically identical to the relations of current algebra [Tre85, Sak73, Adl68]. Of course, this is not surprising as the main idea of current algebra is to derive an algebra of currents which are constructed of the fields and their canonical momenta, which in turn fulfill the canonical commutation relations dictated by classical Poisson brackets. In this respect, current algebra may be considered as the quantized version of an algebra already given at classical level. Quantum effects, in the sense of different algebraic relations, show up as anomalies. To finish our considerations, recall the fact that current algebra and results which can be derived from it merely rely on general symmetry considerations and not on the special form of the Lagrangian.

Fermions and chiral transformations

The preceding results will now be applied to the specific case of free fermions, which is the relevant example for the topic dealt with in this thesis. Introducing additional fields and interactions does not alter the derived relations as long as the considered transformation leaves all additional terms invariant. This is the case for, e. g., the coupling to photon or gluon fields. Therefore, all results remain valid for QCD. Let $\mathscr{L}$ be the free fermion Lagrangian of N_f quark flavors

$$\mathscr{L} = \bar\psi \left(i\hat\partial - M\right)\psi \qquad (A.1.36)$$

where $\hat{\partial} = \gamma_\mu \partial^\mu$, $\psi = (q_1, \ldots, q_{N_f})$ and $M = \text{diag}(m_1, \ldots, m_{N_f})$ a general mass matrix. Canonical conjugate momenta are

$$\pi_\mu = i\bar{\psi}\gamma_\mu = \frac{1}{4}\pi_\nu\gamma^\nu\gamma_\mu. \tag{A.1.37}$$

For a constant-time surface, Eqs. (A.1.29) and (A.1.37) give rise to the following Poisson bracket

$$\left\{\psi_i(x), \pi_j^\mu(y)\right\}_{x_0=y_0} = \delta^{(3)}(\vec{x}-\vec{y})\left(\gamma_0\gamma^\mu\right)_{ij}, \tag{A.1.38}$$

where the i, j are Dirac indices. It is clear that a relation like Eq. (A.1.38) can also be given for any other space-like surface. Having the Poisson brackets at hand one can give the complete algebra of currents without any specific knowledge about the transformation or about symmetry properties w.r.t. this transformation. By virtue of Eqs. (A.1.38) and (A.1.37) the currents associated with infinitesimal internal transformations of the type (A.1.33) fulfill the algebra

$$\left\{j_\mu^\Lambda(x), j_\nu^{\Lambda'}(y)\right\}_t = \delta^{(3)}(\vec{x}-\vec{y})\pi_0\left[\gamma_0\gamma_\mu T, \gamma_0\gamma_\nu T'\right]_- \psi. \tag{A.1.39}$$

Integration w.r.t. $\vec{x}$ or $\vec{y}$ gives the variation of one of the currents under the transformation induced by the generator of the other current. Now, consider the infinitesimal vector and axial transformations in flavor space

$$\Lambda_V : \psi \to \psi' = e^{-i\Theta^a t^a}\psi \approx (1 - i\Theta^a t^a)\psi, \tag{A.1.40a}$$

$$\Lambda_A : \psi \to \psi' = e^{-i\Theta^a t^a \gamma_5}\psi \approx (1 - i\Theta^a t^a \gamma_5)\psi, \tag{A.1.40b}$$

where t^a are the generators of $U(N_f) = U(1) \times SU(N_f)$ acting on the flavor indices of ψ and Θ^a are arbitrary infinitesimal rotation angles. For brevity and because algebraically there is no difference, the U(1) transformation is explicitly included. Later on, these transformations will show up as special cases of a set of more general transformations, which are called chiral transformations. The variation of the Lagrangian density reads

$$\Lambda_V : \delta\mathcal{L} = -i\Theta^a \bar{\psi}\left[t^a, M\right]_- \psi, \tag{A.1.41a}$$

$$\Lambda_A : \delta\mathcal{L} = +i\Theta^a \bar{\psi}\gamma_5\left[t^a, M\right]_+ \psi. \tag{A.1.41b}$$

A Chiral symmetry, currents and order parameters

Note that adding terms $\propto \bar{\psi}\gamma_\mu\psi$, e.g. gauge field coupling terms, does not change Eq. (A.1.41). According to Eq. (A.1.31) the following currents can be associated with (A.1.40)

$$j_\mu^{V,\tau} = \bar{\psi}\gamma_\mu\tau\psi, \tag{A.1.42a}$$

$$j_\mu^{A,\tau} = \bar{\psi}\gamma_5\gamma_\mu\tau\psi. \tag{A.1.42b}$$

Note the difference in sign for Eq. (A.1.42b) compared to the definition Eq. (A.1.31), which can be seen from Eqs. (A.1.37) and (A.1.40b). It is merely convention. The charge obtained from $j_0^{A,\tau}$ is the generator of the transformation which is adjoint to Λ_A; it gives a field variation which is the negative of (A.1.40b). Colloquial phrased, the fields are transformed in the other direction. According to their behavior under Lorentz transformations, (A.1.42a) and (A.1.42b) are currents with internal angular momentum 1. Additionally, one can define the scalar and pseudo-scalar currents with internal angular momentum 0 as

$$j^{S,\tau} = \bar{\psi}\tau\psi, \tag{A.1.42c}$$

$$j^{P,\tau} = i\bar{\psi}\gamma_5\tau\psi. \tag{A.1.42d}$$

Generically, the currents (A.1.42) are denoted as

$$j^{X,\tau} = \bar{\psi}\Gamma^X\tau\psi, \tag{A.1.43}$$

with $\Gamma^X \in \{\mathbb{1}, i\gamma_5, \gamma_5\gamma_\mu, \gamma_\mu\}$ and where we omit the Lorentz index if X also stands for spin-0 currents. The imaginary unit in the definition of the pseudo-scalar current is convention. Due to the special choice of Γ^X the current fulfills $\left(j_\mu^{X,\tau}\right)^\dagger = j_\mu^{X,\tau^\dagger}$, thereby relating particles and antiparticles. According to Eq. (A.1.32) the divergences are given by

$$i\partial^\mu j_\mu^{(V,A),\tau} = \bar{\psi}\left(\gamma_5\right)[\tau, M]_\mp \psi = (-i)^{(S,P)} j^{(S,P),[\tau,M]_\mp}, \tag{A.1.44}$$

where the upper (lower) sign corresponds to the vector (axial-vector) current. For the moment being, τ denotes an arbitrary infinitesimal matrix in flavor space $\tau = \Theta^a t^a$. As the transformations (A.1.40) are infinitesimal, all variations, i.e. (A.1.41) and $\delta^\Lambda\psi$ in Eq. (A.1.40), are proportional to the transformation parameters Θ^a. The same holds true for the corresponding currents, as well as their divergences. Consequently, one can define currents $j_\mu \equiv \Theta^a j_\mu^a$, for which the infinitesimal transformation parame-

A.1 The formalism for classical fields

Table A.1.1: Transformation pattern for the currents (A.1.42) under infinitesimal transformations (A.1.40).

	$j_\mu^{V,\tau}$	$j_\mu^{A,\tau}$	$j^{S,\tau}$	$j^{P,\tau}$
$\delta^{V,\tau'} j^{X,\tau}$	$ij_\mu^{V}[\tau',\tau]_-$	$ij_\mu^{A}[\tau',\tau]_-$	$ij^{S}[\tau',\tau]_-$	$ij^{P}[\tau',\tau]_-$
$\delta^{A,\tau'} j^{X,\tau}$	$-ij_\mu^{A}[\tau',\tau]_-$	$-ij_\mu^{V}[\tau',\tau]_-$	$-j^{P}[\tau',\tau]_+$	$j^{S}[\tau',\tau]_+$

ters cancel each other in Eq. (A.1.32), and Eq. (A.1.44) holds for an arbitrary rotation matrix τ in Eq. (A.1.42). Recall that Eq. (A.1.44) remains unaltered by the introduction of terms which describe the coupling of fermions to gauge fields. Equivalently, Eq. (A.1.44) could be derived using the explicit form of the EoM. However, the EoM of course depend on gauge coupling terms.

The currents (A.1.43) transform under (A.1.40) (infinitesimal transformation) as (see Eq. (A.1.35))

$$\Lambda_{X'}^{\tau'}: \quad j^{X,\tau} \to j^{X,\tau} \pm \frac{i}{2}\bar{\psi}\left([\Gamma^{X'},\Gamma^X]_+ [\tau',\tau]_\mp + [\Gamma^{X'},\Gamma^X]_- [\tau',\tau]_\pm\right)\psi,$$
(A.1.45)

where the upper (lower) sign refers to $\Gamma^{X'} = \mathbb{1}$ ($\Gamma^{X'} = \gamma_5$), which specifies the transformation. The complete transformation pattern for (A.1.42) is listed in Tab. A.1.1. Hence, Λ_A mixes parity partner. Moreover, if no flavor changing currents are considered, i.e. $\tau \propto \mathbb{1}$, Λ_V leaves all currents invariant and Λ_A only transforms spin-0 currents into each other.

The (anti) commutator in Eqs. (A.1.41) and (A.1.44) is given by (no Einstein convention for dotted indices)

$$\left([\tau,M]_\mp\right)_{rs} = \mp\left(m_r \mp m_s\right)\tau_{rs}$$
(A.1.46)

for arbitrary but diagonal mass matrix. Current conservation for the vector (axial-vector) current, therefore, holds if the masses m_r, m_s are equal (zero) if τ_{rs} is nonzero. The diagonal elements of the commutator are always zero. Accordingly, the Lagrangian (A.1.36) is invariant w.r.t. axial transformations if $M = 0$ and w.r.t. the vector transformation if $m_r = m_s$ for $\tau_{rs} \neq 0$. Up and down quarks with their tiny masses of a few MeV may be considered as approximately massless and, hence, represent a two-flavor system which would be invariant w.r.t. axial transformations which are restricted to these two flavors. Consequently, the current (A.1.42b) is conserved

A Chiral symmetry, currents and order parameters

and the Lagrangian (A.1.36) is invariant. On the other hand, vector and axial-vector currents are mixed. Hence, if the ground state of the strong interaction would possess the same symmetry, vector and axial-vector states would be degenerate. This means that the spectral densities (cf. App. B.1) are identical, because an axial transformation may be applied which transforms one spectral function into the other. Obviously, this is not the case, e. g. the mass of the ρ meson (vector state with $\tau = \sigma^3/2$ and σ^i as Pauli matrices) is $m_\rho = 775.5$ MeV, whereas the mass of the chiral partner, the axial-vector a_1, is $m_{a_1} = 1260$ MeV. This phenomenon is a manifestation of the spontaneous breaking of the axial symmetry (symmetry w. r. t. axial transformations (A.1.40b)). The explicit breaking of the axial symmetry due to the finite quark masses should lead to mass differences which are small compared to the meson masses. In the example above the mass difference is of the same order as the meson masses. Thus, the explicit symmetry breaking due to a finite quark mass cannot be responsible for the mass splitting among chiral partners.

Let us consider now a 3-flavor system with a mass matrix of 2 light (i. e. light enough to be considered as approximately massless) and one massive flavor. Then, restrict the transformations to the two light flavors, so that in flavor space the 8 rotation angles ($t^a = \lambda^a/2$, λ^a the Gell-Mann matrices) are $\vec{\Theta} = (\Theta_1, \Theta_2, \Theta_3, 0, \ldots, 0)$:

$$\Theta^a t^a = \frac{\vec{\Theta}\vec{\lambda}}{2} = \frac{1}{2}\begin{pmatrix} \sum_{i=1}^{3} \Theta^i \sigma^i & 0 \\ 0 & 0 \end{pmatrix}. \tag{A.1.47}$$

These transformations clearly leave the Lagrangian with $m_{1,2} \approx 0$ approximately invariant. Specifying now flavor changing currents with

$$\tau = \frac{1}{2}(\lambda^4 + i\lambda^5), \tag{A.1.48}$$

i. e. a current with one light (the first quark in this case) and one heavy quark $j_\mu^{X,\tau} = \bar{q}_1 \gamma_\mu q_3$, the transformations (A.1.45) are given by

$$[t, \tau]_- = [t, \tau]_+ = \frac{1}{2}\begin{pmatrix} 0 & 0 & \Theta_3 \\ 0 & 0 & \Theta_1 + i\Theta_2 \\ 0 & 0 & 0 \end{pmatrix} \tag{A.1.49}$$

and the currents transform as

$$j_\mu^{V,\tau} = \bar{q}_1 \gamma_\mu q_3 \to j_\mu^{V,\tau} - \frac{i}{2}\left(\bar{q}_1 \Gamma \gamma_\mu q_3 \Theta_3 + (\Theta_1 + i\Theta_2)\bar{q}_2 \Gamma \gamma_\mu q_3\right),\tag{A.1.50a}$$

$$j^{S,\tau} = \bar{q}_1 q_3 \to j^{S,\tau} - \frac{i}{2}\left(\bar{q}_1 \Gamma q_3 \Theta_3 + (\Theta_1 + i\Theta_2)\bar{q}_2 \Gamma q_3\right),\tag{A.1.50b}$$

where $\Gamma = \mathbb{1}$ ($\Gamma = \gamma_5$) for the vector (axial-vector) transformation. Although, the heavy quark mass explicitly breaks the invariance of the Lagrangian under general flavor rotations, invariance w.r.t. (A.1.47) (for Λ_V and Λ_A) is preserved. Furthermore, the thus defined currents (A.1.48) are neither conserved nor associated with a symmetry transformation, but they still mix under the transformation (A.1.47). The situation is identically to the massless 2-flavor case and the missing of degenerate parity partner in the heavy-light sector of the meson spectrum is a manifestation of the spontaneous breaking of the axial symmetry in the light quark sector.

Left and right handed currents

To establish the connection to left and right handed chirality, define the left and right handed projectors as

$$P_{L,R} \equiv \frac{1}{2}(1 \mp \gamma_5),\tag{A.1.51}$$

where the convention of [Ynd06] has been used. Then, the free Fermion Lagrangian (A.1.36) in terms of left and right handed spinors $\psi_{L,R} \equiv P_{L,R}\psi$ can be written as

$$\mathscr{L} = \bar{\psi}_L i\partial \psi_L - \bar{\psi}_L M \psi_R + L \leftrightarrow R.\tag{A.1.52}$$

Consider the separate $U(N_f) = U(1) \times SU(N_f)$ transformations of left and right handed fields, also known as the chiral transformations

$$U(N_f)_{L,R} = U(1)_{L,R} \times SU(N_f)_{L,R} : \psi_{L,R} \to \psi'_{L,R} = e^{-i\Theta_{L,R}^a t^a}\psi_{L,R},\tag{A.1.53}$$

where the generators of $SU(N_f)$, t^a for $a = 1,\ldots,N_f^2 - 1$, and of $U(1)$, $t^0 = \mathbb{1}/2$, again act on the flavor indices.

The kinetic terms in Eq. (A.1.52) are invariant w.r.t. (A.1.53). Whereas the mass terms are in general not, because of the independence of the transformation parameters Θ_L and Θ_R.

For infinitesimal transformations the change of the Lagrangian is

$$\delta L = -i\bar{\psi}_L \left(\Theta_L^a t^a M - M \Theta_R^a t^a \right) \psi_R + L \leftrightarrow R \,. \tag{A.1.54}$$

According to Eqs. (A.1.31), (A.1.32) and (A.1.8) the well known currents of left and right handed chirality

$$j_\mu^{(L,R),\tau} = \bar{\psi}_{(L,R)} \gamma_\mu \tau \psi_{(L,R)} \tag{A.1.55}$$

may be defined; their divergences are

$$i\partial^\mu j_\mu^{(L,R),\tau} = \bar{\psi}_{(L,R)} \tau M \psi_{(R,L)} - \bar{\psi}_{(R,L)} M \tau \psi_{(L,R)} \,. \tag{A.1.56}$$

As left and right handed currents are eigenvalues of the chirality operator, $\gamma_5 \psi_{L,R} = \mp \psi_{L,R}$, the actual independent left and right handed transformations can be combined to the vector (axial) transformation for $\Theta_L^a = \Theta_R^a = \Theta_V^a$ ($-\Theta_L^a = \Theta_R^a = \Theta_A^a$):

$$\psi = \psi_L + \psi_R \to \psi' = e^{\mp i \Theta_{V,A}^a t^a} \psi_L + e^{-i \Theta_{V,A}^a t^a} \psi_R = \begin{cases} e^{-i\Theta_V^a t^a} \psi \\ e^{-i\Theta_A^a t^a \gamma_5} \psi \end{cases} . \tag{A.1.57}$$

Thus, vector and axial transformations are special cases of left- and right-handed transformations. From Eq. (A.1.51) the currents (A.1.42a) and (A.1.42b) are given as $j_\mu^{V,\tau} = j_\mu^{L,\tau} + j_\mu^{R,\tau}$ and $j_\mu^{A,\tau} = j_\mu^{L,\tau} - j_\mu^{R,\tau}$.

Consider for a moment the two-flavor case, $N_f = 2$. Then Eq. (A.1.53) by virtue of the SU(2) generators $t^a = \sigma^a/2$, with σ^a again the Pauli isospin matrices, becomes

$$\psi_C \to \psi'_C = \left(\cos \frac{\Theta_C}{2} - i \frac{\vec{\Theta}_C \vec{\sigma}}{\Theta_C} \sin \frac{\Theta_C}{2} \right) \psi_C \tag{A.1.58a}$$

$$\bar{\psi}_C \to \bar{\psi}'_C = \bar{\psi}_C \left(\cos \frac{\Theta_C}{2} + i \frac{\vec{\Theta}_C \vec{\sigma}}{\Theta_C} \sin \frac{\Theta_C}{2} \right) , \tag{A.1.58b}$$

with $\Theta_C \equiv |\vec{\Theta}_C|$ and $C \in \{L, R\}$. The left and right handed currents (A.1.55) transform as

$$\vec{j}_\mu^C \to \cos \Theta_C \vec{j}_\mu^C + \frac{\sin \Theta_C}{\Theta_C} \left(\vec{\Theta}_C \times \vec{j}_\mu^C \right) + 2 \sin^2 \frac{\Theta_C}{2} \frac{\vec{\Theta}_C \circ \vec{\Theta}_C}{\Theta_C^2} \vec{j}_\mu^C , \tag{A.1.59}$$

where $\vec{j}_\mu^C \equiv \vec{j}_\mu^{C,\vec{\sigma}}$ is the left or right handed vector–isospin-vector current with compo-

A.1 The formalism for classical fields

nents j_μ^{C,σ^a} and $\circ$ denotes the dyadic product. Thus a vector current transforms as

$$\vec{j}_\mu^V = \vec{j}_\mu^L + \vec{j}_\mu^R$$
$$\to \cos\Theta_L \left(\vec{j}_\mu^L - \frac{\vec{\Theta}_L \circ \vec{\Theta}_L}{\Theta_L^2} \vec{j}_\mu^L \right) + \frac{\vec{\Theta}_L \circ \vec{\Theta}_L}{\Theta_L^2} \vec{j}_\mu^L + \frac{\sin\Theta_L}{\Theta_L} \vec{\Theta}_L \times \vec{j}_\mu^L + L \to R, \quad (A.1.60)$$

A particular choice is

$$\vec{\Theta}_L = 2\vec{\Theta}_R \equiv 2\vec{\Theta}, \qquad \Theta = \pi. \tag{A.1.61}$$

For the vector current this transformation gives

$$\vec{j}_\mu^V \to \vec{j}_\mu^A + 2\frac{\vec{\Theta} \circ \vec{\Theta}}{\Theta^2} \vec{j}_\mu^R. \tag{A.1.62}$$

This reveals the mixing of vector and axial-vector currents under the chiral transformation (A.1.53). Definition (A.1.61) does not uniquely fix the components of $\vec{\Theta}_C$, and for each component of the vector–isospin-vector or axial-vector–isospin-vector current $\vec{j}_\mu^{(V,A)}$ there is a chiral transformation (A.1.53) which transforms it into its chiral counterpart. Namely, any choice with (A.1.62) and $\Theta^a = 0$ transforms $j_\mu^{V,\sigma^a} \to j_\mu^{A,\sigma^a}$. Thus, the chiral transformations relate parity partners. To be specific and in order to supplement the discussion of Sec. 5.1:

$$SU(2)_L \times SU(2)_L = \left(e^{i\pi\sigma^2} \right)_L \times \left(e^{i\frac{\pi}{2}\sigma^2} \right)_R = \begin{pmatrix} -1 & 0 \\ 0 & -1 \end{pmatrix}_L \times \begin{pmatrix} 0 & -1 \\ 1 & 0 \end{pmatrix}_R \tag{A.1.63}$$

accounts for $j_\mu^{V,\frac{\sigma^3}{2}} \to j_\mu^{A,\frac{\sigma^3}{2}}$.

Analogously, any chiral current (A.1.55) may be transformed by virtue of (A.1.59) into its negative by an appropriate chiral transformation. Thus, if the chiral transformations (A.1.53) are symmetries of the theory, the exchange of chiral partners is a symmetry transformation as well and should not alter observables. Besides the choice (A.1.61) there are other choices possible to transform vector into axial-vector currents. Note that the particular transformation (A.1.61) cannot be a vector transformation (A.1.40a), as the latter does not mix chiral partners.

A Chiral symmetry, currents and order parameters

The choice $\vec{\Theta} \equiv \vec{\Theta}_L = -\vec{\Theta}_R$ is an axial transformation (A.1.40b) and yields

$$\vec{j}_\mu^V \rightarrow \cos\Theta \left[\vec{j}_\mu^V - \frac{\vec{\Theta} \circ \vec{\Theta}}{\Theta^2} \vec{j}_\mu^V \right] + \frac{\vec{\Theta} \circ \vec{\Theta}}{\Theta^2} \vec{j}_\mu^V + \frac{\sin\Theta}{\Theta} \vec{\Theta} \times \vec{j}_\mu^A . \tag{A.1.64}$$

This is the transformation law for vector–isospin-vector currents under finite axial transformations.[22] An analog relation holds true for axial-vector–isospin-vector currents.

Setting $\cos\Theta = 0$, i.e. $\Theta = \frac{\pi}{2}(2n+1)$, $n \in \mathbb{N}$, gives

$$\vec{j}_\mu^V \rightarrow \frac{\vec{\Theta} \circ \vec{\Theta}}{\Theta^2} \vec{j}_\mu^V + \frac{\sin\Theta}{\Theta} \vec{\Theta} \times \vec{j}_\mu^A . \tag{A.1.65}$$

Again we conclude that each component of the vector–isospin-vector current may be transformed into the component of an axial-vector–isospin-vector current by an appropriate chiral transformation, i.e.

$$j_\mu^{V,\sigma^a} \rightarrow j_\mu^{A,\sigma^a} \tag{A.1.66}$$

is possible for all σ^a. The explicit transformation law for finite vector transformations is obtained from $\vec{\Theta} \equiv \vec{\Theta}_L = \vec{\Theta}_R$:

$$\vec{j}_\mu^V \rightarrow \cos\Theta \left[\vec{j}_\mu^V - \frac{\vec{\Theta} \circ \vec{\Theta}}{\Theta^2} \vec{j}_\mu^V \right] + \frac{\vec{\Theta} \circ \vec{\Theta}}{\Theta^2} \vec{j}_\mu^V + \frac{\sin\Theta}{\Theta} \vec{\Theta} \times \vec{j}_\mu^V . \tag{A.1.67}$$

Again choosing $\Theta = \frac{\pi}{2}(2n+1)$, $n \in \mathbb{N}$, yields

$$\vec{j}_\mu^V \rightarrow \frac{\vec{\Theta} \circ \vec{\Theta}}{\Theta^2} \vec{j}_\mu^V + \frac{\sin\Theta}{\Theta} \vec{\Theta} \times \vec{j}_\mu^V . \tag{A.1.68}$$

An analog relation holds true for the axial-vector–isospin-vector current. Equation (A.1.68) tells us that any component of an isospin-vector current can be transformed into any other component by an appropriate choice of $\vec{\Theta}$.

A particular construct of two equal chirality currents is $\vec{j}_\mu^C \vec{j}^{C,\mu}$. According to Eq. (A.1.59) it transforms under arbitrary chiral transformations (A.1.53) as

$$\vec{j}_\mu^C \vec{j}^{C,\mu} \rightarrow \cos^2\Theta_C \vec{j}_\mu^C \vec{j}^{C,\mu} + \frac{\sin^2\Theta_C}{\Theta_C^2} \left(\vec{\Theta}_C \times \vec{j}_\mu^C \right) \left(\vec{\Theta}_C \times \vec{j}^{C,\mu} \right)$$

---
[22]Obviously, the thus obtained transformation is, according to the definition (A.1.40b), an axial transformation with angle $-\Theta_A$.

$$+4\frac{\sin^4\frac{\Theta_C}{2}}{\Theta_C^2}\left(\vec{\Theta}_C\cdot\vec{j}_\mu^C\right)\left(\vec{\Theta}_C\cdot\vec{j}^{C,\mu}\right)+2\cos\Theta_C\frac{\sin\Theta_C}{\Theta_C}\vec{j}_\mu^C\cdot\left(\vec{\Theta}_C\times\vec{j}^{C,\mu}\right)$$

$$+4\cos\Theta_C\frac{\sin^2\frac{\Theta_C}{2}}{\Theta_C^2}\left(\vec{\Theta}_C\cdot\vec{j}_\mu^C\right)\left(\vec{\Theta}_C\cdot\vec{j}^{C,\mu}\right)$$

$$+4\frac{\sin\Theta_C}{\Theta_C}\frac{\sin^2\frac{\Theta_C}{2}}{\Theta_C^2}\left(\vec{\Theta}_C\times\vec{j}_\mu^C\right)\cdot\vec{\Theta}_C\left(\vec{\Theta}_C\cdot\vec{j}^{C,\mu}\right). \tag{A.1.69}$$

The forth and sixth terms are zero due to the vector product. Using $\epsilon^{abc}\epsilon^{cde} = \delta^{ad}\delta^{be} - \delta^{ae}\delta^{bd}$ and $\sin^2 x = (1-\cos 2x)/2$ a straightforward calculation reveals that $\vec{j}_\mu^C\vec{j}^{C,\mu} \to \vec{j}_\mu^C\vec{j}^{C,\mu}$ is invariant under arbitrary chiral transformations (A.1.53). In particular it is invariant w. r. t. arbitrary vector and axial transformations (A.1.40).

For completeness we also rewrite the currents (A.1.43) in terms of left and right handed fields:

$$j^{S,\tau} = \bar{\psi}_L \tau \psi_R + R \leftrightarrow L, \tag{A.1.70a}$$
$$j^{P,\tau} = i\bar{\psi}_L \gamma_5 \tau \psi_R + R \leftrightarrow L, \tag{A.1.70b}$$
$$j_\mu^{V,\tau} = \bar{\psi}_L \gamma_\mu \tau \psi_L + R \leftrightarrow L, \tag{A.1.70c}$$
$$j_\nu^{A,\tau} = \bar{\psi}_L \gamma_5 \gamma_\mu \tau \psi_L + R \leftrightarrow L. \tag{A.1.70d}$$

A.2 The formalism for quantized fields

Introductions to quantum field theory can be found in many textbooks, e. g. [Rom69, Rom65, Wei95, Wei96, Gre96, Pok00, Ynd06, Pes95, Kug97, Mut87, Itz80, Kak93, Bjo65, Hat98, Bro92]. In the following the focus is on quantizing the equal-time Poisson brackets in order to obtain the equal-time current commutation relations (ETCCRs) and to introduce anomalies and order parameters.

Quantization of the Lagrangian theory within Schwinger's action principle is done by the transition from classical equal-time Poisson brackets to quantum mechanical commutators such that the Poisson bracket (A.1.30) is the coefficient of the lowest order term in an expansion of the commutator in terms of $i\hbar$ ($\hbar = 1$ throughout the thesis), i. e.

$$[\mathcal{O},Q]_- = i\{\mathcal{O},Q\} = i\delta\mathcal{O} \tag{A.2.1}$$

A Chiral symmetry, currents and order parameters

is demanded for all local operators $\mathcal{O}$.[23] The quantity $\delta\mathcal{O}$ is the change of $\mathcal{O}$ under the transformation associated with the charge Q (see Eq. (A.1.8)). Explicit evaluation of (A.2.1) for field operators ψ or the operator of the canonical conjugate momentum π_0 in terms of the definition of the generator Q (A.1.8) reveals its equivalence to demanding the canonical commutation relations (or anti-commutation relations, both satisfy Schwinger's action principle) of a field operator and the operator of its canonical conjugate momentum (A.1.4c)

$$i\left[\pi_0(x_0,\vec{x}),\psi(x_0,\vec{y})\right]_\pm = \delta^{(3)}(\vec{x}-\vec{y}). \tag{A.2.2}$$

In particular,

$$i\left[P_\nu,\mathcal{O}(x)\right]_- = \partial_\nu\mathcal{O}(x), \tag{A.2.3}$$

for arbitrary operators $\mathcal{O}(x)$ and the momentum operator P_ν given in Eq. (A.1.11) (see Eq. (A.1.18)), which is equivalent to the EoM in the Heisenberg picture

$$\mathcal{O}(x+a) = e^{iPa}\mathcal{O}(x)e^{-iPa}. \tag{A.2.4}$$

Moreover,

$$i\left[Q^{X,\tau}(x_0),\psi(x_0,\vec{x})\right]_- = \delta^{X,\tau}\psi(x), \tag{A.2.5}$$

where X indicates the type of transformation, e.g. $X \in \{V,A\}$ for (A.1.40), and $\tau = \Theta^a t^a$ specifies the group parameter. Thus, regardless of the symmetry properties of the Lagrangian $\mathscr{L}$ and the conservation properties of (A.1.31), the currents and the corresponding charges fulfill

$$\left[j_0^{X,\tau},j_0^{X',\tau'}\right]_{x_0=y_0} = -\frac{i}{2}\delta^{(3)}(\vec{x}-\vec{y})\pi_0\left(\left[\Gamma^X,\Gamma^{X'}\right]_+\left[\tau,\tau'\right]_-\right)\psi, \tag{A.2.6}$$

with $\Gamma \in \{\mathbb{1},\gamma_5\}$, according to the transformation specified by X, and

$$\left[\Gamma\tau,\Gamma'\tau'\right]_\pm = \frac{1}{2}\left(\left[\Gamma,\Gamma'\right]_+\left[\tau,\tau'\right]_\pm + \left[\Gamma,\Gamma'\right]_-\left[\tau,\tau'\right]_\mp\right) \tag{A.2.7}$$

has been used. The second term in Eq. (A.2.7) is always zero if $\Gamma \in \{\mathbb{1},\gamma_5\}$. All

---
[23]To consider contributions of higher powers of $\hbar$ to the commutators is also known as deformation quantization.

A.2 The formalism for quantized fields

relations given in section A.1 remain valid. Hence, the algebra given by the Poisson brackets remains valid when quantizing the theory using canonical quantization. It consistently follows that the commutators only differ by their parity structure, i. e.

$$\Gamma = \Gamma' : \left[j_0^{X,\tau}, j_0^{X,\tau'} \right]_{x_0 = y_0} = \delta^{(3)}(\vec{x} - \vec{y}) j_0^{V,[\tau,\tau']_-}, \tag{A.2.8a}$$

$$\Gamma \neq \Gamma' : \left[j_0^{X,\tau}, j_0^{X',\tau'} \right]_{x_0 = y_0} = \delta^{(3)}(\vec{x} - \vec{y}) j_0^{A,[\tau,\tau']_-} \tag{A.2.8b}$$

follows.

The ETCCRs (A.2.6) are the starting point for the so called current algebras [Adl68, Tre85, Sak73]. Their development preceded the invention of QCD and provided crucial insights and inspirations for the theory of strong interaction [Cao10]. The algebra (A.2.8) can be extended by including spatial components of the currents – either by definition, general arguments (e. g. Lorentz covariance) or on ground of the canonical quantization within a concrete model, i. e. for a specified Lagrangian $\mathscr{L}$.

Digression: Schwinger terms

In any case, the thus defined algebra cannot be complete. Moreover, there is a striking difference between the algebra of Poisson brackets and commutators of quantum field operators. The latter one is a mapping which is directly related to the multiplication properties of the field operators, the first one not. Therefore, one might suspect that multiplication of field operators may lead to divergences. Investigating the Källén-Lehmann or spectral representation (cf. App. B.1) of the ETC of a temporal and a spatial component, one finds that terms proportional to derivatives of Dirac distributions have to be introduced in order to preserve positivity of the energy spectrum [Adl68, Tre85, Sak73, Rom69]. These terms are called Schwinger terms and were first found to exist in [Got55]. Unfortunately, it is not possible to make general (model independent) arguments about the nature of Schwinger terms, e. g. whether they are c-numbers or not. To understand their existence note that, due to the canonical commutation relations (A.2.2), the product of two field operators at the same space-time point is actually ill-defined, i. e. it is irregular. If, instead of Eq. (A.1.42), the currents would have been defined as the limit of the product of two infinitesimally spatially separated field operators [Sak73]

$$j_\mu^{X,\tau}(x) = \lim_{\epsilon \to 0} \bar{\psi}(x_0, \vec{x} + \vec{\epsilon}) \Gamma_\mu^X \tau \psi(x_0, \vec{x} - \vec{\epsilon}), \tag{A.2.9}$$

A Chiral symmetry, currents and order parameters

then derivatives of Dirac delta distributions naturally arise in the commutation relations. This is the so called point-splitting technique [Sch59]. Other definitions of point-splitting based currents, e. g. an asymmetric splitting, may lead to different prefactors for Schwinger terms [Cha70]. Furthermore, the definition (A.2.9) is not gauge invariant. Define [Tre85, Ynd06]

$$j_\mu^{X,\tau}(x) = \lim_{\epsilon \to 0} \bar{\psi}(x_0, \vec{x}+\vec{\epsilon}) \Gamma_\mu^X \tau e^{ia\int_{\vec{x}-\vec{\epsilon}}^{\vec{x}+\vec{\epsilon}} g \mathscr{A}^a(y) \mathrm{d}y_a} \psi(x_0, \vec{x}-\vec{\epsilon}), \qquad (A.2.10)$$

which is gauge invariant for $a = 1$. Note, that in case of non-Abelian gauge field theories the gauge-invariance ensuring insertion $\left[\exp\left\{ia \int g \mathscr{A}^a\right\}\right]$ has to be path-ordered w. r. t. to the integration [Pes95]. This ensures, that the extra terms which emerge due to the non-commutativity of the gauge fields are properly canceled when switching to other gauges. With this definition also higher derivatives of Dirac distributions can contribute to the commutation relations. Other details and approaches have been worked out in [Hel67, Bou70, Bar69, Fuj87, Bra68].

From Eq. (A.2.10) the divergence of the currents can be reevaluated. In particular, the divergence of the axial-vector $U_A(1)$ current, which refers to the $U_A(1)$ symmetry, gains an anomalous contribution, the well known axial anomaly. Employing the OPE technique (see Sec. 2.2) in conjunction with the background field method in Fock-Schwinger gauge (see App. C.1), the divergence of the axial-current in point-splitting technique (A.2.10) for zero quark masses can be evaluated [Nov84b] and reads

$$\partial^\mu j_\mu^{A,\mathbb{1}} = \frac{\alpha_s}{4\pi} G_{\mu\nu}^A \tilde{G}^{\mu\nu,A}, \qquad (A.2.11)$$

with the dual field-strength tensor $\tilde{G}_{\mu\nu}^A \equiv \epsilon_{\mu\nu}{}^{\alpha\beta} G_{\alpha\beta}^A$. This anomaly breaks the symmetry explicitly. Thus, the axial U(1) symmetry, classically realized for zero quark masses, is not a symmetry of the quantized theory. Moreover, it can be shown that if Schwinger terms exist in the commutator of two currents, their time-ordered product is not Lorentz covariant anymore [Sak73]. Lorentz covariance is restored by adding a so-called "seagull term" to the naive time-ordered product. This also accounts for a formal redefinition of causal current-current correlation functions (see App. B) by means of the Lorentz covariant time-ordered product. The seagull term is completely determined by and can be evaluated from the Schwinger term. Also the non-anomalous Ward identity (cf. App. B.4) may not give the correct result and an anomalous Ward identity can be derived. In order for the non-anomalous Ward identity to be true, the Schwinger terms and divergences of seagull terms must cancel each other. The so-called Feynman conjecture states this cancellation, but there are processes where it fails. An example for such a process is $\pi^0 \to 2\gamma$ [Tre85, Ynd06, Wei96, Itz80], see also [Iof06] for a review on the axial anomaly, where Schwinger terms result in the so-called triangle anomaly.

Another anomaly is generated by the breakdown of scale-invariance due to quantum corrections [Pas84]. Scale invariance of the classical massless theory defines the conserved dilatation current

$$j_\mu^{\text{dil}} = T_{\mu\nu} x^\nu, \quad (A.2.12)$$

with $T_{\mu\nu}$ being the energy-momentum tensor.[24] Finite quark masses break the symmetry in the classical theory, but for small masses it may be considered as an approximate symmetry – similar to the axial transformation. Hence, the divergence of the dilatation current in the classical theory is $\partial^\mu j_\mu^{\text{dil}} = T_\mu^{\ \mu} = \bar{\psi} M \psi$. Quantum corrections, however, yield [Coh92]

$$T_\mu^{\ \mu} = -\frac{1}{8}\left(\frac{11}{3}N_c - \frac{2}{3}N_f\right)\frac{\alpha_s}{\pi} G_{\mu\nu}^A \tilde{G}^{\mu\nu,A} + \bar{\psi} M \psi, \quad (A.2.13)$$

neglecting higher-order corrections in α_s.

The apparent relation between Schwinger terms and the singular behavior of operator products immediately suggests a link to the OPE (cf. Sec. 2). Indeed, it turns out that the most convenient way to evaluate Schwinger terms is by usage of the Bjorken-Johnson-Low (BJL) limit [Bjo66, Joh66] (see also [Tre85]) in combination with the OPE [Shi99, Tre85]. If $\Pi(q)$ is the causal correlator of two field operators A and B (cf. App. B) then the BJL definition of the ETC is [Tre85]

$$\lim_{q_0 \to \infty} q_0 \Pi(q) \equiv i \int d^3x \, e^{-i\vec{q}\vec{x}} \langle [A(0,\vec{x}), B(0)] \rangle. \quad (A.2.14)$$

In this manner, the most singular term in the OPE gives the Schwinger term. As can be seen from the discussion presented so far, there exists an intimate connection between canonical quantization of a field theory, Schwinger terms, anomalies and the OPE technique.

However, the discussion of Schwinger terms and anomalies (see e. g. [Tre85, Wei96, Adl68]), which gives enlightening insights and a deeper understanding of quantum

---

[24]Actually, it is the Belifante energy-momentum tensor, which is obtained from the canonical energy-momentum and angular-momentum tensor by requiring symmetry w. r. t. Lorentz indices of the energy-momentum tensor [Pas84].

field theories, is beyond the scope of this essay.

Equal-time current commutation relations

The ETC of a current and a divergence, which is not treatable within the methods of current algebra, will be needed. Ignoring Schwinger terms the ETCs for the currents (A.1.43) evaluate to

$$\left[j_\mu^{X,\tau}(x_0,\vec{x}),\left(j_\nu^{X',\tau'}(x_0,\vec{y})\right)^\dagger\right]_-$$
$$= \delta^{(3)}(\vec{x}-\vec{y})\left(\Gamma_\mu^X\tau\right)_{ij}\left(\Gamma_\nu^{X'}\tau'\right)_{kl}\left(\bar{\psi}_i(x)(\gamma_0\mathbb{1}_f)_{jk}\psi_l(y)\right.$$
$$\left.-\bar{\psi}_k(y)(\gamma_0\mathbb{1}_f)_{il}\psi_j(x)\right)_{x_0=y_0}$$
$$= \delta^{(3)}(\vec{x}-\vec{y})\psi^\dagger(x)\left(\gamma_0\Gamma_\mu^X\tau\gamma_0\Gamma_\nu^{X'}\tau' - \gamma_0\Gamma_\nu^{X'}\tau'\gamma_0\Gamma_\mu^X\tau\right)\psi(x)$$
$$= \delta^{(3)}(\vec{x}-\vec{y})\psi^\dagger(x)\left[\gamma_0\Gamma_\nu^{X'}\tau'^\dagger,\gamma_0\Gamma_\mu^X\tau\right]_-\psi(x)$$
$$= \frac{1}{2}\delta^{(3)}(\vec{x}-\vec{y})\psi^\dagger(x)\left(\left[\gamma_0\Gamma_\mu^X,\gamma_0\Gamma_\nu^{X'}\right]_+[\tau,\tau'^\dagger]_-\right.$$
$$\left.+\left[\gamma_0\Gamma_\mu^X,\gamma_0\Gamma_\nu^{X'}\right]_-[\tau,\tau'^\dagger]_+\right)\psi(x),\quad\text{(A.2.15)}$$

where $[A,BC]_- = [A,B]_-C + B[A,C]_- = [A,B]_+C - B[A,C]_+$, and Eqs. (A.2.2) as well as (A.2.7) have been used. Technically, Schwinger terms dropped out when going from the first equality to the second [BD71]. Taking the limit $x\to y$ in the above spirit is only allowed if the operator product $\bar{\psi}(x+\epsilon)\psi(x)$ is a smooth function in ϵ at $\epsilon = 0$. As local operator products are highly singular objects, this is clearly not the case. It is clear, however, that the ETCs (A.2.15) must reproduce the transformation properties of the currents given in (A.1.45).

In order to show the consistency, only those commutators need to be considered where at least one operator is the time-component of an axial- or vector current, specified by $\Gamma^{X'}$. This current induces the transformation X'. For $\mu = 0$, Eq. (A.2.15) becomes

$$\left[j_0^{X',\tau'}(x_0,\vec{x}),j_\nu^{X,\tau}(x_0,\vec{y})\right]$$
$$= \frac{1}{2}\delta^{(3)}(\vec{x}-\vec{y})\psi^\dagger(x)\left(\left[\gamma_0\Gamma_0^{X'},\gamma_0\Gamma_\nu^X\right]_+[\tau',\tau]_-\right.$$
$$\left.+\left[\gamma_0\Gamma_0^{X'},\gamma_0\Gamma_\nu^X\right]_-[\tau',\tau]_+\right)\psi(x)$$

$$= \frac{1}{2}\delta^{(3)}(\vec{x}-\vec{y})\psi^\dagger(x) \left(\begin{cases} [\mathbb{1},\gamma_0\Gamma^X_\nu]_+ & :X'=V \\ -[\gamma_5,\gamma_0\Gamma^X_\nu]_+ & :X'=A \end{cases} [\tau',\tau]_- \right.$$

$$\left. + \begin{cases} [\mathbb{1},\gamma_0\Gamma^X_\nu]_- & :X'=V \\ -[\gamma_5,\gamma_0\Gamma^X_\nu]_- & :X'=A \end{cases} [\tau',\tau]_+ \right) \psi(x)$$

$$= \frac{1}{2}\delta^{(3)}(\vec{x}-\vec{y})\psi^\dagger(x) \Bigg($$

$$\begin{cases} \gamma_0[\mathbb{1},\Gamma^X_\nu]_+ & :X'=V \\ -\left(-[\gamma_5,\gamma_0]_+\Gamma^X_\nu - \gamma_0[\gamma_5,\Gamma^X_\nu]_-\right) & :X'=A \end{cases} [\tau',\tau]_-$$

$$+ \begin{cases} \gamma_0[\mathbb{1},\Gamma^X_\nu]_- & :X'=V \\ -\left(-[\gamma_5,\gamma_0]_+\Gamma^X_\nu - \gamma_0[\gamma_5,\Gamma^X_\nu]_+\right) & :X'=A \end{cases} [\tau',\tau]_+ \Bigg)\psi(x)$$

$$= \frac{1}{2}\delta^{(3)}(\vec{x}-\vec{y})\psi^\dagger(x) \left(\begin{cases} \gamma_0[\Gamma_{X'},\Gamma^X_\nu]_+ & :X'=V, \Gamma_{X'}=\mathbb{1} \\ \gamma_0[\Gamma_{X'},\Gamma^X_\nu]_- & :X'=A, \Gamma_{X'}=\gamma_5 \end{cases} [\tau',\tau]_- \right.$$

$$\left. + \begin{cases} \gamma_0[\Gamma_{X'},\Gamma^X_\nu]_- & :X'=V, \Gamma_{X'}=\mathbb{1} \\ \gamma_0[\Gamma_{X'},\Gamma^X_\nu]_+ & :X'=A, \Gamma_{X'}=\gamma_5 \end{cases} [\tau',\tau]_+ \right)\psi(x), \quad \text{(A.2.16)}$$

where $[A,BC]_+ = [A,B]_-C + B[A,C]_+ = [A,B]_+C - B[A,C]_-$ and $[A,BC]_- = [A,B]_+C - B[A,C]_+$ have been used. In case of the axial transformation a global sign has to be added, because the charge of the axial current defined in (A.1.42b) is the negative generator of Λ_A (A.1.40b).[25] Collecting commutators and anticommutators in Dirac space, multiplying with i and integrating w. r. t. $\vec{x}$ gives

$$i\left[Q^{X',\tau'}(x_0), j^{X,\tau}_\nu(x_0,\vec{y})\right]_- = \delta^{X',\tau'} j^{X,\tau}_\nu(x_0,\vec{y})$$

$$= \pm \frac{i}{2}\bar{\psi}\left([\Gamma^{X'},\Gamma^X]_+[\tau',\tau]_\mp + [\Gamma^{X'},\Gamma^X]_-[\tau',\tau]_\pm\right)\psi, \quad \text{(A.2.17)}$$

where upper (lower) signs refer to the vector (axial) transformation, proving the consistency.

Commutators involving currents of different parity transform as pseudo tensors.

---

[25] Cf. discussion after Eq. (A.1.43).

The opposite is the case for commutators involving currents of the same parity:

$$\left[j_\mu^{(V,A),\tau}(x),\left(j_0^{(V,A),\tau'}(y)\right)^\dagger\right]_{x_0=y_0} = \delta^{(3)}(\vec{x}-\vec{y})j_\mu^{V,[\tau,\tau'^\dagger]_-}(x), \qquad (A.2.18a)$$

$$\left[j_0^{(V,A),\tau}(x),i\partial^\nu\left(j_\nu^{(V,A),\tau'}(y)\right)^\dagger\right]_{x_0=y_0} = \mp\delta^{(3)}(\vec{x}-\vec{y})j^{S,[\tau,[M,\tau'^\dagger]_\mp]_\mp}(x). \qquad (A.2.18b)$$

Only commutators involving currents with the same parity are considered throughout the thesis. Consequently the r. h. s. of Eq. (A.2.18) must be of vectorial or scalar character, respectively.

Order parameters

In quantum field theory, a symmetry is spontaneously broken, if the ground state $|\Omega\rangle$ is not invariant under the corresponding transformation, i. e. $Q|\Omega\rangle \neq 0$ or, equivalent, $e^{iQ\Theta}|\Omega\rangle \neq |\Omega\rangle$. This is also called the Nambu-Goldstone realization of the symmetry, in contrast to the Wigner-Weyl realization where one has $Q|\Omega\rangle = 0$. Using Eq. (A.2.1) the ground state expectation value $\langle\Omega|\delta\mathcal{O}|\Omega\rangle$ is an order parameter of symmetry breaking or restoration:

$$\langle\Omega|i\,[Q,\mathcal{O}]_-|\Omega\rangle = \langle\Omega|\delta\mathcal{O}|\Omega\rangle. \qquad (A.2.19)$$

If $|\Omega\rangle$ is symmetric, i. e. in the Wigner-Weyl phase, one has $\langle\Omega|i\,[Q,\mathcal{O}]|\Omega\rangle = 0$, and $\langle\Omega|\delta\mathcal{O}|\Omega\rangle$ vanishes. Conversely, $\langle\Omega|\delta\mathcal{O}|\Omega\rangle \neq 0$ means that the symmetry must be spontaneously broken. Note that the vanishing of Eq. (A.2.19) does not necessarily require the Wigner-Weyl realization, i. e. symmetry restoration. In fact, the r. h. s. of Eq. (A.2.19) might be zero due to other symmetries such as parity invariance or flavor symmetries (cf. [Tho08a]). The ground state is assumed to be Lorentz invariant and invariant w. r. t. parity transformations and time reversal. Owing to

$$\langle\Omega|\mathcal{O}|\Omega\rangle = \langle\Omega|\mathcal{U}^\dagger\mathcal{O}\mathcal{U}|\Omega\rangle, \qquad (A.2.20)$$

with $\mathcal{U}$ being time reversal, parity or Lorentz transformation, order parameters must also be Lorentz invariant and invariant w. r. t. parity transformations and time reversal.[26] Concerning the charge operator Q, defined in Eq. (A.1.8) for the classical theory, it must be noted that the integral is ill-defined in the Nambu-Goldstone

---

[26] Note that the time inversion operator is anti-unitary, see Eq. (B.2.3). Therefore, Eq. (A.2.20) must be modified for the time inversion operator: $\langle\Omega|\mathcal{O}|\Omega\rangle = \langle\Omega|\mathcal{U}^{-1}\mathcal{O}\mathcal{U}|\Omega\rangle^\dagger$. Hence, for $\mathcal{U}^{-1}\mathcal{O}\mathcal{U} = \mathcal{O}$ the corresponding condensate must be real.

A.2 The formalism for quantized fields

phase: it diverges. However, due to the canonical commutation relations (A.2.2) any commutator of a current with a local Heisenberg field operator is local, i. e. nonzero only in a finite domain, and the space-integral is thus well-defined. Therefore, in the Nambu-Goldstone phase Eq. (A.2.19) is understood to be the space-integral of the ETC of the current, which generates the charge, and the operator $\mathcal{O}$ [Kug97].

The most prominent example of such an order parameter can be deduced from Eq. (A.1.45) in a two-flavor system for the axial transformation of the pseudo-scalar current which can be taken from Tab. A.1.1. Then Eq. (A.2.1) tells us that

$$\langle\Omega|\left[Q^{A,\tau}(x_0), j^{P,\tau'}(x)\right]_-|\Omega\rangle = -i\langle\Omega|j^{S,[\tau,\tau']}_+(x)|\Omega\rangle \tag{A.2.21}$$

is an order parameter of spontaneous chiral symmetry breaking. Choosing $\tau \to \sigma^a/2$ and $\tau' \to \sigma^b/2$ in Eq. (A.2.21), σ being again the Pauli matrices, gives the famous chiral condensate:

$$\langle\Omega|\left[Q^{A,\frac{\sigma^a}{2}}(x_0), j^{P,\frac{\sigma^b}{2}}(x)\right]_-|\Omega\rangle = -i\frac{\delta^{ab}}{2}\langle\Omega|\bar{\psi}(x)\psi(x)|\Omega\rangle. \tag{A.2.22}$$

The formalism presented so far in this section already allows to derive the Gell-Mann–Oakes–Renner relation [GM68]. Extracting the coordinate dependence of the matrix element $\langle\Omega|j_\mu^{A,\sigma^b}(x)|\pi^c(q)\rangle$, with $|\pi^c(q)\rangle$ being a one-pion state with momentum q, by virtue of Eq. (A.2.4)[27] and projecting the Lorentz structure, one obtains

$$\langle\Omega|j_\mu^{A,\frac{\sigma^b}{2}}(x)|\pi^c(q)\rangle = e^{-iqx}i\delta^{bc}q_\mu f_\pi(q^2), \tag{A.2.23}$$

where the pion decay constant is defined as (no Einstein convention for dotted indices)

$$if_\pi(q^2) \equiv \frac{1}{3m_\pi^2}\langle\Omega|q^\mu j_\mu^{A,\frac{\sigma^a}{2}}(0)|\pi^a(q)\rangle = \frac{1}{m_\pi^2}\langle\Omega|q^\mu j_\mu^{A,\frac{\sigma^a}{2}}(0)|\pi^{\dot{a}}(q)\rangle. \tag{A.2.24}$$

This definition may be justified by applying $i\partial^\mu$ to Eq. (A.2.23) and making use of Eq. (A.1.44). Thereby the pseudo-scalar current is introduced, which has the quantum numbers of the pion [Ynd06]

$$\langle\Omega|i\partial^\mu j_\mu^{A,\frac{\sigma^b}{2}}(x)|\pi^c(q)\rangle = -i\langle\Omega|j^{P,\left[\frac{\sigma^b}{2},M\right]}_+(x)|\pi^c(q)\rangle = e^{-iqx}i\delta^{bc}m_\pi^2 f_\pi(q^2). \tag{A.2.25}$$

---

[27]The ground state is assumed to be translational invariant.

For a two-flavor system one has $\left[\sigma^b, M\right]_+ = (m_u+m_d)\sigma^b + (m_u-m_d)\delta^{b3}\mathbb{1}$, and thus Eq. (A.2.21) becomes

$$\langle\Omega|\left[Q^{A,\frac{\sigma^a}{2}}(x_0), j^{P,\left[\frac{\sigma^b}{2},M\right]_+}(x)\right]_-|\Omega\rangle$$
$$= -i\delta^{ab}\frac{m_u+m_d}{2}\langle\bar\psi\psi\rangle - i\delta^{b3}\frac{m_u-m_d}{2}\langle\bar\psi\sigma^a\psi\rangle. \quad (A.2.26)$$

Inserting a complete set of covariantly normalized one-pion states,

$$\mathbb{1} = \sum_c \int \frac{d^3p}{2(2\pi)^3 E_p}|\pi^c(p)\rangle\langle\pi^c(p)|, \quad (A.2.27)$$

into Eq. (A.2.26), using Eqs. (A.2.23) and (A.2.25) and performing the trace w. r. t. flavor indices, reveals the Gell-Mann–Oakes–Renner relation

$$f_\pi^2 m_\pi^2 = -\frac{m_u+m_d}{2}\langle\bar\psi\psi\rangle - \frac{m_u-m_d}{2}\langle\bar\psi\sigma^3\psi\rangle = -\langle m_u\bar uu + m_d\bar dd\rangle. \quad (A.2.28)$$

Employing the standard values $f_\pi = 92.4\,\text{MeV}$, $m_\pi = 139.6\,\text{MeV}$, $m_u = 4\,\text{MeV}$ and $m_d = 7\,\text{MeV}$ [Nak10] gives a chiral condensate of $\langle\bar qq\rangle = -(247\,\text{MeV})^3$, where we defined the averaged chiral condensate as $\langle\bar qq\rangle = \langle\bar\psi M\psi\rangle/\text{Tr}M$.

Another example can be obtained for a two-flavor system by considering the operator [Tho08a]

$$J_{\mu\nu}^{\tau,\tau'} \equiv j_\mu^{A,\tau} j_\nu^{V,\tau'}. \quad (A.2.29)$$

It transforms under axial transformations as (cf. Tab. A.1.1)

$$i\left[Q^{\tilde\tau}, J_{\mu\nu}^{\tau,\tau'}\right] = -i\left(j_\mu^{V,[\tilde\tau,\tau]_-} j_\nu^{V,\tau'} + j_\mu^{A,\tau} j_\nu^{A,[\tilde\tau,\tau']_-}\right). \quad (A.2.30)$$

Inserting the Pauli matrices $\tau = \sigma^a/2$, $\tau' = \sigma^b/2$, $\tilde\tau = \sigma^c/2$, results in

$$i\left[Q^{\sigma^c/2}, J_{\mu\nu}^{\sigma^a/2,\sigma^b/2}\right] = -\frac{i}{2}\left(\epsilon^{cax} j_\mu^{V,\sigma^x} j_\nu^{V,\sigma^b} + \epsilon^{cbx} j_\mu^{A,\sigma^a} j_\nu^{A,\sigma^x}\right) \quad (A.2.31)$$

and contracting with ϵ^{abc} yields

$$i\epsilon^{abc}\left[Q^{\sigma^c/2}, J_{\mu\nu}^{\sigma^a/2,\sigma^b/2}\right] = -i\left(\vec j_\mu^V \vec j_\nu^V - \vec j_\mu^A \vec j_\nu^A\right), \quad (A.2.32)$$

where the notation of vector–isospin-vector and axial-vector–isospin-vector currents

A.2 The formalism for quantized fields

of App. A.1 is used. Equation (A.2.32) clearly identifies $\left(\vec{j}_\mu^V \vec{j}_\nu^V - \vec{j}_\mu^A \vec{j}_\nu^A\right)$ as an order parameter of chiral symmetry breaking.

As has been discussed below Eq. (A.1.68) in App. A.1, the components of an isospin-vector current may be transformed into each other by a finite vector transformation. If the ground state is symmetric w. r. t. the quark flavors it follows that expectation values are invariant w. r. t. isospin transformations, because the components of an isospin-vector current can be transformed into each other by an appropriate finite vector transformation (cf. App. A.1). Consequently, each component $j_\mu^{V,\sigma^a} j_\nu^{V,\sigma^a} - j_\mu^{A,\sigma^a} j_\nu^{A,\sigma^a}$ is an order parameter.

Let us summarize the symmetry properties of the Lagrangian (A.1.36) under chiral transformations. The classical chiral $U_L(N_f) \times U_R(N_f)$ symmetry is broken explicitly by nonzero quark masses down to a $U_V(N_f)$ symmetry in case of equal quark masses and to a $U_V(1)$ in case of unequal quark masses. If the current quark masses and their differences are small, the $U_L(N_f) \times U_R(N_f)$ is an approximate symmetry. While the exact $U_V(1)$ symmetry is always satisfied (it reflects the invariance w. r. t. phase transformations and corresponds to the baryon number conservation) the approximate $U_A(1)$ symmetry of the quantized theory is broken explicitly by the axial anomaly. Therefore, $j_\mu^{V,1}$ is conserved in the classical as well as in the quantized theory, whereas $j_\mu^{A,1}$ is only partially (i. e. approximately) conserved in the classical theory and not conserved in the quantized theory. The approximate $SU_A(N_f)$ symmetry is spontaneously or dynamically broken. Contrary to the $U_A(1)$ current, the $SU_A(N_f)$ currents are partially conserved if the quark masses are approximately equal. This is referred to as partial conservation of the axial-vector current. The $SU_A(N_f)$ transformations relate parity partners to each other. The spontaneous breakdown of this symmetry results in the non-degeneracy of parity partner. For obvious reasons, the approximate $SU_V(N_f)$ symmetry is called flavor symmetry. It implicates conservation of the vector–isospin-vector current. Contrarily to the $SU_A(N_f)$ symmetry, the flavor symmetry is not broken spontaneously, which is signaled by the approximate degeneracy within the isospin triplet of, e. g., the ρ mesons. The tiny mass differences of a few MeV can be traced back to the explicit breaking of the flavor symmetry by unequal quark masses. One might argue that the mass splitting of chiral partners caused by the explicit chiral symmetry breaking by finite quark masses must be of the same order.

A.3 Brief survey on Quantum Chromodynamics

The chapter is finished by a short basic introduction to QCD. The focus is on the essential equations and relations which are needed in this thesis. For a detailed essay on QCD, we recommend [Pas84, Mut87]. The Lagrange density of classical chromodynamics follows from the requirement of invariance w. r. t. local color gauge transformations and reads

$$\mathscr{L} = \bar{\psi}\left(i\hat{D} - M\right)\psi - \frac{1}{4}G^A_{\mu\nu}G^{\mu\nu}_A, \tag{A.3.1}$$

where we have introduced the following notation

$$D^{ab}_\mu(x) = \partial_\mu \mathbb{1}^{ab} - ig\mathscr{A}^{ab}_\mu(x) : \quad \text{covariant derivative}, \tag{A.3.2a}$$

$$\mathscr{A}^\mu_{ab} = A^{A\mu} t^A_{ab} : \quad \text{gluon fields}, \tag{A.3.2b}$$

$$\mathscr{G}_{\mu\nu} = G^A_{\mu\nu} t^A = \frac{i}{g}\left[D_\mu, D_\nu\right] : \quad \text{gluon field strength tensor}, \tag{A.3.2c}$$

$$\psi^a_i : \quad \text{quark field operators}, \tag{A.3.2d}$$

$$t^A_{ab} = \frac{1}{2}\lambda^A : \quad \text{generators of SU}(N_c), \tag{A.3.2e}$$

$$\lambda^A : \quad \text{Gell-Mann matrices in case of } N_c = 3. \tag{A.3.2f}$$

Dirac indices are denoted by Latin letters $i, j, k, \cdots$, Lorentz indices by Greek letters $\mu, \nu, \kappa, \cdots$ and color indices by Latin letters $a, b, c, \cdots$. The generators t^A satisfy

$$\left[t^A, t^B\right] = if^{ABC}t^C, \quad tr\left(t^A\right) = 0, \quad tr\left(t^A t^B\right) = \frac{\delta^{AB}}{2}, \tag{A.3.3}$$

where f^{ABC} are the structure constants of the SU(N_c) algebra. From (A.3.2c) one can show that

$$G^A_{\mu\nu} = \partial_\mu A^A_\nu - \partial_\nu A^A_\mu + gf^{ABC} A^B_\mu A^C_\nu, \tag{A.3.4}$$

and

$$G^A_{\mu\nu} = -G^A_{\nu\mu}. \tag{A.3.5}$$

The last term in Eq. (A.3.4) accounts for the gluon self-interaction, i. e. gluon-gluon, triple-gluon and quartic-gluon interactions. N_c denotes the number of colors ($N_c = 3$

A.3 Brief survey on Quantum Chromodynamics

for QCD), and the index A refers to the generators of SU(N_c). The quark field spinor ψ denotes the flavor multiplet, $\psi = (u, d, s, c, b, t)^T$. In case of $N_f = 2$ it is called isodoublet. The mass matrix M is diagonal with the following entries [Nak10]

$$m_u = 1.5 \text{ to } 4.0 \, \text{MeV}, \tag{A.3.6}$$

$$m_d = 3 \text{ to } 7 \, \text{MeV}, \tag{A.3.7}$$

$$m_s = 95 \pm 25 \, \text{MeV}, \tag{A.3.8}$$

$$m_c = 1.25 \pm 0.09 \, \text{GeV}, \tag{A.3.9}$$

$$m_b = 4.20 \pm 0.07 \, \text{GeV} \quad (\overline{\text{MS}} \text{ mass}), \tag{A.3.10}$$

$$\text{or} \quad 4.70 \pm 0.07 \, \text{GeV} \quad (\text{1S mass}), \tag{A.3.11}$$

$$m_t = 174.2 \pm 3.3 \, \text{GeV} \quad (\text{direct observation of top events}), \tag{A.3.12}$$

$$\text{or} \quad 172.3^{+10.2}_{-7.6} \, \text{GeV} \quad (\text{Standard Model electroweak fit}). \tag{A.3.13}$$

Because quarks are confined within hadrons and can not be observed as isolated free particles, the determination of their masses depends on the theoretical framework used. Therefore, different values for the bottom and top quark masses are quoted. The difference is of no importance for our calculations as the top quark is not considered here; for the bottom quarks the difference is of minor impact.

The coupling strength of strong interaction, denoted by g, is defined as

$$g = \sqrt{4\pi\alpha_s}, \tag{A.3.14}$$

$$\alpha_s(q^2) = \frac{4\pi}{(33 - 2N_f) \ln\left(\frac{-q^2}{\Lambda_{\text{QCD}}^2}\right)}, \tag{A.3.15}$$

with N_f being the number of active quark flavors and Λ_{QCD}^2 the renormalization scale parameter of QCD, usually determined to reproduce $\alpha_s(M_Z) \approx 0.19$ with $M_Z = 91.1876 \pm 0.0021 \, \text{GeV}$ being the Z boson mass. The running coupling in Eq. (A.3.15) is the one-loop result.

The Lagrange density (A.3.1) emerges from the requirement of invariance under the local gauge color transformation

$$\psi_i^a(x) \to \psi_i'^a(x) = \left[e^{-igt^A \Theta_A(x)}\right]^{ab} \psi_i^b(x). \tag{A.3.16}$$

A Chiral symmetry, currents and order parameters

By this one obtains the following transformation laws for color transformations

$$D_\mu \psi_i(x) \to D'_\mu \psi'_i(x) = e^{-igt^A\Theta_A(x)} D_\mu \psi_i(x), \tag{A.3.17}$$

$$D_\mu \to D'_\mu = e^{-igt^A\Theta_A(x)} D_\mu e^{igt^A\Theta_A(x)}, \tag{A.3.18}$$

$$\mathcal{A}_\mu \to \mathcal{A}'_\mu = e^{-igt^A\Theta_A} \mathcal{A}_\mu e^{igt^A\Theta_A(x)} + \left[\partial_\mu e^{-igt^A\Theta_A(x)}\right] e^{igt^A\Theta_A(x)}, \tag{A.3.19}$$

$$\mathcal{G}_{\mu\nu}(x) \to \mathcal{G}'_{\mu\nu} = e^{-igt^A\Theta_A} \mathcal{G}_{\mu\nu}(x) e^{igt^A\Theta_A(x)}. \tag{A.3.20}$$

The explicit EoM for the quark and gluon fields follow from (A.3.1) as

$$\hat{D} q = -i m_q q, \tag{A.3.21}$$

$$\bar{q} \overleftarrow{\hat{D}} = i m_q \bar{q}, \tag{A.3.22}$$

$$\left[D^\mu, \mathcal{G}_{\mu\nu}\right](x) = -g t^A \sum_{n=u,d,s,c} \bar{n}(x) t^A \gamma_\nu n(x), \tag{A.3.23}$$

where summation over A is understood and we have defined

$$\gamma^\mu D_\mu = \hat{D}, \tag{A.3.24}$$

$$\sigma^{\mu\nu} \mathcal{G}_{\mu\nu} = \sigma \mathcal{G} \tag{A.3.25}$$

with spin matrices

$$\sigma_{\mu\nu} = \frac{i}{2}[\gamma_\mu, \gamma_\nu] = i(\gamma_\mu \gamma_\nu - g_{\mu\nu}). \tag{A.3.26}$$

Exploiting these relations, one can show that the gluon field strength tensor fulfills the following useful relations

$$D^2 = \hat{D}\hat{D} + \frac{1}{2} g \sigma \mathcal{G}, \tag{A.3.27a}$$

$$D^2 q = \left(\frac{1}{2} g \sigma \mathcal{G} - m^2\right) q, \tag{A.3.27b}$$

where we have defined

$$D^2 = g^{\mu\nu} D_\mu D_\nu. \tag{A.3.28}$$

Splitting up the Lagrange density into a free part $\mathcal{L}_0$ and a part containing the

A.3 Brief survey on Quantum Chromodynamics

interaction, we define the interaction Lagrange density as

$$\mathscr{L}_{\text{int}} = \mathscr{L} - \mathscr{L}_0. \tag{A.3.29}$$

The free, i. e. non-interacting, Lagrange density reads

$$\mathscr{L}_0 = \bar{\psi}\left(i\hat{\partial} - M\right)\psi - \frac{1}{4}\left(\partial_\mu A^A_\nu - \partial_\nu A^B_\mu\right)\left(\partial^\mu A^{A,\nu} - \partial^\nu A^{B,\mu}\right). \tag{A.3.30}$$

However, canonical quantization of the Lagrangian (A.3.1) requires the introduction of gauge fixing terms in order to restore Lorentz covariance. Because the canonical conjugate momentum to the time-component of the gauge field is zero, the canonical commutation relations cannot be fulfilled and Lorentz covariance is lost. Therefore a gauge fixing term is introduced which eliminates the unphysical degrees of freedom of the massless gauge fields. But as the resulting Fock space of states has negative norm and unitarity is violated, Faddeev-Popov ghosts need to be introduced as well in order to preserve unitarity in the Fock space of physical states, i. e. the probabilistic interpretation. The resulting quantized Lagrangian is not gauge invariant anymore. In [Bec74, Bec75] the Becchi-Rouet-Stora transformation has been derived, which leaves the Lagrangian invariant and may therefore serve as replacement of the local gauge invariance requirement. The introduction of the Faddeev-Popov ghosts causes an additional complication of the coupled set of DSEs of the theory. Apart from quark-gluon, gluon-gluon, triple- and quartic-gluon interactions, also ghost-gluon interactions have to be considered (see Sec. 6.1.2). The QCD Lagrangian reads [Pas84, Mut87]

$$\mathscr{L} = \bar{\psi}\left(i\hat{D} - M\right)\psi - \frac{1}{4}G^A_{\mu\nu}G_A^{\mu\nu}$$
$$- \frac{1}{2a}\left(\partial^\mu A^A_\mu\right)\left(\partial_\nu A^\nu_A\right) - \left(\partial^\mu \bar{\phi}_A\right)\left(\partial_\mu \phi^A\right) + g f_{ABC}\left(\partial^\mu \bar{\phi}^A\right)\phi^B A^C_\mu, \tag{A.3.31}$$

where ϕ denotes the ghost field.

B Correlation functions

The central objects within a QSR analysis, performed in the chapters 3, 4 and 5, are current-current correlation functions. A detailed knowledge about analytic properties, relations among different current-current correlation functions and symmetries is mandatory. In the following, these points will be investigated and a comprehensive collection of the relations needed throughout this work is given. Special emphasis is put on nonzero densities and temperatures. In-medium current-current correlation functions are of particular relevance in solid-state physics, and a lot of the material collected here can be found in standard many-body textbooks such as e. g. [Neg88, Fet71, Nol05]. We generalize this treatment to the case of non-contracted Lorentz indices of spin-1 currents.

Different correlation functions of two current operators (A.1.43) with different analytic properties may be defined, such as the causal correlator

$$\Pi_{\mu\nu}^{X,\tau}(q) = i \int d^4x \, e^{iqx} \langle T \left[j_\mu^{X,\tau}(x) j_\nu^{X,\tau^\dagger}(0) \right] \rangle \,, \tag{B.0.1a}$$

the retarded and advanced correlators

$$R_{\mu\nu}^{X,\tau}(q) = i \int d^4x \, e^{iqx} \Theta(x_0) \langle \left[j_\mu^{X,\tau}(x), j_\nu^{X,\tau^\dagger}(0) \right] \rangle \,, \tag{B.0.1b}$$

$$A_{\mu\nu}^{X,\tau}(q) = i \int d^4x \, e^{iqx} \Theta(-x_0) \langle \left[j_\nu^{X,\tau^\dagger}(0), j_\mu^{X,\tau}(x) \right] \rangle \,, \tag{B.0.1c}$$

and the spectral density

$$\rho_{\mu\nu}^{X,\tau}(q) = -\int d^4x \, e^{iqx} \langle \left[j_\mu^{X,\tau}(x), j_\nu^{X,\tau^\dagger}(0) \right] \rangle \,. \tag{B.0.1d}$$

T denotes time-ordering. In vacuum $\langle \mathcal{O} \rangle$ denotes the ground state expectation value of the operator $\mathcal{O}$. In the medium, i. e. at finite density and/or temperature, it denotes

B Correlation functions

the Gibbs average, which is defined as

$$\langle \mathcal{O} \rangle = \frac{1}{Z} \sum_n \langle n | e^{-\beta(H-\mu N)} \mathcal{O} | n \rangle , \qquad (B.0.2)$$

where H is the Hamiltonian, N is some additive quantum number, e. g. the particle number, and μ the corresponding chemical potential. The set $\{|n\rangle\}$ stands for a complete set of eigenstates of $\bar{P}_0 = H - \mu N$, i. e. $\bar{P}_0 |n\rangle = (E_n - \mu N_n)|n\rangle$. The partition function is denoted by $Z = \text{Tr} \exp\{-\beta(H-\mu N)\}$. The inverse temperature $\beta = 1/k_B T$ is the Legendre transform conjugate to the total energy of the system [Zia09]. Strictly speaking it is the slope of the Lagrangian as a function of energy and thus its derivative w. r. t. energy. Likewise, $-\beta\mu$ is the conjugate to the total particle number. See also [Gol50] for details and background of the Legendre transform. The space-time dependence of the currents is given by Eq. (A.2.4)

$$j_\mu^{X,\tau}(x) = e^{i\bar{P}x} j_\mu^{X,\tau}(0) e^{-i\bar{P}x} , \qquad (B.0.3)$$

with $\bar{P}^\mu = (H - \mu N, \vec{P})$ the generator of translations. All considerations are for vector or axial-vector currents with uncontracted indices. The cases of contracted indices or of scalar and pseudo-scalar currents are special cases thereof.

B.1 Källén-Lehmann representation and analytic properties

With these conventions the Källén-Lehmann representations [Käl52, Leh54] of causal, retarded and advanced correlators for the bosonic currents defined in Eq. (A.1.43) are obtained by expanding each expectation value of the Gibbs averages in (B.0.1) over the complete set of states $\{|n\rangle\}$, often referred to as inserting a representation of the unit operator, and using the integral representation of the Heaviside function [Neg88]:

$$\begin{Bmatrix} \Pi_{\mu\nu}^{X,\tau}(q) \\ R_{\mu\nu}^{X,\tau}(q) \\ A_{\mu\nu}^{X,\tau}(q) \end{Bmatrix} = -\frac{(2\pi)^3}{Z} \lim_{\epsilon \to 0^+} \sum_{n,m} \delta^{(3)}(\vec{q} - \vec{p}_n + \vec{p}_m) \langle m|j_\mu^{X,\tau}|n\rangle \langle n|j_\nu^{X,\tau^\dagger}|m\rangle$$

B.1 Kállén-Lehmann representation and analytic properties

$$\times \left(\frac{e^{-\beta(E_m - \mu N_m)}}{q_0 - (E_n - E_m - \mu) \begin{Bmatrix} + \\ + \\ - \end{Bmatrix} i\epsilon} - \frac{e^{-\beta(E_n - \mu N_n)}}{q_0 - (E_n - E_m - \mu) \begin{Bmatrix} - \\ + \\ - \end{Bmatrix} i\epsilon} \right). $$

(B.1.1)

Hence, the retarded correlator has poles below the real axis and the advanced correlator has poles above the real axis. The causal correlator (2.0.1) is analytic neither in the upper nor in the lower energy plane. We assumed $\mu(N) = \mu(N+1) \equiv \mu$, i.e. the chemical potential is independent of the particle number, which is justified in a grand canonical ensemble. Here and in the following we insist on explicitly indicating the limiting procedure in order to prevent an erroneously interchanging of limits. The spectral density becomes

$$\rho_{\mu\nu}^{X,\tau}(q)$$
$$= -\frac{(2\pi)^4}{Z} \sum_{n,m} \delta^{(4)}\left(q - \bar{p}_n + \bar{p}_m\right) \langle m|j_\mu^{X,\tau}|n\rangle \langle n|j_\nu^{X,\tau^\dagger}|m\rangle$$
$$\times \left(e^{-\beta(E_m - \mu N_m)} - e^{-\beta(E_n - \mu N_n)} \right)$$
$$= -\frac{(2\pi)^4}{Z} \sum_{n,m} \delta^{(4)}\left(q - \bar{p}_n + \bar{p}_m\right) \langle m|j_\mu^{X,\tau}|n\rangle \langle n|j_\nu^{X,\tau^\dagger}|m\rangle$$
$$\times e^{-\beta(E_m - \mu N_m)} \left(1 - e^{-\beta q_0}\right), \quad \text{(B.1.2)}$$

where we have used that only states with particle number $N_n = N_m + 1$ contribute and hence $\bar{p}_{n,0} - \bar{p}_{m,0} = E_n - E_m - \mu$. Note that in general $\operatorname{Im} \rho_{\mu\nu}^{X,\tau}(q) \neq 0$ due to the uncontracted Lorentz indices of the currents. In terms of the spectral density the correlators can be written as

$$\begin{Bmatrix} R_{\mu\nu}^{X,\tau}(q) \\ A_{\mu\nu}^{X,\tau}(q) \end{Bmatrix} = \lim_{\epsilon \to 0^+} \int_{-\infty}^{+\infty} \frac{d\omega}{2\pi} \frac{\rho_{\mu\nu}^{X,\tau}(\omega, \vec{q})}{q_0 - \omega \pm i\epsilon}$$
$$= \mp \frac{i}{2} \rho_{\mu\nu}^{X,\tau}(q) + \text{p.v.} \int_{-\infty}^{+\infty} \frac{d\omega}{2\pi} \frac{\rho_{\mu\nu}^{X,\tau}(\omega, \vec{q})}{q_0 - \omega}, \quad \text{(B.1.3)}$$

where we have used the Sokhatsky-Weierstrass theorem and "p.v." denotes the Cauchy principle value of the integral. From Eq. (B.1.3), $R_{\mu\nu}^{X,\tau}(q)^* = A_{\mu\nu}^{X,\tau}(q)$ follows for $q_0 \in \mathbb{R}$. Using $1 - e^{-x} = (1 + e^{-x}) \tanh(x/2)$, real and imaginary parts of the correlators thus

B Correlation functions

read

$$\operatorname{Re}\left\{\begin{array}{c}\Pi_{\mu\nu}^{X,\tau}(q)\\ R_{\mu\nu}^{X,\tau}(q)\\ A_{\mu\nu}^{X,\tau}(q)\end{array}\right\} = \frac{1}{2}\left\{\begin{array}{c}-\tanh^{-1}\left(\frac{\beta q_0}{2}\right)\\ +\\ -\end{array}\right\}\operatorname{Im}\rho_{\mu\nu}^{X,\tau}(q) + \text{p.v.}\int_{-\infty}^{+\infty}\frac{d\omega}{2\pi}\frac{\operatorname{Re}\rho_{\mu\nu}^{X,\tau}(\omega,\vec{q})}{q_0-\omega}$$

(B.1.4)

and

$$\operatorname{Im}\left\{\begin{array}{c}\Pi_{\mu\nu}^{X,\tau}(q)\\ R_{\mu\nu}^{X,\tau}(q)\\ A_{\mu\nu}^{X,\tau}(q)\end{array}\right\} = \frac{1}{2}\left\{\begin{array}{c}-\tanh^{-1}\left(\frac{\beta q_0}{2}\right)\\ -\\ +\end{array}\right\}\operatorname{Re}\rho_{\mu\nu}^{X,\tau}(q) + \text{p.v.}\int_{-\infty}^{+\infty}\frac{d\omega}{2\pi}\frac{\operatorname{Im}\rho_{\mu\nu}^{X,\tau}(\omega,\vec{q})}{q_0-\omega}.$$

(B.1.5)

Finally, we remark that without further assumptions it is not possible to give a dispersion relation expressing the real parts of the two-point functions by their imaginary parts only, i.e. time and parity reversal invariance and translational invariance, due to the nonzero imaginary part of the spectral density. Eliminating the real part of the spectral density gives

$$\operatorname{Re}\left\{\begin{array}{c}\Pi_{\mu\nu}^{X,\tau}(q)\\ R_{\mu\nu}^{X,\tau}(q)\\ A_{\mu\nu}^{X,\tau}(q)\end{array}\right\} = \frac{1}{2}\left\{\begin{array}{c}-\tanh^{-1}\left(\frac{\beta q_0}{2}\right)\\ +\\ -\end{array}\right\}\operatorname{Im}\rho_{\mu\nu}^{X,\tau}(q)$$

$$+ \text{p.v.}\int_{-\infty}^{+\infty}\frac{d\omega}{\pi}\frac{1}{q_0-\omega}\left\{\begin{array}{c}-\tanh\left(\frac{\beta\omega}{2}\right)\\ -\\ +\end{array}\right\}$$

$$\times\left[\operatorname{Im}\left\{\begin{array}{c}\Pi_{\mu\nu}^{X,\tau}(\omega,\vec{q})\\ R_{\mu\nu}^{X,\tau}(\omega,\vec{q})\\ A_{\mu\nu}^{X,\tau}(\omega,\vec{q})\end{array}\right\} - \text{p.v.}\int_{-\infty}^{+\infty}\frac{d\omega'}{2\pi}\frac{\operatorname{Im}\rho_{\mu\nu}^{X,\tau}(\omega',\vec{q})}{\omega-\omega'}\right]$$

(B.1.6)

whereas eliminating the imaginary part we end up with

$$\operatorname{Im}\left\{\begin{array}{c}\Pi_{\mu\nu}^{X,\tau}(q)\\ R_{\mu\nu}^{X,\tau}(q)\\ A_{\mu\nu}^{X,\tau}(q)\end{array}\right\} = \frac{1}{2}\left\{\begin{array}{c}-\tanh^{-1}\left(\frac{\beta q_0}{2}\right)\\ +\\ -\end{array}\right\}\operatorname{Re}\rho_{\mu\nu}^{X,\tau}(q)$$

B.1 Källén-Lehmann representation and analytic properties

$$+ \text{p.v.} \int_{-\infty}^{+\infty} \frac{d\omega}{\pi} \frac{1}{q_0 - \omega} \begin{Bmatrix} -\tanh\left(\frac{\beta\omega}{2}\right) \\ - \\ + \end{Bmatrix}$$

$$\times \left[\text{Re} \begin{Bmatrix} \Pi_{\mu\nu}^{X,\tau}(\omega, \vec{q}) \\ R_{\mu\nu}^{X,\tau}(\omega, \vec{q}) \\ A_{\mu\nu}^{X,\tau}(\omega, \vec{q}) \end{Bmatrix} - \text{p.v.} \int_{-\infty}^{+\infty} \frac{d\omega'}{2\pi} \frac{\text{Re}\rho_{\mu\nu}^{X,\tau}(\omega', \vec{q})}{\omega - \omega'} \right]. \quad (B.1.7)$$

If the spectral function is real, Eqs. (B.1.6) and (B.1.7) reduce to the well-known dispersion relations which relate real and imaginary parts of correlators.

Considering the imaginary part of the spectral density we find

$$\text{Im}\rho_{\mu\nu}^{X,\tau}(q) = \frac{1}{2i} \left(\rho_{\mu\nu}^{X,\tau}(q) - \rho_{\mu\nu}^{X,\tau}(q)^* \right)$$
$$\propto \frac{1}{i} \left(\langle m | j_\mu^{X,\tau} | n \rangle \langle n | j_\nu^{X,\tau^\dagger} | m \rangle - \langle m | j_\nu^{X,\tau} | n \rangle \langle n | j_\mu^{X,\tau^\dagger} | m \rangle \right), \quad (B.1.8)$$

which is antisymmetric in its Lorentz indices. Similarly, the real part is symmetric in (μ, ν). From the spectral representations we read off that the propagator and the correlators are symmetric if and only if the spectral density is real, i.e. $\langle m | j_\mu^{X,\tau} | n \rangle \langle n | j_\nu^{X,\tau^\dagger} | m \rangle \in \mathbb{R}$. Assuming time and parity reversal invariance and translational invariance, the spectral density is symmetric w.r.t. its Lorentz indices and, therefore, real. Thus, all correlators are symmetric w.r.t. their Lorentz indices. This will be shown in App. B.2.[28]

Relating real and imaginary parts of the causal correlator to real and imaginary

---

[28] Assuming translational invariance of the Gibbs average, relating real and imaginary parts of the spectral density to its symmetric and antisymmetric contributions can directly be done from the definition:

$$\left(\rho_{\mu\nu}^{X,\tau}(q)\right)^* = -\int d^4x \, e^{-iqx} \langle \left[j_\nu^{X,\tau}(0), j_\mu^{X,\tau^\dagger}(x) \right] \rangle = \rho_{\mu\nu}^{X,\tau^\dagger}(-q)$$
$$= -\int d^4x \, e^{-iqx} \langle \left[j_\nu^{X,\tau}(-x), j_\mu^{X,\tau^\dagger}(0) \right] \rangle$$
$$= -\int d^4x \, e^{iqx} \langle \left[j_\nu^{X,\tau}(x), j_\mu^{X,\tau^\dagger}(0) \right] \rangle$$
$$= \rho_{\nu\mu}^{X,\tau}(q).$$

Hence, the real (imaginary) part equals the symmetric (antisymmetric) part:

$$\begin{Bmatrix} \text{Re} \\ \text{Im} \end{Bmatrix} \rho_{\mu\nu}^{X,\tau}(q) = \frac{1}{2} \left(\rho_{\mu\nu}^{X,\tau}(q) + \begin{Bmatrix} + \\ - \end{Bmatrix} \rho_{\nu\mu}^{X,\tau}(q) \right)$$

as stated above.

B Correlation functions

parts of advanced and retarded correlator we obtain

$$\begin{Bmatrix} \text{Re} \\ \text{Im} \end{Bmatrix} \Pi_{\mu\nu}^{X,\tau}(q) = \frac{1}{2} \begin{Bmatrix} \text{Re} \\ \text{Im} \end{Bmatrix} (R+A)_{\mu\nu}^{X,\tau}(q)$$
$$+ \frac{1}{2} \begin{Bmatrix} - \\ + \end{Bmatrix} \tanh^{-1}\left(\frac{\beta q_0}{2}\right) \begin{Bmatrix} \text{Re} \\ \text{Im} \end{Bmatrix} (R-A)_{\mu\nu}^{X,\tau}(q). \quad \text{(B.1.9)}$$

Hence, the causal correlator in terms of retarded and advanced correlators is given by

$$\begin{aligned}
\Pi_{\mu\nu}^{X,\tau}(q) &= \frac{1}{2}(R+A)_{\mu\nu}^{X,\tau}(q) - \frac{1}{2}\tanh^{-1}\left(\frac{\beta q_0}{2}\right)(R^* - A^*)_{\mu\nu}^{X,\tau}(q) \\
&= \frac{R_{\mu\nu}^{X,\tau}(q)}{1-e^{\beta q_0}} + \frac{A_{\mu\nu}^{X,\tau}(q)}{1-e^{-\beta q_0}} - i\tanh^{-1}\left(\frac{\beta q_0}{2}\right)\text{Re}\rho_{\mu\nu}^{X,\tau}(q) \\
&= \frac{R_{\mu\nu}^{X,\tau}(q)}{1-e^{-\beta q_0}} + \frac{A_{\mu\nu}^{X,\tau}(q)}{1-e^{\beta q_0}} - \tanh^{-1}\left(\frac{\beta q_0}{2}\right)\text{Im}\rho_{\mu\nu}^{X,\tau}(q), \quad \text{(B.1.10)}
\end{aligned}$$

which distinguishes from the well-known expression by the imaginary part of the spectral density. It features overlapping poles from the contributions of the retarded and advanced correlator above and below the real axis, which complicate applications of contour integrals in the complex plane.

The zero temperature limit, $\beta \to \infty$, yields

$$\Pi_{\mu\nu}^{X,\tau}(q) = \Theta(q_0)R_{\mu\nu}^{X,\tau}(q) + \Theta(-q_0)A_{\mu\nu}^{X,\tau}(q) - \text{sign}(q_0)\text{Im}\rho_{\mu\nu}^{X,\tau}(q), \quad \text{(B.1.11)}$$

recovering non-overlapping pole contributions. Only for real, i. e. symmetric spectral densities, the uncontracted causal correlator is given by retarded and advanced correlators only.

B.2 Symmetry constraints

Throughout this thesis, invariance w. r. t. time reversal and parity transformations and translational invariance of the ground state or the medium are assumed. In this appendix subsection the inferences thereof for the correlators and the spectral density are derived. Concerning these transformations, retarded and advanced correlators transform identically, thus only the retarded correlator is discussed in detail. The same relations hold for the advanced correlator. All results are summarized in Tab. B.2.1.

Within a nuclear medium, charge conjugation is not a symmetry operation. How-

B.2 Symmetry constraints

ever, for the sake of completeness it is included. For spinor fields, the charge conjugation $\mathscr{C}$ reads as $\mathscr{C}\psi_\alpha\mathscr{C}^{-1} = -\mathscr{C}\psi_\alpha\mathscr{C} = C_{\alpha\beta}\bar{\psi}_\beta$, where $C = i\gamma_0\gamma_2$ is the spinor representation of the unitary charge conjugation operator $\mathscr{C}$ in the canonical representation of the Dirac matrices [Pes95]. For the currents defined in Eq. (A.1.43), the charge conjugation thus results in

$$
\begin{aligned}
\mathscr{C} j_\mu^{X,\tau}(x)\mathscr{C} &= (C\psi)_{i,a}\left(\Gamma_\mu^X\right)_{ij}\tau_{ab}\left(\bar{\psi}C\right)_{j,b} \\
&= (-1)^X \bar{\psi}\Gamma_\mu^X \tau^T \psi \\
&= (-1)^X j_\mu^{X,\tau^T}(x),
\end{aligned}
\qquad (B.2.1)
$$

meaning that $\mathscr{C} j_\mu^{X,\tau}(x)\mathscr{C} = (-1)^X \left(j_\mu^{X,\tau}(x)\right)^\dagger$ if $\tau = \tau^*$ is real. If the ground state or the medium is symmetric w.r.t. charge conjugation $\mathscr{C}|\Omega\rangle = |\Omega\rangle$, the correlators defined in Eq. (B.0.1) transform as

$$
\begin{aligned}
G_{\mu\nu}^{X,\tau}(q) &= i\int d^4x\, e^{iqx} f(x_0)\langle j_\mu^{X,\tau}(x) j_\nu^{X,\tau^\dagger}(0)\rangle \\
&= i\int d^4x\, e^{iqx} f(x_0)\langle j_\mu^{X,\tau^T}(x) j_\nu^{X,\tau^*}(0)\rangle \\
&= G_{\mu\nu}^{X,\tau^T}(q),
\end{aligned}
\qquad (B.2.2)
$$

where f is an arbitrary function, e.g. the Heaviside function. As any of the correlators defined in Eq. (B.0.1) can be built up from suitable functions $G_{\mu\nu}^{X,\tau}(q)$, all correlators transform in the same way. The effect of charge conjugation is to transpose the flavor matrix τ.

Before proceeding with the investigation of the symmetry properties of the correlator functions, recall that the time reversal operator $\mathscr{T}$ is an antiunitary operator [Kug97], i.e.

$$
(\mathscr{T}\psi, \mathscr{T}\phi) = (\psi, \phi)^* = (\phi, \psi), \qquad (B.2.3)
$$

where an appropriate notation for the scalar product has been used in order to underline the difference to a unitary operator. Requiring invariance of a state $|\phi\rangle = \mathscr{T}|\phi\rangle$, the expectation value of two current operators becomes

$$
\begin{aligned}
\langle\phi| j_\mu^{X,\tau}(x) j_\nu^{X,\tau^\dagger}(0)|\phi\rangle &= \langle\mathscr{T}\phi| j_\mu^{X,\tau}(x) j_\nu^{X,\tau^\dagger}(0)\mathscr{T}|\phi\rangle \\
&= \langle\mathscr{T}\phi| \mathscr{T}\mathscr{T}^{-1} j_\mu^{X,\tau}(x)\mathscr{T}\mathscr{T}^{-1} j_\nu^{X,\tau^\dagger}(0)\mathscr{T}|\phi\rangle
\end{aligned}
$$

B Correlation functions

$$\begin{aligned}
&= \langle \phi | \mathcal{T}^{-1} j_\mu^{X,\tau}(x) \mathcal{T} \mathcal{T}^{-1} j_\nu^{X,\tau\dagger}(0) \mathcal{T} | \phi \rangle^* \\
&= (-1)^{\dot{\mu}}(-1)^{\dot{\nu}} \langle \phi | j_\mu^{X,\tau}(-x_0,\vec{x}) j_\nu^{X,\tau\dagger}(0) | \phi \rangle^* \\
&= (-1)^{\dot{\mu}}(-1)^{\dot{\nu}} \langle \phi | j_\nu^{X,\tau}(0) j_\mu^{X,\tau\dagger}(-x_0,\vec{x}) | \phi \rangle .
\end{aligned} \quad \text{(B.2.4)}$$

A dotted index is not summed and

$$(-1)^{\dot{\mu}} = \begin{cases} +1 & : \mu = 0 \\ -1 & : \mu = 1,2,3 \end{cases}. \quad \text{(B.2.5)}$$

The implications of the ground state or Gibbs average symmetries for each correlation function are now derived and summarized in Tab. B.2.1.

Spectral density

- parity reversal $\mathcal{P}$:

$$\begin{aligned}
\rho_{\mu\nu}^{X,\tau}(q) &= -\int d^4x\, e^{iqx} \langle [j_\mu^{X,\tau}(x), j_\nu^{X,\tau\dagger}(0)] \rangle \\
&= -(-1)^{\dot{\mu}}(-1)^{\dot{\nu}} \int d^4x\, e^{iqx} \langle [j_\mu^{X,\tau}(x_0,-\vec{x}), j_\nu^{X,\tau\dagger}(0)] \rangle \\
&= -(-1)^{\dot{\mu}}(-1)^{\dot{\nu}} \int d^4x\, e^{i(q_0 x_0 + \vec{q}\vec{x})} \langle [j_\mu^{X,\tau}(x_0,\vec{x}), j_\nu^{X,\tau\dagger}(0)] \rangle \\
&= (-1)^{\dot{\mu}}(-1)^{\dot{\nu}} \rho_{\mu\nu}^{X,\tau}(q_0,-\vec{q}) .
\end{aligned} \quad \text{(B.2.6)}$$

- time reversal $\mathcal{T}$:

$$\begin{aligned}
\rho_{\mu\nu}^{X,\tau}(q) &= -\int d^4x\, e^{iqx} \langle [j_\mu^{X,\tau}(x), j_\nu^{X,\tau\dagger}(0)] \rangle \\
&= -(-1)^{\dot{\mu}}(-1)^{\dot{\nu}} \int d^4x\, e^{iqx} \langle [j_\nu^{X,\tau}(0), j_\mu^{X,\tau\dagger}(-x_0,\vec{x})] \rangle \\
&= (-1)^{\dot{\mu}}(-1)^{\dot{\nu}} \int d^4x\, e^{i(-q_0 x_0 - \vec{q}\vec{x})} \langle [j_\mu^{X,\tau\dagger}(x_0,\vec{x}), j_\nu^{X,\tau}(0)] \rangle \\
&= -(-1)^{\dot{\mu}}(-1)^{\dot{\nu}} \rho_{\mu\nu}^{X,\tau\dagger}(-q_0,\vec{q}) .
\end{aligned} \quad \text{(B.2.7)}$$

B.2 Symmetry constraints

- translational invariance $\mathscr{S}$:

$$\begin{aligned}\rho_{\mu\nu}^{X,\tau}(q) &= -\int d^4x\, e^{iqx} \langle [j_\mu^{X,\tau}(x), j_\nu^{X,\tau^\dagger}(0)]\rangle \\ &= \int d^4x\, e^{iqx} \langle [j_\nu^{X,\tau^\dagger}(-x), j_\mu^{X,\tau}(0)]\rangle \\ &= -\rho_{\nu\mu}^{X,\tau^\dagger}(-q).\end{aligned} \qquad (B.2.8)$$

Retarded correlator

- parity reversal $\mathscr{P}$:

$$\begin{aligned}R_{\mu\nu}^{X,\tau}(q) &= (-1)^\mu(-1)^\nu i\int d^4x\, e^{iqx}\Theta(x_0)\langle[j_\mu^{X,\tau}(x_0,-\vec{x}), j_\nu^{X,\tau^\dagger}(0)]\rangle \\ &= (-1)^\mu(-1)^\nu i\int d^4x\, e^{i(q_0 x_0+\vec{q}\vec{x})}\Theta(x_0)\langle[j_\mu^{X,\tau}(x_0,\vec{x}), j_\nu^{X,\tau^\dagger}(0)]\rangle \\ &= (-1)^\mu(-1)^\nu R_{\mu\nu}^{X,\tau}(q_0,-\vec{q}).\end{aligned} \qquad (B.2.9)$$

- time reversal $\mathscr{T}$:

$$\begin{aligned}R_{\mu\nu}^{X,\tau}(q) &= (-1)^\mu(-1)^\nu i\int d^4x\, e^{iqx}\Theta(x_0)\langle[j_\nu^{X,\tau}(0), j_\mu^{X,\tau^\dagger}(-x_0,\vec{q})]\rangle \\ &= (-1)^\mu(-1)^\nu i\int d^4x\, e^{i(-q_0 x_0-\vec{q}\vec{x})}\Theta(-x_0)\langle[j_\nu^{X,\tau}(0), j_\mu^{X,\tau^\dagger}(x)]\rangle \\ &= (-1)^\mu(-1)^\nu A_{\mu\nu}^{X,\tau}(-q_0,\vec{q}).\end{aligned} \qquad (B.2.10)$$

- translational invariance $\mathscr{S}$:

$$\begin{aligned}R_{\mu\nu}^{X,\tau}(q) &= \int d^4x\, e^{iqx}\Theta(x_0)\langle[j_\mu^{X,\tau}(0), j_\nu^{X,\tau^\dagger}(-x)]\rangle \\ &= \int d^4x\, e^{-iqx}\Theta(-x_0)\langle[j_\mu^{X,\tau}(0), j_\nu^{X,\tau^\dagger}(x)]\rangle \\ &= A_{\nu\mu}^{X,\tau^\dagger}(-q).\end{aligned} \qquad (B.2.11)$$

Note the interrelation between retarded and advanced correlator in case of time reversal and translational invariance.

B Correlation functions

Causal correlator

- translational invariance $\mathscr{S}$:

$$\begin{aligned}
\Pi_{\mu\nu}^{X,\tau}(q) &= i \int d^4x e^{iqx} \langle T\left[j_\mu^{X,\tau}(x) j_\nu^{X,\tau^\dagger}(0)\right]\rangle \\
&= i \int d^4x e^{iqx} \langle T\left[j_\mu^{X,\tau}(0) j_\nu^{X,\tau^\dagger}(-x)\right]\rangle \\
&= i \int d^4x e^{-iqx} \langle \Theta(-x_0) j_\mu^{X,\tau}(0) j_\nu^\dagger(x) + \Theta(x_0) j_\nu^{X,\tau^\dagger}(x) j_\mu(0)\rangle \\
&= i \int d^4x e^{-iqx} \langle T\left[j_\nu^{X,\tau^\dagger}(x) j_\mu^{X,\tau}(0)\right]\rangle \\
&= \Pi_{\nu\mu}^{X,\tau^\dagger}(-q).
\end{aligned} \qquad (B.2.12)$$

- parity reversal $\mathscr{P}$:

$$\begin{aligned}
\Pi_{\mu\nu}^{X,\tau}(q) &= (-1)^\mu (-1)^\nu i \int d^4x e^{iqx} \langle T\left[j_\mu^{X,\tau}(x_0, -\vec{x}) j_\nu^{X,\tau^\dagger}(0)\right]\rangle \\
&= (-1)^\mu (-1)^\nu i \int d^4x e^{i(q_0 x_0 + \vec{q}\vec{x})} \langle T\left[j_\mu^{X,\tau}(x) j_\nu^{X,\tau^\dagger}(0)\right]\rangle \\
&= (-1)^\mu (-1)^\nu \Pi_{\mu\nu}^{X,\tau}(q_0, -\vec{q}).
\end{aligned} \qquad (B.2.13)$$

- time reversal $\mathscr{T}$:

$$\begin{aligned}
\Pi_{\mu\nu}^{X,\tau}(q) &= (-1)^\mu (-1)^\nu i \int d^4x e^{iqx} \langle \Theta(x_0) j_\nu^{X,\tau}(0) j_\mu^{X,\tau^\dagger}(-x_0, \vec{x}) \\
&\qquad\qquad\qquad + \Theta(-x_0) j_\mu^{X,\tau^\dagger}(-x_0, \vec{x}) j_\nu^{X,\tau}(0)\rangle \\
&= (-1)^\mu (-1)^\nu i \int d^4x e^{i(-q_0 x_0 - \vec{q}\vec{x})} \langle \Theta(-x_0) j_\nu^{X,\tau}(0) j_\mu^{X,\tau^\dagger}(x) \\
&\qquad\qquad\qquad + \Theta(x_0) j_\mu^{X,\tau^\dagger}(x) j_\nu^{X,\tau}(0)\rangle \\
&= (-1)^\mu (-1)^\nu i \int d^4x e^{i(-q_0 x_0 - \vec{q}\vec{x})} \langle T\left[j_\mu^{X,\tau^\dagger}(x) j_\nu^{X,\tau}(0)\right]\rangle \\
&= (-1)^\mu (-1)^\nu \Pi_{\mu\nu}^{X,\tau^\dagger}(-q_0, \vec{q}).
\end{aligned} \qquad (B.2.14)$$

If $\tau = \tau^*$ is real, i.e. $\tau^\dagger = \tau^T$, and provided the state in (B.2.12) is symmetric w.r.t. charge conjugation $\mathscr{C}|\Omega\rangle = |\Omega\rangle$, then

Table B.2.1: Symmetry properties of the correlation functions for assumed symmetries of the ground state in vacuum or of the set $\{|n\rangle\}$ in the medium. With reference to (B.2.5) we denote $(-1)^{\mu\nu} \equiv (-1)^{\mu}(-1)^{\nu}$.

	$\rho_{\mu\nu}^{X,\tau}(q)$	$R_{\mu\nu}^{X,\tau}(q)$	$\Pi_{\mu\nu}^{X,\tau}(q)$
$\mathscr{S}$	$-\rho_{\nu\mu}^{X,\tau\dagger}(-q)$	$A_{\nu\mu}^{X,\tau\dagger}(-q)$	$\Pi_{\nu\mu}^{X,\tau\dagger}(-q)$
$\mathscr{P}$	$(-1)^{\mu\nu}\rho_{\mu\nu}^{X,\tau}(q_0,-\vec{q})$	$(-1)^{\mu\nu}R_{\mu\nu}^{X,\tau}(q_0,-\vec{q})$	$(-1)^{\mu\nu}\Pi_{\mu\nu}^{X,\tau}(q_0,-\vec{q})$
$\mathscr{T}$	$-(-1)^{\mu\nu}\rho_{\mu\nu}^{X,\tau\dagger}(-q_0,\vec{q})$	$A_{\mu\nu}^{X,\tau\dagger}(-q_0,\vec{q})$	$(-1)^{\mu\nu}\Pi_{\mu\nu}^{X,\tau\dagger}(-q_0,\vec{q})$
$\mathscr{C}$	$\rho_{\mu\nu}^{X,\tau}(q)$	$R_{\mu\nu}^{X,\tau}(q)$	$\Pi_{\mu\nu}^{X,\tau}(q)$
$\mathscr{PT}$	$-\rho_{\mu\nu}^{X,\tau\dagger}(-q)$	$A_{\mu\nu}^{X,\tau\dagger}(-q)$	$\Pi_{\mu\nu}^{X,\tau\dagger}(-q)$
$\mathscr{SP}$	$-(-1)^{\mu\nu}\rho_{\nu\mu}^{X,\tau\dagger}(-q_0,\vec{q})$	$A_{\nu\mu}^{X,\tau\dagger}(-q_0,\vec{q})$	$(-1)^{\mu\nu}\Pi_{\nu\mu}^{X,\tau\dagger}(-q_0,\vec{q})$
$\mathscr{ST}$	$(-1)^{\mu\nu}\rho_{\nu\mu}^{X,\tau}(q_0,-\vec{q})$	$(-1)^{\mu\nu}R_{\nu\mu}^{X,\tau}(q_0,-\vec{q})$	$(-1)^{\mu\nu}\Pi_{\nu\mu}^{X,\tau}(q_0,-\vec{q})$
$\mathscr{PTS}$	$\rho_{\nu\mu}^{X,\tau}(q)$	$R_{\nu\mu}^{X,\tau}(q)$	$\Pi_{\nu\mu}^{X,\tau}(q)$

$$\Pi_{\mu\nu}(q) = \Pi_{\nu\mu}(-q). \tag{B.2.15}$$

B.3 Decomposition

Instead of treating directly tensorial quantities it is more convenient to deal with Lorentz invariants. Thereby, independent contributions to the current-current correlators are disentangled. In general, a rank-2 tensor in four-dimensional Minkowski space can be decomposed into 6 independent algebraic invariants. Four of them belong to the symmetric part, $\tilde{A}, \ldots, \tilde{D}$, and two to the antisymmetric part, E and F. The current-current correlation functions depend on the external momentum q and on the medium's four-velocity, which is encoded in the Gibbs average.

A system at nonzero temperature and/or (baryon) density is characterized by a heat-bath vector v_μ which can be normalized by $v^2 = 1$. The heat bath is at rest in a reference frame with $\vec{v} = 0$. Defining the medium velocity v_μ in relativistic field theory is constrained by the tight relation between mass and energy. To employ the heat flux among different fluid elements, as is usually done in non-relativistic problems, suffers from this ambiguity, cf. [Ran09]. For a system with a conserved current j_μ^B, e.g., the baryon current, one can use $v_\mu \propto j_\mu^B$ (Eckart choice) which locks the velocity to the charge flow [Ran09]. For a thermal system without conserved currents one must lock the flow with energy-momentum flow (Landau choice, cf. [Ran09]).

B Correlation functions

Table B.3.1: Contraction table of the projectors defined in Eq. (B.3.3).

	$P^{L\mu\nu}$	$P^{T\mu\nu}$	$L^{\mu\nu}$	$T^{\mu\nu}$
$P^L_{\mu\nu}$	1	0	0	0
$P^T_{\mu\nu}$	0	3	1	2
$L_{\mu\nu}$	0	1	1	0
$T_{\mu\nu}$	0	2	0	2

In a strongly interacting medium, the coefficients $\tilde{A}, \ldots, \tilde{D}, E, F$ are functions of q^2, v^2 and $q \cdot v$. The most general decomposition reads

$$\Pi_{\mu\nu}(q) = g_{\mu\nu}\tilde{A} + q_\mu q_\nu \tilde{B} + \frac{v_\mu v_\nu}{v^2}\tilde{C}$$
$$+ (q_\mu v_\nu + v_\mu q_\nu)\tilde{D} + (q_\mu v_\nu - v_\mu q_\nu)E + \epsilon^{\mu\nu\alpha\beta}q_\alpha v_\beta F. \quad \text{(B.3.1)}$$

However, it is more appropriate to introduce a decomposition, with coefficients $A, \ldots, D$, which distinguishes transversal and longitudinal states relative to the four-momentum q. Additionally, the 4-transversal part can be decomposed into a part which is transversal and one that is longitudinal to the three-momentum $\vec{q}$, cf. e.g. [Kap06]. In the frame where the medium is at rest, the three-momentum can be expressed as

$$\bar{v}_\mu \equiv v_\mu - \frac{v \cdot q}{q^2} q_\mu. \quad \text{(B.3.2)}$$

Introducing the four-longitudinal, the four-transversal, the three-longitudinal and three-transversal projectors as

$$P^L_{\mu\nu} \equiv \frac{q_\mu q_\nu}{q^2}, \quad \text{(B.3.3a)}$$

$$P^T_{\mu\nu} \equiv g_{\mu\nu} - \frac{q_\mu q_\nu}{q^2} = g_{\mu\nu} - P^L_{\mu\nu}, \quad \text{(B.3.3b)}$$

$$L_{\mu\nu} \equiv \frac{v_\mu v_\nu}{v^2}, \quad \text{(B.3.3c)}$$

$$T_{\mu\nu} \equiv g_{\mu\nu} - \frac{q_\mu q_\nu}{q^2} - \frac{v_\mu v_\nu}{v^2} = P^T_{\mu\nu} - L_{\mu\nu}, \quad \text{(B.3.3d)}$$

the properties given in Tab. B.3.1 can be confirmed. The decomposition thus reads

$$\Pi_{\mu\nu}(q) = P^T_{\mu\nu}\Pi_T + P^L_{\mu\nu}\Pi_L + L_{\mu\nu}\tilde{\Pi}^{(3)}_L + (q_\mu \bar{v}_\nu + \bar{v}_\mu q_\nu)D$$

B.3 Decomposition

$$+ (q_\mu \bar{v}_\nu - \bar{v}_\mu q_\nu) E + \epsilon^{\mu\nu\alpha\beta} q_\alpha \bar{v}_\beta F \,. \quad \text{(B.3.4)}$$

The coefficients Π_T and $\tilde{\Pi}_L^{(3)}$ carry the information about the four-transversal degrees of freedom referring to vector or axial-vector components which can be decomposed into three-momentum transversal and three-momentum longitudinal states [Gal91]. The coefficient Π_L is related to the four-longitudinal states referring to a scalar or pseudo-scalar component. The coefficient D encodes the mixing between three-momentum longitudinal (axial-) vector states and (pseudo-) scalar states occurring due to broken rotational invariance for an excitation (hadron) moving with nonzero velocity in the medium [Wol98, Chi77]. Thus, the spin is no longer a conserved quantum number.

Introducing three-transversal part $\Pi_T^{(3)} \equiv \frac{1}{2} T^{\mu\nu} \Pi_{\mu\nu}$ and three-longitudinal part $\Pi_L^{(3)} \equiv L^{\mu\nu} \Pi_{\mu\nu}$ the invariants are given by

$$\Pi_T = \Pi_T^{(3)}, \quad \text{(B.3.5a)}$$

$$\Pi_L = P_{\mu\nu}^L \Pi^{\mu\nu}, \quad \text{(B.3.5b)}$$

$$\Pi_L^{(3)} = \Pi_T + \tilde{\Pi}_L^{(3)}, \quad \text{(B.3.5c)}$$

$$D = \frac{1}{2q^2 \bar{v}^2} \left(q^\mu \bar{v}^\nu + \bar{v}^\mu q^\nu \right) \Pi_{\mu\nu}(q), \quad \text{(B.3.5d)}$$

$$E = \frac{1}{2} \frac{q_\alpha \bar{v}_\beta - q_\beta \bar{v}_\alpha}{q^2 \bar{v}^2} \Pi_{\mu\nu}(q), \quad \text{(B.3.5e)}$$

$$F = -\frac{1}{2} \frac{\epsilon^{\mu\nu\alpha\beta} q_\alpha \bar{v}_\beta}{q^2 \bar{v}^2} \Pi_{\mu\nu}(q). \quad \text{(B.3.5f)}$$

Equations (B.3.5a) and (B.3.5c) reflect the fact that the three-longitudinal part has to be subtracted from the four-transversal part in order to obtain the three-transversal part, see (B.3.3d). If the correlator is symmetric w.r.t. its Lorentz indices, one has $E = F = 0$. In cases of parity violation, F gives a non-vanishing contribution [Gre92].

Antisymmetric contributions to the correlator exist for parity violating processes as, for example, in weak interaction processes. These are not considered here. In the following, let the correlator be symmetric. In the rest frame of the medium $v = (1, \vec{0})$ and for the mesons at rest $q = (q_0, \vec{q} = 0)$, one has $\bar{v}_\mu = 0$ and the correlator is decomposed solely into four-longitudinal and four-transversal parts. The decomposition is therefore given by

$$\Pi_{\mu\nu}(q) = P_{\mu\nu}^T \Pi_T(q) + P_{\mu\nu}^L \Pi_L(q), \quad \text{(B.3.6)}$$

B Correlation functions

with $\Pi_T(q) = \frac{1}{3} P^T_{\mu\nu} \Pi^{\mu\nu}(q)$ and $\Pi_L(q) = P^L_{\mu\nu} \Pi^{\mu\nu}(q)$, both written explicitly covariant.

B.4 Non-anomalous Ward identities

The currents j^S and j^P are supposed to carry the quantum numbers of scalar and pseudo-scalar mesons, respectively, meaning that $j^{(S,P)}|\Omega\rangle$ is a state with the respective quantum numbers. The issue is more complicated for j^V_μ and j^A_μ. In vacuum, current conservation leads to the transversality of the causal correlator. Thus, according to Eq. (B.3.3), the OPE has to be performed merely for the trace of the correlator. Furthermore, in this case only spin-1 contributions enter. In a medium and for non-conserved currents the situation is more involved. In this section, the non-anomalous Ward identity relating spin-1 and spin-0 correlators are derived and conditions for the transversality of spin-1 correlators are given.

Consider the causal correlator of two mesonic (bosonic) Heisenberg field operators A_μ and B_ν (to be identified with $j^{(V,A),\tau}_\mu$ and $j^{(V,A),\tau^\dagger}_\nu$)

$$\Pi_{\mu\nu}(q) = i \int d^4x \, e^{iqx} \langle T[A_\mu(x)B_\nu(0)] \rangle, \quad (B.4.1)$$

with the same conventions as in Eqs. (B.0.1). In a symmetric medium (symmetric w. r. t. the participating flavors) Π_L vanishes for A_μ or B_ν being conserved as will be seen later on.

To relate the four-longitudinal part of the (axial-) vector current to the (pseudo-) scalar current consider the correlator

$$\Pi_{\mu\nu}(q,p) = i^2 \int d^4x \, d^4y \, e^{iqx} e^{-ipy} \langle T[A_\mu(x)B_\nu(y)] \rangle. \quad (B.4.2)$$

Extracting the coordinate dependence of the second operator, using the translational invariance of the medium or the ground state and substituting $x \to x+y$ one arrives at

$$\begin{aligned} \Pi_{\mu\nu}(q,p) &= i^2 \int d^4x \, d^4y \, e^{iqx} e^{-i(p-q)y} \langle T[A_\mu(x)B_\nu(0)] \rangle \\ &= i^2 \int d^4x \, e^{iqx} (2\pi)^4 \delta^{(4)}(p-q) \langle T[A_\mu(x)B_\nu(0)] \rangle \\ &= i(2\pi)^4 \delta^{(4)}(p-q) \Pi_{\mu\nu}(q) \end{aligned} \quad (B.4.3)$$

B.4 Non-anomalous Ward identities

and hence

$$q^\mu p^\nu \Pi_{\mu\nu}(q,p) = i(2\pi)^4 \delta^{(4)}(p-q) q^\mu q^\nu \Pi_{\mu\nu}(q), \qquad (B.4.4)$$

where we set $p^\nu = q^\nu$ due to Dirac's delta distribution on the right hand side. The partial derivatives of the time-ordered product read

$$\begin{Bmatrix} i\partial^\mu_x \\ i\partial^\nu_y \end{Bmatrix} T\left[A_\mu(x)B_\nu(y)\right] = \begin{Bmatrix} T\left[i\partial^\mu_x A_\mu(x)B_\nu(y)\right] \\ T\left[A_\mu(x)i\partial^\nu_y B_\nu(y)\right] \end{Bmatrix}$$
$$+ \begin{Bmatrix} + \\ - \end{Bmatrix} i\delta(x_0 - y_0) \begin{Bmatrix} \left[A_0(x), B_\nu(y)\right] \\ \left[A_\mu(x), B_0(y)\right] \end{Bmatrix} \quad (B.4.5)$$

and

$$i\partial^\mu_x i\partial^\nu_y T\left[A_\mu(x)B_\nu(y)\right] = T\left[i\partial^\mu_x A_\mu(x) i\partial^\nu_y B_\nu(y)\right]$$
$$- i^2 \left(\partial^0_x \delta(x_0 - y_0)\right) \left[A_0(x), B_0(y)\right]$$
$$- i\delta(x_0 - y_0) \left(\left[i\partial^\mu_x A_\mu(x), B_0(y)\right] - \left[A_0(x), i\partial^\nu_y B_\nu(y)\right]\right). \quad (B.4.6)$$

The transversality therefore reads

$$q^\mu \Pi_{\mu\nu}(q) = i \int d^4x \left(-i\partial^\mu e^{iqx}\right) \langle T\left[A_\mu(x)B_\nu(0)\right]\rangle$$
$$= i^2 \int d^4x \, e^{iqx} \langle \partial^\mu T\left[A_\mu(x)B_\nu(0)\right]\rangle$$
$$= i \int d^4x \, e^{iqx} \langle T\left[i\partial^\mu A_\mu(x)B_\nu(0)\right] + i\delta(x_0)\left[A_0(x), B_\nu(0)\right]\rangle \quad (B.4.7)$$

and, using translational invariance of the Gibbs average,

$$q^\nu \Pi_{\mu\nu}(q) = i \int d^4x \left(-i\partial^\nu_x e^{iqx}\right) \langle T\left[A_\mu(0)B_\nu(-x)\right]\rangle$$
$$= i^2 \int d^4x \, e^{iqx} \langle \partial^\nu_x T\left[A_\mu(0)B_\nu(-x)\right]\rangle$$
$$= i \int d^4x \, e^{iqx} \langle T\left[A_\mu(0) i\partial^\nu_x B_\nu(-x)\right] + i\delta(x_0)\left[A_\mu(0), B_0(-x)\right]\rangle$$
$$= i \int d^4x \, e^{iqx} \langle T\left[A_\mu(0)\left(-i\partial^\nu_{-x} B_\nu(-x)\right)\right] + i\delta(x_0)\left[A_\mu(0), B_0(-x)\right]\rangle$$

B Correlation functions

$$= i \int d^4x \, e^{iqx} \langle -T \left[A_\mu(x) i \partial^\nu B_\nu(0) \right] + i \delta(x_0) \left[A_\mu(x), B_0(0) \right] \rangle \quad \text{(B.4.8)}$$

where an additional sign in the first term occurs. We also find

$$q^\mu p^\nu \Pi_{\mu\nu}(q, p)$$
$$= i^2 \int d^4x \, d^4y \left(-i \partial_x^\mu e^{iqx} \right) \left(i \partial_y^\nu e^{-ipy} \right) \langle T \left[A_\mu(x) B_\nu(y) \right] \rangle$$
$$= -i^2 \int d^4x \, d^4y \, e^{iqx} e^{-ipy} i \partial_x^\mu i \partial_y^\nu \langle T \left[A_\mu(x) B_\nu(y) \right] \rangle$$
$$= -i^2 \int d^4x \, d^4y \, e^{iqx} e^{-ipy} \langle T \left[i \partial_x^\mu A_\mu(x) i \partial_y^\nu B_\nu(y) \right]$$
$$\qquad - i^2 \left(\partial_x^0 \delta(x_0 - y_0) \right) \left[A_0(x), B_0(y) \right]$$
$$\qquad - i \delta(x_0 - y_0) \left(\left[i \partial_x^\mu A_\mu(x), B_0(y) \right] - \left[A_0(x), i \partial_y^\nu B_\nu(y) \right] \right) \rangle$$
$$= -i^2 \int d^4x \, d^4y \, e^{iqx} e^{-ipy} \langle T \left[i \partial_x^\mu A_\mu(x) i \partial_y^\nu B_\nu(y) \right]$$
$$\qquad + i^2 \delta(x_0 - y_0) e^{-iqx} \partial_x^0 e^{iqx} \left[A_0(x), B_0(y) \right]$$
$$\qquad - i \delta(x_0 - y_0) \left(\left[i \partial_x^\mu A_\mu(x), B_0(y) \right] - \left[A_0(x), i \partial_y^\nu B_\nu(y) \right] \right) \rangle$$
$$= -i^2 \int d^4x \, d^4y \, e^{iqx} e^{-ipy} \langle T \left[i \partial_x^\mu A_\mu(x) i \partial_y^\nu B_\nu(y) \right]$$
$$\qquad + i^2 \delta(x_0 - y_0) \left(i q_0 + \partial_x^0 \right) \left[A_0(x), B_0(y) \right]$$
$$\qquad - i \delta(x_0 - y_0) \left(\left[i \partial_x^\mu A_\mu(x), B_0(y) \right] - \left[A_0(x), i \partial_y^\nu B_\nu(y) \right] \right) \rangle . \quad \text{(B.4.9)}$$

In the second line we assumed vanishing currents (or expectation values thereof) at the surface of the integration domain $x = \pm \infty$. Using $\partial_x^0 A_0(x) = \partial_x^\mu A_\mu(x) - \partial_x^i A_i(x)$ gives

$$q^\mu p^\nu \Pi_{\mu\nu}(q, p)$$
$$= -i^2 \int d^4x \, d^4y \, e^{iqx} e^{-ipy} \langle T \left[i \partial_x^\mu A_\mu(x) i \partial_y^\nu B_\nu(y) \right]$$
$$+ i \delta(x_0 - y_0) \left(-q_0 \left[A_0(x), B_0(y) \right] + \left[i \partial_x^\mu A_\mu(x), B_0(y) \right] - i \partial_x^i \left[A_i(x), B_0(y) \right] \right)$$
$$\qquad - i \delta(x_0 - y_0) \left(\left[i \partial_x^\mu A_\mu(x), B_0(y) \right] - \left[A_0(x), i \partial_y^\nu B_\nu(y) \right] \right) \rangle$$
$$= -i^2 \int d^4x \, d^4y \, e^{iqx} e^{-ipy} \langle T \left[i \partial_x^\mu A_\mu(x) i \partial_y^\nu B_\nu(y) \right]$$

B.4 Non-anomalous Ward identities

$$+i\delta(x_0-y_0)\left(-q_0\left[A_0(x),B_0(y)\right]+\left[A_0(x),i\partial_y^\nu B_\nu(y)\right]-i\partial_x^i\left[A_i(x),B_0(y)\right]\right)\rangle$$

$$=-i^2\int d^4x\,d^4y\,e^{iqx}e^{-ipy}\langle T\left[i\partial_x^\mu A_\mu(x)i\partial_y^\nu B_\nu(y)\right]$$

$$+i\delta(x_0-y_0)\left(-q_0\left[A_0(x),B_0(y)\right]+\left[A_0(x),i\partial_y^\nu B_\nu(y)\right]\right.$$

$$\left.+i(-iq^i)\left[A_i(x),B_0(y)\right]\right)\rangle$$

$$=-i^2\int d^4x\,d^4y\,e^{iqx}e^{-ipy}\langle T\left[i\partial_x^\mu A_\mu(x)i\partial_y^\nu B_\nu(y)\right]$$

$$+i\delta(x_0-y_0)\left(-q^\mu\left[A_\mu(x),B_0(y)\right]_{x_0=y_0}+\left[A_0(x),i\partial_y^\nu B_\nu(y)\right]_{x_0=y_0}\right)\rangle. \quad \text{(B.4.10)}$$

The first and third terms on the right hand side of this equation vanish if $\partial^\nu B_\nu = 0$ holds.

Identifying A_μ and B_ν with the currents (A.1.41a) and (A.1.41b), inserting the divergences (A.1.44), the current-current commutators (A.2.18) and ignoring Schwinger terms gives

$$q^\mu p^\nu \Pi_{\mu\nu}^{(V,A)}(q,p|\tau,\tau')$$
$$=-i^2\int d^4x\,d^4y\,e^{iqx}e^{-ipy}\langle -T\left[j^{(S,P),[\tau,M]_\mp}(x)j^{(S,P),[\tau',M]_\mp^\dagger}(y)\right]$$
$$+i\delta^{(4)}(x-y)\left(-q^\mu j_\mu^{V,[\tau,\tau'^\dagger]}(x)\mp j^{S,[\tau,[M,\tau'^\dagger]_\mp]_\mp}(x)\right)\rangle$$
$$=\Pi^{(S,P)}\left(q,p\left|[\tau,M]_\mp,[\tau',M]_\mp\right.\right)$$
$$+i(2\pi)^4\delta^{(4)}(q-p)\langle -q^\mu j_\mu^{V,[\tau,\tau'^\dagger]}\mp j^{S,[\tau,[M,\tau'^\dagger]_\mp]_\mp}\rangle. \quad \text{(B.4.11)}$$

In the last step we used translational invariance of the ground state or the medium. Comparing Eqs. (B.4.4) and (B.4.11) leads to the non-anomalous Ward identity which relates the causal correlator of spin-0 currents to the longitudinal parts of the spin-1 current-current correlator:

$$q^\mu q^\nu \Pi_{\mu\nu}^{(V,A)}(q|\tau,\tau')$$
$$=\Pi^{(S,P)}\left(q\left|[\tau,M]_\mp,[\tau',M]_\mp\right.\right)-\langle q^\mu j_\mu^{V,[\tau,\tau'^\dagger]}\rangle \mp \langle j^{S,[\tau,[M,\tau'^\dagger]_\mp]_\mp}\rangle. \quad \text{(B.4.12)}$$

Accordingly, following the same steps as in the previous case and assuming current

B Correlation functions

conservation, Eq. (B.4.8) results in

$$q^\mu \Pi_{\mu\nu}^{(V,A)}(q|\tau,\tau') = -\langle j_\nu^V(0|[\tau,\tau'^\dagger])\rangle. \tag{B.4.13}$$

Consequently, the causal correlator is transversal if the current is conserved and the expectation value of the vector-current is zero. Current conservation alone is not a sufficient condition for transversality.

Following the same steps, the non-anomalous Ward identities for advanced and retarded correlator can be derived. A careful evaluation for the retarded correlator yields

$$\begin{aligned}
q^\mu q^\nu R_{\mu\nu}^{(V,A)}(q|\tau,\tau') \\
&= q^\nu i \int d^4x \left[(-i\partial^\mu) e^{iqx}\right] \Theta(x_0) \langle [j_\mu^{(V,A),\tau}(x), j_\nu^{(V,A),\tau'^\dagger}(0)] \rangle \\
&= q^\nu i \int d^4x \, e^{iqx} i\partial^\mu \Theta(x_0) \langle [j_\mu^{(V,A),\tau}(x), j_\nu^{(V,A),\tau'^\dagger}(0)] \rangle \\
&= q^\nu i \int d^4x \, e^{iqx} \left\{ i\delta(-x_0) \langle [j_0^{(V,A),\tau}(x), j_\nu^{(V,A),\tau'^\dagger}(0)] \rangle \right. \\
&\qquad\qquad \left. + \Theta(x_0) \langle [i\partial^\mu j_\mu^{(V,A),\tau}(x), j_\nu^{(V,A),\tau'^\dagger}(0)] \rangle \right\} \\
&= q^\nu i \int d^4x \, e^{iqx} \left\{ i\delta(x_0) \langle [j_0^{(V,A),\tau}(x), j_\nu^{(V,A),\tau'^\dagger}(0)] \rangle \right\} \\
&\qquad + i \int d^4x \, e^{iqx} i\partial^\nu \Theta(x_0) \langle [j_\mu^{(V,A),\tau}(x), j_\nu^{(V,A),\tau'^\dagger}(0)] \rangle \\
&= i \int d^4x \, e^{iqx} q^\nu \left\{ i\delta(x_0) \langle -\delta^{(3)}(\vec{x}) j_\nu^{V,[\tau'^\dagger,\tau]_-}(x) \rangle \right\} \\
&\qquad + i \int d^4x \, e^{iqx} \left\{ i\delta(x_0) \langle [i\partial^\mu j_\mu^{(V,A),\tau}(x), j_0^{(V,A),\tau'^\dagger}(0)] \rangle \right. \\
&\qquad\qquad \left. + \Theta(x_0) \langle [i\partial^\mu j_\mu^{(V,A),\tau}(x), i\partial^\nu j_\nu^{(V,A),\tau'^\dagger}(0)] \rangle \right\} \\
&= i^2 q^\nu \langle j_\nu^{V,[\tau,\tau'^\dagger]_-}(0) \rangle + i \int d^4x \, e^{iqx} \left\{ i\delta(x_0) \langle \mp \delta^{(3)}(\vec{x}) j_\nu^{S,[\tau'^\dagger,[M,\tau]_\mp]_\mp}(x) \rangle \right. \\
&\qquad\qquad \left. + \Theta(x_0) \langle [i\partial^\mu j_\mu^{(V,A),\tau}(x), i\partial^\nu j_\nu^{(V,A),\tau'^\dagger}(0)] \rangle \right\} \\
&= -q^\nu \langle j_\nu^{V,[\tau,\tau'^\dagger]}_- \rangle \pm \langle j_\nu^{S,[\tau'^\dagger,[M,\tau]_\mp]_\mp} \rangle + R^{(S,P)}\left(q \,\big|\, [\tau,M]_\mp, [\tau',M]_\mp \right)
\end{aligned} \tag{B.4.14}$$

B.4 Non-anomalous Ward identities

and analogously for the advanced correlator

$$q^\mu q^\nu A^{(V,A)}_{\mu\nu}(q|\tau,\tau')$$
$$= i \int d^4x \left[(-i\partial^\mu)(-i\partial^\nu)e^{iqx}\right]\Theta(-x_0)\langle\left[j^{(V,A),\tau'}_\nu(0), j^{(V,A),\tau}_\mu(x)\right]\rangle$$
$$= -q^\nu \langle j^{V,[\tau,\tau']}_\nu \rangle \mp \langle j^{S,[\tau',[M,\tau]_\mp]}_\nu \rangle + A^{(S,P)}\left(q\left|[\tau,M]_\mp,[\tau',M]_\mp\right.\right). \quad\text{(B.4.15)}$$

We turn now to an $N_f = 2$ flavor system. Let heavy-light meson currents be given by $\tau = \tau' = (\sigma^1 + i\sigma^2)/2$, where $\sigma^i = 2t^i$ are the Pauli matrices, which gives (4.1.1). The diagonal mass matrix can be written as $M = (C_V\sigma_3 + C_A\mathbb{1})/2$ and the commutators fulfill $[\tau,M]_\mp = \mp C_{(V,A)}\tau$, $[\tau,\tau^\dagger]_\mp = \tau^{(V,A)}$, $[\tau,[M,\tau^\dagger]_\mp]_\mp = \mp C_{(V,A)}\tau^{(V,A)}$, where we have defined $C_V = m_1 - m_2$, $C_A = m_1 + m_2$ and $\tau^V = \sigma^3$, $\tau^A = \mathbb{1}$. Note that q_1 is attributed to a light-quark field (e.g. up or down quarks) and q_2 to a heavy-quark field (e.g. charm or bottom quarks). With these relations the longitudinal part of (B.3.6) is given by

$$q^2 \Pi^{(V,A)}_L(q) = q^\mu q^\nu \Pi^{(V,A)}_{\mu\nu}(q) = C^2_{(V,A)}\Pi^{(S,P)}(q) - \langle\bar\psi \hat q \sigma^3 \psi\rangle + C_{(V,A)}\langle\bar\psi \tau^{(V,A)}\psi\rangle \quad\text{(B.4.16)}$$

with $\hat q = \gamma_\mu q^\mu$. (Equation (B.4.16) also holds in a three-flavor system with two light quarks and one massive quark.)

The statement of Eq. (B.4.13) can now be specified. Π_L is zero if the current $j^{X,\tau}_\nu(y)$ is conserved, i.e. $m_1 = m_2$ for vector currents or $m_1 + m_2 = 0$ for axial-vector currents (first and third term on the right hand side of Eq. (B.4.16) vanish), and the difference of light and heavy net quark currents being zero, i.e. $\langle\bar\psi \hat q \sigma^3 \psi\rangle = 0$, which is only true in a medium which is symmetric w.r.t. the quark flavors of the meson current. If Π_L is zero, the transversal projection $\Pi^{(V,A)}_T(q)$ is proportional to the trace of the correlator $\Pi^{(V,A)}_T(q) = -g^{\mu\nu}\Pi^{(V,A)}_{\mu\nu}(q)/3$. Otherwise the trace $\Pi^{(V,A)} \equiv g^{\mu\nu}\Pi^{(V,A)}_{\mu\nu}(q)$ contains pieces of (axial-) vectors and (pseudo-) scalars. The pure (axial-) vector information is encoded in $\Pi^{(V,A)}_T(q)$ for which one obtains

$$3\Pi^{(V,A)}_T = \frac{C^2_{(V,A)}}{q^2}\Pi^{(S,P)} + \frac{1}{q^2}\langle\bar\psi \hat q \sigma^3 \psi\rangle + \frac{C_{(V,A)}}{q^2}\langle\bar\psi \tau^{(V,A)}\psi\rangle - g^{\mu\nu}\Pi^{(V,A)}_{\mu\nu} \quad\text{(B.4.17)}$$

which relates $\Pi^{(V,A)}_T$ to the trace of the correlator and $\Pi^{(S,P)}$. For m_1 and m_2 being arbitrary quark masses, the chiral difference $\Pi^{P-S}(q)$ and the sum $\Pi^{S+P}(q)$ enter the

B Correlation functions

chiral difference $q^\mu q^\nu \Pi_{\mu\nu}^{V-A}(q)$ and, hence, Π_T^{V-A}. The expectation value of the quark current cancels out in any case:

$$\Pi_T^{V-A} = -\frac{1}{3}\left(\frac{m_1^2 + m_2^2}{q^2}\Pi^{P-S} + 2\frac{m_1 m_2}{q^2}\Pi^{P+S}\right.$$
$$\left. +\frac{2}{q^2}\left(m_1\langle\bar{q}_2 q_2\rangle + m_2\langle\bar{q}_1 q_1\rangle\right) + \frac{\Pi^{V-A}}{q^2}\right). \quad (B.4.18)$$

If one quark mass is zero, $m_1 \to 0$, one obtains the relation

$$q^\mu q^\nu \Pi_{\mu\nu}^{V-A}(q) = -m_2^2 \Pi^{P-S}(q) - 2m_2 \langle\bar{q}_1 q_1\rangle \quad (B.4.19)$$

and therefore

$$\Pi_T^{V-A}(q) = -\frac{m_2^2}{3q^2}\Pi^{P-S}(q) - \frac{1}{3}\Pi^{V-A}(q) - \frac{2}{3}\frac{m_2}{q^2}\langle\bar{q}_1 q_1\rangle. \quad (B.4.20)$$

The last term contains the interesting combination of light-quark condensate and heavy-quark mass.

B.5 Subtracted dispersion relations

In this section, the proofs of Eqs. (2.1.2) and (2.1.7) are given. As pointed out in Sec. 2.1, the derivation is done for the causal correlator. Due to Lorentz covariance of the current-current correlator and due to the Theorem of Hall and Wightman (cf. for example [Rom69]), $\Pi(q)$ is merely a function of all possible scalar products of the Lorentz vectors it depends on, i.e. q for the vacuum case and q and v, for the in-medium case, where v stands for the medium four-velocity. If we consider vacuum sum rules, $\Pi(q)$ therefore only depends on q^2, $\Pi(q) = \Pi(q^2)$.

However, considering in-medium sum rules, the ground state also depends on the medium four-velocity v_μ. Hence one has $\Pi(q) = \Pi(q, v) = \Pi(q^2, v^2, qv)$. For fixed medium velocity the current-current correlation function remains a function of q_0 and $\vec{q}$, and we can not make further restrictions to the pole structure. Thus, $\Pi(q)$ is analytic in the complex energy plane apart from the real q_0-axis.

In the following subsections we relate the values of the current-current correlation function for complex values of q_0 (or q^2) to its values on the real axis (positive real axis), i.e. to its pole structure and hence, due to (B.1.1), to the excitation energies of the considered particle. This enables us to relate hadronic observables to properties in

B.5 Subtracted dispersion relations

the domain of large Euclidean momenta, determined by the quark structure. Due to the additional dependence of the current-current correlation function on the medium velocity these relations differ in vacuum and medium.

B.5.1 Vacuum dispersion relations

By the analytic structure of the correlation function, Cauchy's theorem enables us to give an integral representation for it, called dispersion relation [Fur92, Sug61]. Therefore, we use the analyticity of $\Pi(q^2)$ for the vacuum case in the area surrounded by the contour exhibited in the left diagram of Fig. 2.1.1.

For q^2 off the positive real axis, $q^2 \notin \mathbb{R}^+ \cup \{0\}$, one gets the identity

$$\Pi(q^2) = \frac{1}{2\pi i} \int_\Gamma \frac{\Pi(s)}{s-q^2} ds \quad \text{(B.5.1)}$$

$$= \frac{1}{2\pi i} \int_0^{+\infty} \frac{\Pi(s+i\epsilon)}{s-q^2} ds + \frac{1}{2\pi i} \int_{+\infty}^0 \frac{\Pi(s-i\epsilon)}{s-q^2} ds$$

$$+ \frac{1}{2\pi i} \oint_\epsilon \frac{\Pi(s)}{s-q^2} ds + \frac{1}{2\pi i} \oint_\infty \frac{\Pi(s)}{s-q^2} ds, \quad \text{(B.5.2)}$$

where the last integral is for the integration over the outer circle for the radius tending to infinity, and the third one is for integration on the semicircle at the origin. The limit $\eta \to 0$ is to be understood. Thus, the integral over the semicircle becomes zero.

One can now show that the contribution of the integral over the infinitely large circle is a finite polynomial in q^2 if and only if $|\Pi(q^2)| \leq |q^2|^N$ for $|q^2| \to \infty$, where $N \in \mathbb{N}$ is a finite and fixed number [Sug61]. Because $|q^2| < |s|$, if s goes along the outer circle, one can write

$$\frac{1}{s-q^2} = \sum_{n=0}^{\infty} \frac{1}{s}\left(\frac{q^2}{s}\right)^n. \quad \text{(B.5.3)}$$

By making use of the boundary condition for $\Pi(q^2)$ the integral over the outer circle reads

$$\frac{1}{2\pi i} \oint_\infty \frac{\Pi(s)}{s-q^2} ds = \sum_{n=0}^{\infty} \frac{(q^2)^n}{2\pi i} \oint_\infty \frac{\Pi(s)}{s^{n+1}} ds = \sum_{n=0}^{\infty} \frac{(q^2)^n}{2\pi} \int_0^{2\pi} \frac{\Pi(s)}{s^n} d\phi$$

$$= \sum_{n=0}^{N} \frac{(q^2)^n}{2\pi} \int_0^{2\pi} \frac{\Pi(s)}{s^n} d\phi = \sum_{n=0}^{N} a_n (q^2)^n \quad \text{(B.5.4)}$$

B Correlation functions

which is a finite polynomial in q^2. This polynomial is not equal to the so-called subtractions, which we will introduce later on. The exact dispersion relation in vacuum then reads

$$\Pi(q^2) = \frac{1}{\pi} \int_0^\infty \frac{\Delta \Pi^{\text{vac}}(s)}{s - q^2} ds + \sum_{n=0}^{N} a_n q_0^n \tag{B.5.5}$$

with $\Delta\Pi^{\text{vac}}$ defined in Eq. (2.1.3). If $\Pi(q^2)$ does not vanish fast enough when $|q^2|$ approaches infinity, the boundary condition is essential and adequate to an elimination of the contribution of the infinite circle by subtracting a finite polynomial in q^2. It is important to note that the coefficients a_n are not proportional to the derivatives of $\Pi(q^2)$ for $q^2 = 0$ even if $\Pi(q^2)$ is analytic in an open circle around the origin (unless the circle is the infinite circle itself, which would require infinite excitation energies in (B.1.1)) [Sug61]. Because of the pole structure along the real axis, the current-current correlation function is not analytic inside the infinite circle and Cauchy's theorem is therefore not applicable to the integral over the infinite circle. Instead, in [Sug61], the authors show that the contribution from the infinite circle can be expressed by the boundary values of $\Pi(q^2)$ along the real axis.

However, there are several methods to get rid of the polynomial contributions. One possibility is to take the $(N+1)$-st derivative of (B.5.5). The polynomial is canceled and one gets

$$\left(\frac{d}{dq^2}\right)^{N+1} \Pi(q^2) = \frac{(N+1)!}{\pi} \int_0^\infty \frac{\Delta\Pi^{\text{vac}}(s)}{(s-q^2)^{N+2}} ds \,. \tag{B.5.6}$$

Of course, we also could have done this right at the beginning in (B.5.1) and by the boundary condition the integral over the outer circle would be zero. This method is called the method of power moments $M_n(q^2)$; the n-th moment is given by

$$M_n(q^2) = \frac{1}{n!} \left(\frac{d}{dq^2}\right)^n \Pi(q^2) \,. \tag{B.5.7}$$

We mention this method just for completeness. It will not be considered anymore throughout this thesis.

However, if $\Pi(q^2)$ is also analytic in an open (finite) circle around the origin we can give a dispersion relation which has no polynomial contribution of the infinite circle. In this case, the lower limit for the integration along the positive real axis

B.5 Subtracted dispersion relations

effectively starts from a lower boundary threshold s_0,[29] because the integrations above and below the positive real axis cancel each other up to the threshold due to analyticity. Now, the dispersion relation can be obtained from (B.5.1) by subtracting a polynomial of degree $N-1$, with coefficients $\propto \Pi^{(n)}(0)$. Here, the derivatives are solely calculated from (B.5.1), which means that we use the contour given in the left panel of Fig. 2.1.1.

In order to clarify the difference between the subtractions and the polynomial coefficients a_n in (B.5.5), we emphasize that the derivatives $\Pi^{(n)}(0)$ used as coefficients for the subtractions are not calculated from an integral over the infinite circle, in contrast to the polynomial coefficients a_n, i. e.

$$\Pi(q^2) - \sum_{n=0}^{N-1} \frac{\Pi^{(n)}(0)}{n!}(q^2)^n = \frac{1}{2\pi i}\int_\Gamma \frac{\Pi(s)}{s-q^2}ds - \sum_{n=0}^{N-1}\frac{(q^2)^n}{2\pi i}\int_\Gamma \frac{\Pi(s)}{s^{n+1}}ds. \quad (B.5.8)$$

Again by using the standard expression for geometric series

$$\frac{1}{s-q^2} - \sum_{n=0}^{N-1}\frac{(q^2)^n}{s^{n+1}} = \sum_{n=N}^{\infty}\frac{1}{s}\left(\frac{q^2}{s}\right)^n = \frac{1}{s}\left(\frac{q^2}{s}\right)^N\sum_{n=0}^{\infty}\left(\frac{q^2}{s}\right)^n = \left(\frac{q^2}{s}\right)^N\frac{1}{s-q^2}, \quad (B.5.9)$$

one ends up with

$$\Pi(q^2) - \sum_{n=0}^{N-1}\frac{\Pi^{(n)}(0)}{n!}(q^2)^n = \frac{1}{2\pi i}\int_\Gamma \left(\frac{q^2}{s}\right)^N \frac{\Pi(s)}{s-q^2}ds. \quad (B.5.10)$$

The integral over the infinite circle vanishes due to the boundary condition of the current-current correlation function:

$$\frac{1}{2\pi i}\oint_\infty \left(\frac{q^2}{s}\right)^N \frac{\Pi(s)}{s-q^2}ds = \sum_{n=0}^{\infty}\frac{(q^2)^{n+N}}{2\pi i}\oint_\infty \frac{\Pi(s)}{s^{N+n+1}}ds = 0. \quad (B.5.11)$$

The N times subtracted dispersion relation in vacuum finally reads

$$\Pi(q^2) - \sum_{n=0}^{N-1}\frac{\Pi^{(n)}(0)}{n!}(q^2)^n = \frac{1}{\pi}\int_0^\infty \left(\frac{q^2}{s}\right)^N \frac{\Delta\Pi^{\text{vac}}(s)}{s-q^2}ds$$

$$= \frac{1}{\pi}\int_{s_0}^\infty \left(\frac{q^2}{s}\right)^N \frac{\Delta\Pi^{\text{vac}}(s)}{s-q^2}ds, \quad (B.5.12)$$

---

[29] The quantity s_0 corresponds to the lowest-lying excitation energy.

B Correlation functions

where the coefficients of the polynomial are well known functions. The N-th derivative of (B.5.12) eliminates the polynomial.

B.5.2 In-medium dispersion relations

The in-medium case proceeds in a similar way, but with the difference that the dispersion relation may only be given in the complex q_0 plane and the real axis is explicitly excluded. Using the integration contour given in the right panel of Fig. 2.1.1 and the analyticity of $\Pi(q_0, \vec{q})$ in the area surrounded by the integration contour we get

$$\Pi(q_0, \vec{q}) = \frac{1}{2\pi i} \int_\Gamma \frac{\Pi(\omega, \vec{q})}{\omega - q_0} d\omega$$

$$= \frac{1}{2\pi i} \int_{-\infty}^{+\infty} \frac{\Pi(\omega + i\epsilon, \vec{q})}{\omega - q_0} d\omega + \frac{1}{2\pi i} \int_{+\infty}^{-\infty} \frac{\Pi(\omega - i\epsilon, \vec{q})}{\omega - q_0} d\omega$$

$$+ \frac{1}{2\pi i} \oint_\infty \frac{\Pi(\omega, \vec{q})}{\omega - q_0} d\omega, \qquad (B.5.13)$$

for fixed $\vec{q}$. Following the same arguments as for the vacuum case, the dispersion relation for the in-medium case is

$$\Pi(q_0, \vec{q}) = \frac{1}{\pi} \int_{-\infty}^{+\infty} \frac{\Delta\Pi(\omega, \vec{q})}{\omega - q_0} d\omega + \sum_{n=0}^{N} a_n q_0^n, \qquad (B.5.14)$$

where the polynomial again corresponds to the contribution of the infinite circle and $\Delta\Pi(s, \vec{q})$ is defined in (2.1.6). If $\Pi(q_0, \vec{q})$ is analytic in an open (finite) area at the origin, the integration along the cuts at the real axis effectively starts from thresholds s_0^+, s_0^-.[30] Again, a subtracted dispersion relation may be derived which does not contain polynomial contributions arising from the integral over the infinite circle:

$$\Pi(q_0, \vec{q}) - \sum_{n=0}^{N-1} \frac{\Pi^{(n)}(q_0 = 0, \vec{q})}{n!} q_0^n = \frac{1}{2\pi i} \int_\Gamma \Pi(\omega, \vec{q}) \left(\sum_{n=0}^{\infty} - \sum_{n=0}^{N-1} \right) \left(\frac{q_0}{\omega} \right)^n \frac{1}{\omega} d\omega$$

$$= \frac{1}{2\pi i} \int_\Gamma \left(\frac{q_0}{\omega} \right)^N \frac{\Pi(\omega, \vec{q})}{\omega - q_0} d\omega. \qquad (B.5.15)$$

---

[30] This corresponds to an energy gap between the lowest-lying particle state and the lowest-lying antiparticle excitation.

B.5 Subtracted dispersion relations

The integral over the infinite circle is zero again due to the boundary condition, giving the N-fold subtracted dispersion relation in q_0:

$$\Pi(q_0,\vec{q}) - \sum_{n=0}^{N-1} \frac{\Pi^{(n)}(0,\vec{q})}{n!}(q_0)^n$$

$$= \frac{1}{\pi}\int_{-\infty}^{+\infty}\left(\frac{q_0}{\omega}\right)^N \frac{\Delta\Pi(\omega,\vec{q})}{\omega-q_0}d\omega$$

$$= \frac{1}{\pi}\int_{s_0^+}^{+\infty}\left(\frac{q_0}{\omega}\right)^N \frac{\Delta\Pi(\omega,\vec{q})}{\omega-q_0}d\omega + \frac{1}{\pi}\int_{-\infty}^{s_0^-}\left(\frac{q_0}{\omega}\right)^N \frac{\Delta\Pi(\omega,\vec{q})}{\omega-q_0}d\omega. \quad (B.5.16)$$

For purposes which will become evident later on, it is more convenient to work with a dispersion relation in q_0^2 rather than a dispersion relation in q_0. Therefore, the current-current correlation function $\Pi(q_0,\vec{q})$ is split up into its symmetric and anti-symmetric part w. r. t. q_0

$$\Pi(q_0,\vec{q}) = \frac{1}{2}\left(\Pi(q_0,\vec{q}) + \Pi(-q_0,\vec{q})\right) + \frac{1}{2}\left(\Pi(q_0,\vec{q}) - \Pi(-q_0,\vec{q})\right). \quad (B.5.17)$$

Even (e) and odd (o) contributions may be defined as

$$\Pi^e(q_0,\vec{q}) = \frac{1}{2}\left(\Pi(q_0,\vec{q}) + \Pi(-q_0,\vec{q})\right) = \Pi^e(-q_0,\vec{q}), \quad (B.5.18a)$$

$$\Pi^o(q_0,\vec{q}) = \frac{1}{2q_0}\left(\Pi(q_0,\vec{q}) - \Pi(-q_0,\vec{q})\right) = \Pi^o(-q_0,\vec{q}), \quad (B.5.18b)$$

where the energy factor in front of $\Pi^o(q_0,\vec{q})$ is convention. Summation and subtraction of the dispersion relation (B.5.16) for $\Pi(q_0,\vec{q})$ and $\Pi(-q_0,\vec{q})$ result in the following separate dispersion relations for the even part $\Pi^e(q_0,\vec{q})$ of the current-current correlation function

$$\Pi^e(q_0,\vec{q}) - \frac{1}{2}\sum_{n=0}^{N-1}\frac{\Pi^{(n)}(0,\vec{q})}{n!}(q_0)^n(1+(-1)^n)$$

$$= \frac{1}{2\pi}\int_{-\infty}^{+\infty}d\omega\,\Delta\Pi(\omega,\vec{q})\frac{q_0^N}{\omega^{N-1}}\frac{\left(1+(-1)^N\right)+\frac{q_0}{\omega}\left(1-(-1)^N\right)}{\omega^2-q_0^2} \quad (B.5.19)$$

and for the odd part of the correlation function

$$\Pi^o(q_0,\vec{q}) - \frac{1}{2}\sum_{n=0}^{N-1}\frac{\Pi^{(n)}(0,\vec{q})}{n!}(q_0)^{n-1}(1-(-1)^n)$$

B Correlation functions

$$= \frac{1}{2\pi} \int_{-\infty}^{+\infty} d\omega \, \Delta\Pi(\omega,\vec{q}\,) \frac{q_0^{N-1}}{\omega^{N-1}} \frac{\left(1-(-1)^N\right) + \frac{q_0}{\omega}\left(1+(-1)^N\right)}{\omega^2 - q_0^2}. \quad \text{(B.5.20)}$$

Both functions, $\Pi^e(q_0,\vec{q}\,)$ and $\Pi^o(q_0,\vec{q}\,)$, only depend on q_0^2, in line with their definition (B.5.18). Hence, one may write

$$\Pi(q_0,\vec{q}\,) = \Pi^e(q_0^2,\vec{q}\,) + q_0 \Pi^o(q_0^2,\vec{q}\,). \quad \text{(B.5.21)}$$

The not subtracted dispersion relations can be obtained from these expressions by setting $N = 0$. At this point we remark the similarity of the vacuum dispersion relation and the even in-medium dispersion relation, which can be seen by substituting $\omega^2 = t$ in (B.5.19). This is not the case for the odd in-medium dispersion relation (B.5.20).

C In-medium operator product expansion for pseudo-scalar heavy-light mesons

This appendix starts with the preparation of basic ingredients for the evaluation of an OPE in the heavy-light meson sector. The Fock-Schwinger gauge and its application to the background field method are presented. Furthermore, several essential relations for the quark propagator are derived. The Borel transformation is introduced and applied to the subtracted dispersion relation obtained in App. B.5. The projection of ground state expectation values or Gibbs averages of quantum field operators onto invariants is performed for all quantities which are met within the evaluation of an OPE for heavy-light mesons up to and including mass dimension 5. For heavy-light pseudo-scalar mesons the in-medium OPE evaluation in terms of normal ordered condensates is given in detail. The relation between normal and non-normal ordered condensates is derived in general and evaluated explicitly for the condensates which are required for the in-medium OPE of heavy-light pseudo-scalar mesons. The cancellation of infrared divergences is demonstrated and recurrence relations for the Wilson coefficients are derived. We explain our strategy of analyzing the QSR and show the discrepancy. Finally, the trace anomaly of QCD (A.2.13) is applied to the evaluation of condensates at high densities and temperatures. The results of App. C.5 are extensively used in Secs. 3, 4 and D.

C.1 Fock-Schwinger gauge and background field method

In this section, we discuss the Fock-Schwinger gauge, which enables us to give simple expressions for gluon fields and quark propagators. These are used to develop techniques for the calculation of the OPE. This section follows closely [Nov84b].

C In-medium OPE for heavy-light mesons

The Fock-Schwinger gauge reads

$$\left(x^\mu - x_0^\mu\right)\mathscr{A}_\mu(x) = 0. \tag{C.1.1}$$

Usually one chooses $x_0^\mu = 0$. Obviously, this gauge is not translation invariant. The invariance is broken due to the special role of x_0. Therefore, a shift in space-time would destroy the gauge condition. But, as we will see, this condition is crucial to the techniques used in the calculations performed afterwards. Hence, we must restrict ourselves to one selected frame. By Eq. (C.1.1) we can write

$$0 = \partial_\mu\left(y^\nu A_\nu^A(y)\right) = A_\mu^A(y) + y^\nu \partial_\mu A_\nu^A(y). \tag{C.1.2}$$

Furthermore, using Eq. (A.3.4) one obtains

$$y^\nu \partial_\mu A_\nu^A = y^\nu G_{\mu\nu}^A + y^\nu \partial_\nu A_\mu^A, \tag{C.1.3}$$

where the non-Abelian term in Eq. (A.3.4) vanishes due to the gauge condition (C.1.1). Together with Eq. (C.1.2) one ends up with

$$A_\mu^A(y) + y^\nu \partial_\nu A_\mu^A = y^\nu G_{\nu\mu}^A. \tag{C.1.4}$$

Substituting $y^\nu = \alpha x^\nu$ and inserting (C.1.4) again leads to

$$\frac{\mathrm{d}}{\mathrm{d}\alpha}\left(\alpha A_\mu(\alpha x)\right) = \alpha x^\nu G_{\nu\mu}^A(\alpha x). \tag{C.1.5}$$

Integration over α from 0 to 1 finally yields

$$A_\mu^A(x) = \int_0^1 \mathrm{d}\alpha\, \alpha x^\nu G_{\nu\mu}^A(\alpha x). \tag{C.1.6}$$

We observe that the Fock-Schwinger gauge (C.1.1) enables us to express the gluon fields in terms of the gluon field strength tensor. Expanding the gluon field $\mathscr{A}_\mu$ in Eq. (C.1.1) in x yields

$$x^\mu \left(\sum_{n=0}^\infty \frac{1}{n!} x^{\alpha_1} \ldots x^{\alpha_n} \partial_{\alpha_1} \ldots \partial_{\alpha_n} \mathscr{A}_\mu(0)\right) = 0 \tag{C.1.7}$$

C.1 Fock-Schwinger gauge and background field method

for all x. Therefore

$$x^\mu x^{\alpha_1}\ldots x^{\alpha_n}\partial_{\alpha_1}\ldots\partial_{\alpha_n}\mathscr{A}_\mu(0) = 0\,. \tag{C.1.8}$$

As a direct consequence of the last equation one can show that

$$x^{\alpha_1}\ldots x^{\alpha_n}\left(\partial_{\alpha_1}\ldots\partial_{\alpha_n}\mathscr{G}_{\mu\nu}\right)_{x=0} = x^{\alpha_1}\ldots x^{\alpha_n}\left(D_{\alpha_1}\ldots D_{\alpha_n}\mathscr{G}_{\mu\nu}\right)_{x=0}\,. \tag{C.1.9}$$

Thus, the gluon field strength tensor in Eq. (C.1.6) may be expanded for small x

$$\begin{aligned}\mathscr{A}_\mu(x) &= \int_0^1 d\alpha\, \alpha x^\nu \mathscr{G}_{\nu\mu}(\alpha x) \\
&= \int_0^1 d\alpha\, \alpha x^\nu \sum_{n=0}^\infty \frac{\alpha^n}{n!} x^{\alpha_1}\ldots x^{\alpha_n}\left(\partial_{\alpha_1}\ldots\partial_{\alpha_n}\mathscr{G}_{\nu\mu}\right)_{x=0} \\
&= \int_0^1 d\alpha\, \alpha x^\nu \sum_{n=0}^\infty \frac{\alpha^n}{n!} x^{\alpha_1}\ldots x^{\alpha_n}\left(D_{\alpha_1}\ldots D_{\alpha_n}\mathscr{G}_{\nu\mu}\right)_{x=0}\,.\end{aligned} \tag{C.1.10}$$

Integration over α yields the covariant expansion for the gluon fields in terms of the gluon field strength tensor only:

$$\mathscr{A}_\mu(x) = \sum_{n=0}^\infty \frac{x^\nu}{n!(n+2)} x^{\alpha_1}\ldots x^{\alpha_n}\left(D_{\alpha_1}\ldots D_{\alpha_n}\mathscr{G}_{\nu\mu}\right)_{x=0}\,. \tag{C.1.11}$$

The Fock-Schwinger gauge also provides a covariant expansion for the quark fields:

$$\Psi(x) = \sum_{n=0}^\infty \frac{1}{n!} x^{\alpha_1}\ldots x^{\alpha_n}\left(\vec{D}_{\alpha_1}\ldots\vec{D}_{\alpha_n}\Psi\right)_{x=0}\,, \tag{C.1.12a}$$

$$\bar\Psi(x) = \sum_{k=0}^\infty \frac{1}{k!} x^{\alpha_1}\ldots x^{\alpha_k}\left(\bar\Psi \overleftarrow{D}_{\alpha_k}\ldots\overleftarrow{D}_{\alpha_1}\right)_{x=0}\,. \tag{C.1.12b}$$

The method used to simulate the non-perturbative vacuum and medium effects of the physical ground state is the background field method. A propagating quark interacts with the virtual particles emerging from the ground state. These particles can be modeled effectively by a quantized vector field which is shifted by a classical weak gluonic background field. The interaction is described by a propagation within this field. As it is a weak field, it is possible to give a perturbative expansion of the quark propagator in terms of the coupling strength:

C In-medium OPE for heavy-light mesons

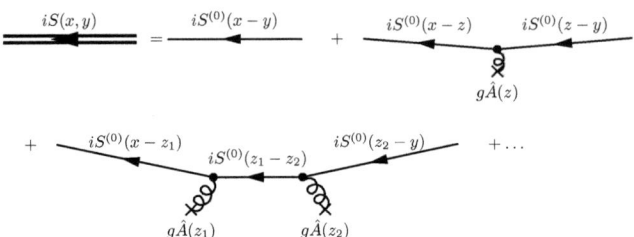

Figure C.1.1: Perturbative quark propagator in a weak glounic background field. The gluons are supposed to emerge from the ground state.

$$iS(x,y) = iS^{(0)}(x-y)$$
$$+ \sum_{n=1}^{\infty} \int iS^{(0)}(x-z_1) ig\hat{A}(z_1) iS^{(0)}(z_1-z_2) \ldots ig\hat{A}(z_n) iS^{(0)}(z_n-y) \, dz_1 \ldots dz_n \,,$$
(C.1.13)

where $S^{(0)}(x-y)$ denotes the free propagator satisfying

$$(i\hat{\partial}_x - m) S^{(0)}(x-y) = \delta^{(4)}(x-y). \tag{C.1.14}$$

The expansion is depicted in Fig. C.1.1. If we think of the gluons as emerging from the ground state, the fields can be understood as operators acting on the Fock space and creating an additional gluon. Thus we use a classical expansion for the quark propagator.

The Fourier transformed free propagator reads

$$S^{(0)}(p) = \int d^4x \, e^{ipx} S^{(0)}(x) = \frac{\hat{p}+m}{p^2-m^2}. \tag{C.1.15}$$

Using Eq. (C.1.14) one can show that (C.1.13) satisfies

$$(i\hat{\partial}_x + g\hat{A}(x) - m) S(x,y) = \delta^{(4)}(x-y), \tag{C.1.16}$$

confirming that $S(x)$ is indeed the propagator of a quark in a weak gluonic background field. In order to ensure the gauge condition (C.1.1), one has to permit shifts in space-time. Hence, one is not allowed to perform the transformation $x \to x' = x - y$. This means that, as a result of the gauge, the propagator does not depend solely on the differences of the coordinates, i.e. $x-y$. Thus, $S(x,y)$ and $S(x-y,0)$ are different

C.1 Fock-Schwinger gauge and background field method

quantities. Let us introduce the Fourier transformed expressions

$$S(p) = \int d^4x\, e^{ipx} S(x,0)\,, \tag{C.1.17}$$

$$\tilde{S}(p) = \int d^4x\, e^{-ipx} S(0,x)\,, \tag{C.1.18}$$

$$A_\mu(p) = \int d^4x\, e^{ipx} A_\mu(x)\,, \tag{C.1.19}$$

where we also introduce the quantity $\tilde{S}(p)$ in order to ensure the gauge condition. We first calculate the Fourier transform of $A_\mu(x)$ giving

$$A_\mu(p) = \sum_{n=0}^{\infty} (-i)^{n+1} \frac{(2\pi)^4}{n!(n+2)} \left(\vec{D}_{\alpha_1} \ldots \vec{D}_{\alpha_n} \mathscr{G}_{\rho\mu}(0) \right) \left(\partial^\rho \partial^{\alpha_1} \ldots \partial^{\alpha_n} \delta^{(4)}(p) \right), \tag{C.1.20}$$

where ∂ denotes a derivative with respect to the momentum. The transformed expression of Eq. (C.1.13) thus reads

$$S(p) = S^{(0)}(p) + \sum_{n=1}^{\infty} (-g)^n S^{(0)}(p)$$
$$\times \int \frac{d^4k_1}{(2\pi)^4} \ldots \frac{d^4k_n}{(2\pi)^4} \hat{A}(p-k_1) S^{(0)}(k_1) \hat{A}(k_1-k_2) \ldots S^{(0)}(k_n)\,. \tag{C.1.21}$$

By integration by parts we observe for an arbitrary function $f(k)$

$$\int \frac{d^4k}{(2\pi)^4} g \hat{A}(p-k) f(k) = (\gamma \tilde{A}) f(p)\,, \tag{C.1.22}$$

where we have defined

$$\tilde{A}_\mu = \sum_{n=0}^{\infty} \tilde{A}_\mu^{(n)} \tag{C.1.23}$$

with

$$\tilde{A}_\mu^{(0)} = i \frac{g}{2} \mathscr{G}_{\mu\nu} \partial^\nu\,,$$
$$\tilde{A}_\mu^{(n)} = -\frac{(-i)^{n+1} g}{n!(n+2)} \left(D_{\alpha_1} \ldots D_{\alpha_n} \mathscr{G}_{\mu\nu}(0) \right) \partial^\nu \partial^{\alpha_1} \ldots \partial^{\alpha_n} \tag{C.1.24}$$

C In-medium OPE for heavy-light mesons

which is here a derivative operator. This enables us to give a more comfortable expression for the perturbative quark propagator. Successive integration by parts of the momentum integrals in Eq. (C.1.21) and using Eq. (C.1.22) yield

$$S(p) = \sum_{n=0}^{\infty} S^{(n)}(p) \tag{C.1.25}$$

with

$$\begin{aligned} S^{(n)}(p) &= (-1)S^{(n-1)}(p)\left(\gamma\tilde{A}\right)S^{(0)}(p) \\ &= (-1)S^{(0)}(p)\left(\gamma\tilde{A}\right)S^{(n-1)}(p) \\ &= (-1)^n S^{(0)}(p)\underbrace{\left(\gamma\tilde{A}\right)S^{(0)}(p) \times \ldots \times S^{(0)}(p)\left(\gamma\tilde{A}\right)}_{n}S^{(0)}(p). \end{aligned} \tag{C.1.26}$$

Here the derivatives contained in $(\gamma\tilde{A})$ act on all functions to the right of them. Moreover, performing the same steps for $\tilde{S}(p)$, one can show that $S(p) = \tilde{S}(p)$. Although (C.1.13) is not translational invariant, the Fourier transform of $S(x,0)$ and $S(0,x)$ are identical. Equation (C.1.26) reveals that the n-th order propagator term $S^{(n)}$ can be obtained by applying the operator $-S^{(0)}\left(\gamma\tilde{A}\right)$ to the $(n-1)$-st term $S^{(n-1)}$. Therefore, Eq. (C.1.26) is a formal solution of the Dyson-Schwinger like equation

$$S = S^{(0)} - S^{(0)}\gamma\tilde{A}S \tag{C.1.27}$$

which may formally be written as

$$S^{-1} = \left[S^{(0)-1} + \gamma\tilde{A}\right]^{-1}. \tag{C.1.28}$$

The corresponding expression for $\tilde{S}(p)$ in terms of $\tilde{A}_\mu$ reads

$$\tilde{S}(p) = S^{(0)}(p) + \sum_{n=1}^{\infty}(-1)^n S^{(0)}(p)\underbrace{\left(\gamma\overleftarrow{\tilde{A}}\right)S^{(0)}(p) \times \ldots \times S^{(0)}(p)\left(\gamma\overleftarrow{\tilde{A}}\right)}_{n}S^{(0)}(p) \tag{C.1.29}$$

which is equal to (C.1.25) because the partial derivatives commute. Employing a very useful formula for further calculations

$$\frac{\partial}{\partial p_\rho}S^{(0)}(p) = -S^{(0)}(p)\gamma^\rho S^{(0)}(p), \tag{C.1.30}$$

C.1 Fock-Schwinger gauge and background field method

which can be confirmed by a direct calculation, we write out the first terms of (C.1.26) in lowest order of the gluon field $\tilde{A}_\mu$

$$S^{(0)}(p) = \frac{\hat{p}+m}{p^2-m^2}, \tag{C.1.31}$$

$$S^{(1)}(p) = i\frac{g}{2}\mathcal{G}_{\mu\nu}(0)S^{(0)}(p)\gamma^\mu S^{(0)}(p)\gamma^\nu S^{(0)}(p), \tag{C.1.32}$$

$$S^{(2)}(p) = \left(i\frac{g}{2}\right)^2 \mathcal{G}_{\mu\nu}(0)\mathcal{G}_{\kappa\lambda}(0)T^{\mu\nu\kappa\lambda}(p), \tag{C.1.33}$$

where we have defined

$$\begin{aligned}T^{\mu\nu\kappa\lambda}(p) \equiv\; & S^{(0)}(p)\gamma^\mu S^{(0)}(p)\gamma^\nu S^{(0)}(p)\gamma^\kappa S^{(0)}(p)\gamma^\lambda S^{(0)}(p) \\ & + S^{(0)}(p)\gamma^\mu S^{(0)}(p)\gamma^\kappa S^{(0)}(p)\gamma^\nu S^{(0)}(p)\gamma^\lambda S^{(0)}(p) \\ & + S^{(0)}(p)\gamma^\mu S^{(0)}(p)\gamma^\kappa S^{(0)}(p)\gamma^\lambda S^{(0)}(p)\gamma^\nu S^{(0)}(p). \end{aligned} \tag{C.1.34}$$

The OPE of the current-current correlator may now be obtained by applying Wick's theorem to the time ordered operator product $T\left[j(x)j^\dagger(0)\right]$ in terms of quark fields. While pure quark condensates directly enter by virtue of Wick's theorem, gluonic operators and gluonic condensates enter via quark-gluon interactions of the current-quarks with the background field. On the one hand, the perturbative expansion of the quark propagator contributes and, on the other hand, the covariant expansion of the quark fields (C.1.12) together with the field equations (A.3.23) of chromodynamics contribute.

Having established the above formalism, it is easy to relate the mass dimension of a power correction to the expansion of the quark propagator. Note that the mass dimension of a condensate is composed of the operators of the condensate and of prefactors which originate from the EoM for the quark fields, i.e. mass factors. Therefore, one has to separate operators which act on the Wilson coefficient and contributions to the condensate in the above treatment. The differential operator $\tilde{A}$ has mass dimension 1 (in all orders, of course). However, the partial derivatives entering each term of Eq. (C.1.23) act on the propagators of contracted, i.e. not condensed, quark fields and, hence, only contribute to the Wilson coefficients. Thus, each term $\tilde{A}^{(n)}$ of the expansion of the gluon field contributes with $n+2$ to the mass dimension. Analogously, the covariant derivatives of the quark-field expansion (C.1.12) contributes to the condensate, whereas the coordinate factors act as partial derivatives in the Wilson coefficient after the Fourier transformation has been performed. The following

213

C In-medium OPE for heavy-light mesons

prescription can be established:

$$\dim_m = \frac{3}{2}n + \sum_{i=1}^{n} m_i + \sum_{p=1}^{P} k_p (2 + l_p), \tag{C.1.35}$$

where n is the number of not contracted quark field operators, m_i is the order of the quark field i w.r.t. the expansion (C.1.12), k_p is the order of quark propagator p w.r.t. the expansion (C.1.25), i.e. it is the number of gluon insertions (or interactions) of the respective quark, and l_p is the order of the gluon field w.r.t. the expansion (C.1.23). P is the total number of quark propagators. Given a mass dimension, the required orders of expansions can be determined from Eq. (C.1.35).

If we restrict ourselves to small distances in Eq. (2.0.1a), a restriction to large momenta is required. Nevertheless, unlike the expectation that perturbative methods are adequate in this energy regime, in particular for the example of D mesons, non-perturbative effects will prohibit us from a pure perturbative treatment even in the high energy regime. Details about this effect and about calculating the OPE by applying the background field method can be found in App. C.4 for the example of the D meson.

C.2 Borel transformed sum rules

In Sec. 2.3 we mentioned the possibility to cast the integrals in Eqs. (2.1.2) and (2.1.7) into the Laplace transform of the dispersion integral and, hence, enhancing the contribution of the lowest resonance by an exponential weight. The according transformation is the Borel transformation [Wid46], which is an algebraic form of the inverse Laplace transform. Considering the OPE, which is an asymptotic expansion, i.e. a divergent series, the Borel transformation transforms the series into its Borel sum by suppressing the coefficients factorially [Wei96]. Therefore, it improves the convergence of the expansion and is thus a suitable tool for summing the actually divergent series.

The Borel transform of a function $f(Q^2)$ is defined as [Fur92]

$$\mathscr{B}[f](M^2) \equiv \lim_{\substack{n \to \infty \\ Q^2 = nM^2}} \frac{(Q^2)^{n+1}}{n!} \left(-\frac{\mathrm{d}}{\mathrm{d}Q^2} \right)^n f(Q^2), \tag{C.2.1}$$

C.2 Borel transformed sum rules

with $Q^2 \geq 0$. Sometimes [Nar02] a slightly different definition is used

$$\mathscr{B}'[f](M^2) \equiv \lim_{\substack{n \to \infty \\ Q^2 = nM^2}} \frac{(Q^2)^{n+1}}{n!} \left(-\frac{d}{dQ^2}\right)^{n+1} f(Q^2), \tag{C.2.2}$$

and the connection between both is given by

$$\mathscr{B}[f](M^2) = M^2 \mathscr{B}'[f](M^2). \tag{C.2.3}$$

In order to spell out the Borel transformed sum rules, we have to give the transformed expressions for typical functions which appear in the OPE. For simple functions f, e. g. $f(Q^2) = (Q^2)^n \ln Q^2$ or $f(Q^2) = 1/(Q^2 + m^2)^n$, the transformed expressions can be calculated directly from Eq. (C.2.1) by using general rules for the n-th derivative, e. g. the Leibniz rule

$$\left(\frac{d}{dx}\right)^n (uv) = \sum_{m=0}^{n} \binom{n}{m} \left(\frac{d^m}{dx^m} u\right) \left(\frac{d^{n-m}}{dx^{n-m}} v\right). \tag{C.2.4}$$

As can be seen from (C.2.1) all functions which are merely finite polynomials in Q^2 vanish under Borel transformation. For more complex functions, direct calculations are rather difficult. Instead, one can use the connection between the Borel transformation and the Laplace transformation [Wid46]. The Laplace transform of a function $g(t)$ is defined as [Bro01]

$$\mathscr{L}[g](s) \equiv \int_0^\infty e^{-st} g(t) dt. \tag{C.2.5}$$

Here $g(t)$ is assumed to be piecewise smooth for $t \geq 0$ and to be bounded by $e^{\alpha t}$ for $t \to \infty$; $g(t) \leq e^{\alpha t}$ for some $\alpha > 0$. The integral converges for Re $s > \alpha$ and is an analytic function in s in that area.

The beneficial point is that an inverse operation to the Laplace transformation is

C In-medium OPE for heavy-light mesons

given by Post's inversion formula (see chapter VII.6 in [Wid46])[31]

$$g(t) = \lim_{n\to\infty} \frac{1}{n!} \left(\frac{n}{t}\right)^{n+1} \left[-\mathscr{L}[g]\left(\frac{n}{t}\right)\right]^{(n)}, \tag{C.2.6}$$

where $[\ldots]^{(n)}$ denotes the n-th derivative. Setting $t = 1/M^2$ and $n/t = nM^2 = Q^2$ one obtains

$$g\left(\frac{1}{M^2}\right) = \lim_{\substack{n\to\infty \\ Q^2=nM^2}} \frac{(Q^2)^{n+1}}{n!} \left(-\frac{d}{dQ^2}\right)^n \mathscr{L}[g](Q^2). \tag{C.2.7}$$

Comparison with (C.2.1) shows that the Borel transform obeys

$$\mathscr{B}[f](M^2) = g(1/M^2), \tag{C.2.8}$$

if the function f is the Laplace transform of g [Raf81]. This gives explicit expressions of Borel transforms for a broad range of functions once they are given as the Laplace transforms of other known functions. Moreover, we can now safely use well known properties of the Laplace transformation to simplify many calculations. For example, the momentum shift

$$f(s-a) \xrightarrow{\mathscr{L}^{-1}} e^{at} g(t) \tag{C.2.9}$$

gives for the Borel transformation

$$\mathscr{B}\left[f(Q^2+m^2)\right](M^2) = e^{-m^2/M^2} \mathscr{B}\left[f(Q^2)\right](M^2). \tag{C.2.10}$$

It is this property of the Borel transform which introduces the exponential damping and transforms the dispersion integral into a Laplace transform.

Although, by the above criteria, the Borel transformation can be performed for a broad range of functions, we merely need the transform of one type of functions

---

[31] Another representation of the inversion of the Laplace transform is given by the Bromwich integral

$$g(t) = \frac{1}{2\pi i} \int_{c-i\infty}^{c+i\infty} e^{pt} \mathscr{L}[g](p) \, dp$$

for $t > 0$ and c greater than the real part of all singularities of $\mathscr{L}[g](p)$.

during this work, namely

$$f(s) = \frac{1}{s^{\alpha+1}} \frac{1}{(\ln s)^{\beta+1}}. \tag{C.2.11}$$

If $\text{Re}\,\alpha > -1$ and $\text{Re}\,s > 1$, then $f(s)$ is the Laplace transform of the transcendental function $\mu(t, \beta, \alpha)$ [Erd55b]

$$\int_0^\infty e^{-st} \mu(t, \beta, \alpha)\,dt = \frac{1}{s^{\alpha+1}} \frac{1}{(\ln s)^{\beta+1}}, \tag{C.2.12}$$

where $\mu(t, \beta, \alpha)$ is defined by the following integral representation

$$\mu(x, \beta, \alpha) = \int_0^\infty \frac{x^{\alpha+1} t^\beta \,dt}{\Gamma(\beta+1)\Gamma(\alpha+t+1)}. \tag{C.2.13}$$

By repeated integration by parts one can cast this into a form that is more convenient for further calculations:

$$\mu(x, -n, \alpha) = (-1)^{n-1} \left(\frac{d}{du}\right)^{n-1} \left.\frac{x^{\alpha+u}}{\Gamma(\alpha+u+1)}\right|_{u=0} \tag{C.2.14}$$

with $n = 1, 2, \cdots$. In particular, one obtains

$$\mu(x, -1, \alpha) = \frac{x^\alpha}{\Gamma(\alpha+1)}, \tag{C.2.15}$$

$$\mu(x, -2, \alpha) = \frac{x^\alpha}{\Gamma(\alpha+1)} \left[-\ln x + \frac{\Gamma'(\alpha+1)}{\Gamma(\alpha+1)}\right] \tag{C.2.16}$$

with

$$\Gamma'(n) = -(n-1)!\left[\frac{1}{n} + \gamma_E - \sum_{k=1}^n \frac{1}{k}\right] \tag{C.2.17}$$

and γ_E being the Euler constant defined as

$$\gamma_E = \lim_{n\to\infty}\left(\sum_{k=1}^n \frac{1}{k} - \ln n\right). \tag{C.2.18}$$

From Eq. (C.2.12) we are now able to represent the Borel transform of (C.2.11) without performing too many elaborate calculations by means of the initial definition

C In-medium OPE for heavy-light mesons

of the Borel transformation (C.2.1)

$$\mathscr{B}\left[\frac{1}{(Q^2)^{\alpha+1}}\frac{1}{(\ln Q^2)^{\beta+1}}\right](M^2) = \mu\left(\frac{1}{M^2},\beta,\alpha\right). \tag{C.2.19}$$

In particular, for $\beta = -1$ and $\alpha = n-1$ the result reads

$$\mathscr{B}\left[\frac{1}{(Q^2)^n}\right](M^2) = \mu\left(\frac{1}{M^2},-1,n-1\right) = \frac{1}{(n-1)!}\frac{1}{(M^2)^{n-1}}, \tag{C.2.20}$$

and for $\beta = -2$ and $\alpha = n-1$ we get

$$\mathscr{B}\left[\frac{1}{(Q^2)^n}\ln Q^2\right](M^2) = \mu\left(\frac{1}{M^2},-2,n-1\right)$$
$$= \frac{1}{(n-1)!}\frac{1}{(M^2)^{n-1}}\left[\ln M^2 - \frac{1}{n} - \gamma_E + \sum_{k=1}^{n}\frac{1}{k}\right]. \tag{C.2.21}$$

Note, due to the factorial prefactor in Eq. (C.2.20), the Borel transform of an expansion in $x = Q^{-2}$ is indeed the corresponding Borel sum.

The Borel transformation may now be applied to the subtracted dispersion relations (B.5.12), as well as to (B.5.19) and (B.5.20). As the OPE is only valid for large space like momenta, i.e. $q^2 \ll 0$, one may substitute $q^2 = -Q^2$. Furthermore, the limiting procedure in the definition of the Borel transformation is not in conflict with the OPE. Indeed, the limit in Eq. (C.2.1) can be interpreted as approaching infinite momenta $Q^2(n)$ by virtue of different slopes M^2. For the dispersion relation in vacuum we get

$$\Pi_{\text{OPE}}(Q^2) - \sum_{n=0}^{N-1}\frac{\Pi_{\text{ph}}^{(n)}(0)}{n!}(-Q^2)^n$$
$$= \frac{1}{\pi}\int_0^{s_0}\left(-\frac{Q^2}{s}\right)^N\frac{\Delta\Pi_{\text{ph}}(s)}{s+Q^2}ds + \frac{1}{\pi}\int_{s_0}^{+\infty}\left(-\frac{Q^2}{s}\right)^N\frac{\Delta\Pi_{\text{OPE}}(s)}{s+Q^2}ds. \tag{C.2.22}$$

For the in-medium dispersion relations we introduce $q_0^2 = -Q^2$. This is equivalent to considering imaginary values of the energy q_0. For the even part one obtains

C.2 Borel transformed sum rules

$$\Pi_{\text{OPE}}^{e}(Q^2,\vec{q}) - \frac{1}{2}\sum_{n=0}^{N-1}\frac{\Pi_{\text{ph}}^{(n)}(0,\vec{q})}{n!}(Q^2)^{n/2}(1+(-1)^n)$$

$$= \left[\frac{1}{\pi}\int_{s_0^-}^{s_0^+}d\omega\frac{\Delta\Pi_{\text{ph}}(\omega,\vec{q})}{\omega^2+Q^2} + \frac{1}{\pi}\left(\int_{-\infty}^{s_0^-} + \int_{s_0^+}^{+\infty}\right)d\omega\frac{\Delta\Pi_{\text{OPE}}(\omega,\vec{q})}{\omega^2+Q^2}\right]$$

$$\times \frac{1}{2}\left(\frac{(-Q^2)^{N/2}}{\omega^{N-1}}(1+(-1)^N) + \frac{(-Q^2)^{(N+1)/2}}{\omega^N}(1-(-1)^N)\right), \quad \text{(C.2.23a)}$$

while the odd part reads

$$\Pi_{\text{OPE}}^{o}(Q^2,\vec{q}) - \frac{1}{2}\sum_{n=0}^{N-1}\frac{\Pi_{\text{ph}}^{(n)}(0,\vec{q})}{n!}(Q^2)^{(n-1)/2}(1-(-1)^n)$$

$$= \left[\frac{1}{\pi}\int_{s_0^-}^{s_0^+}d\omega\frac{\Delta\Pi_{\text{ph}}(\omega,\vec{q})}{\omega^2+Q^2} + \frac{1}{\pi}\left(\int_{-\infty}^{s_0^-} + \int_{s_0^+}^{+\infty}\right)d\omega\frac{\Delta\Pi_{\text{OPE}}(\omega,\vec{q})}{\omega^2+Q^2}\right]$$

$$\times \frac{1}{2}\left(\frac{(-Q^2)^{(N-1)/2}}{\omega^{N-1}}(1-(-1)^N) + \frac{(-Q^2)^{N/2}}{\omega^N}(1+(-1)^N)\right). \quad \text{(C.2.23b)}$$

Application of the Borel transformation (C.2.1) to the vacuum dispersion relation (C.2.22) yields

$$\mathscr{B}\left[\left(-\frac{Q^2}{s}\right)^N\frac{1}{s+Q^2}\right](M^2) = \frac{e^{-s/M^2}}{(-s)^N}\mathscr{B}\left[\frac{(Q^2-s)^N}{Q^2}\right](M^2)$$

$$= \frac{e^{-s/M^2}}{(-s)^N}\mathscr{B}\left[\frac{(-s)^N}{Q^2}\right](M^2) = e^{-s/M^2}, \quad \text{(C.2.24)}$$

where we have used that only the term $\propto 1/Q^2$ survives the Borel transformation. Hence, one finally ends up with

$$\mathscr{B}\left[\Pi_{\text{OPE}}(Q^2)\right](M^2) = \frac{1}{\pi}\int_0^{s_0}e^{-s/M^2}\Delta\Pi_{\text{ph}}(s)\,ds + \frac{1}{\pi}\int_{s_0}^{+\infty}e^{-s/M^2}\Delta\Pi_{\text{OPE}}(s)\,ds. \quad \text{(C.2.25)}$$

In the same way, we obtain for the in-medium dispersion relation (C.2.23a) by

C In-medium OPE for heavy-light mesons

applying the Borel transformation (C.2.1) with respect to Q^2

$$\mathscr{B}\left[\frac{(-Q^2)^{N/2}}{\omega^{N-1}}\frac{1}{\omega^2+Q^2}\right](M^2)$$
$$= (-1)^{N/2}\frac{e^{-\omega^2/M^2}}{\omega^{N-1}}\mathscr{B}\left[\frac{(Q^2-\omega^2)^{N/2}}{Q^2}\right](M^2)$$
$$= (-1)^{N/2}\frac{e^{-\omega^2/M^2}}{\omega^{N-1}}\mathscr{B}\left[\frac{(-\omega^2)^{N/2}}{Q^2}\right](M^2) = (-1)^N \omega e^{-\omega^2/M^2}, \quad \text{(C.2.26)}$$

where we have used that, due to the factor $(1+(-1)^N)$ in the first term of the last line in (C.2.23a), $N/2$ can only have integer values. The same holds true for $(N+1)/2$ in the second term. Thus, we arrive at

$$\mathscr{B}\left[\Pi^e_{\text{OPE}}(Q^2,\vec{q}\,)\right](M^2)$$
$$= \left[\frac{1}{\pi}\int_{s_0^-}^{s_0^+} d\omega\,\Delta\Pi_{\text{ph}}(\omega,\vec{q}\,) + \frac{1}{\pi}\left(\int_{-\infty}^{s_0^-} + \int_{s_0^+}^{+\infty}\right) d\omega\,\Delta\Pi_{\text{OPE}}(\omega,\vec{q}\,)\right]\omega e^{-\omega^2/M^2}. \tag{C.2.27a}$$

For the odd part we use (C.2.24) and obtain

$$\mathscr{B}\left[\Pi^o_{\text{OPE}}(Q^2,\vec{q}\,)\right](M^2)$$
$$= \left[\frac{1}{\pi}\int_{s_0^-}^{s_0^+} d\omega\,\Delta\Pi_{\text{ph}}(\omega,\vec{q}\,) + \frac{1}{\pi}\left(\int_{-\infty}^{s_0^-} + \int_{s_0^+}^{+\infty}\right) d\omega\,\Delta\Pi_{\text{OPE}}(\omega,\vec{q}\,)\right] e^{-\omega^2/M^2}. \tag{C.2.27b}$$

As a byproduct we find that the subtraction terms in Eqs. (2.1.2) and (2.1.7) as well as the contribution from the infinite outer circle in Eqs. (B.5.4) and (B.5.14) for the non-subtracted dispersion relations are eliminated under Borel transformation since they are polynomials in Q^2.

As discussed in Sec. 2, we are interested in the low-lying strength encoded in $\int_{s_0^-}^{s_0^+} d\omega\,\omega \times \Delta\Pi\, e^{-\omega^2/M^2}$ and $\int_{s_0^-}^{s_0^+} d\omega\,\Delta\Pi\, e^{-\omega^2/M^2}$, whereas the continuum parts encoded in $\pm\int_{s_0^\pm}^{\pm\infty} d\omega \times \omega\Delta\Pi\, e^{-\omega^2/M^2}$ and $\pm\int_{s_0^\pm}^{\pm\infty} d\omega\,\Delta\Pi\, e^{-\omega^2/M^2}$ will be merged into the perturbative OPE part $\Pi_{\text{per}}(\omega)$ (see Eq. (3.1.1)) according to the semi-local duality hypothesis;[32] $s_0^\pm$ are the corresponding continuum thresholds.

---

[32] The semi-local duality hypothesis states that the integral of the continuum is well approximated by

In Sec. 2.1 and App. B.5 we review the method of subtracted dispersion relations. They enable us to suppress polynomial contributions to the dispersion relations from the integral over the infinite circle and ensure the convergence of the dispersion integral. From the definition of the Borel transformation, we see that all functions vanish which are merely polynomials in Q^2. Hence, by applying (C.2.1) to the dispersion relations (B.5.14) or (B.5.5) one gets rid of the polynomial contribution from the outer circle integral. Thus, we do not need to consider subtracted dispersion relations and could have started from the non-subtracted dispersion relation.

C.3 Projection of color, Dirac and Lorentz indices

Condensates are expectation values or Gibbs averages of quantum field operators. They reflect basic properties of the QCD ground state or the strongly interacting medium and are assumed to be color singlets, Lorentz invariants and invariants under parity transformations and time reversal. Expectation values or Gibbs averages which are not invariant under parity transformations and time reversal are supposed to be zero. In the following description we adopt the method described in [Jin93] and list the projections that are needed throughout the thesis.

Up to mass dimension 5 we meet the following structures

$$\langle : \bar{q}_i^a q_j^b : \rangle, \quad \langle : (\bar{q}_i D_\mu)^a q_j^b : \rangle, \quad \langle : (\bar{q}_i D_\mu D_\nu)^a q_j^b : \rangle,$$
$$\langle : G_{\mu\nu}^A G_{\kappa\lambda}^B : \rangle, \quad \langle : \bar{q}_i^a \mathscr{G}_{\mu\nu}^{ab} q_j^b : \rangle. \tag{C.3.1}$$

The meaning of various indices is explained in App. A.3. The last structure is invariant under color rotations. The generators of $U(N_c)$ form a complete set of matrices in color space. Dirac indices are projected onto elements of the Clifford algebra. The projection of Lorentz indices accounts for different condensates in vacuum and medium.

In the following, a list of the in-medium projections up to mass dimension 5 is given. Terms that violate time reversal or parity invariance are omitted. The EoM (A.3.27) have been used whenever possible.

(*i*) Condensates of mass dimension 3:

$$\langle : \bar{q}q : \rangle = \langle : \bar{q}q : \rangle, \tag{C.3.2}$$

---

an integral of the perturbative part.

C In-medium OPE for heavy-light mesons

$$\langle:\bar{q}\gamma_\mu q:\rangle = \langle:\bar{q}\hat{v}q:\rangle \frac{v_\mu}{v^2}. \tag{C.3.3}$$

A condensate of the type $\langle:\bar{q}\sigma_{\mu\nu}q:\rangle$ occurs neither in vacuum nor in a medium since there is no possibility to create an antisymmetric structure in the Lorentz indices μ, ν. Condensates of the type $\langle:\bar{q}\gamma_5\gamma_\mu q:\rangle$, $\langle:\bar{q}\gamma_5 q:\rangle$ can not be projected onto structures that are invariant under parity transformations.

(ii) Condensates of mass dimension 4:

$$\langle:\bar{q}D_\mu q:\rangle = -\langle:\bar{q}\hat{v}q:\rangle i\frac{m_q v_\mu}{v^2}, \tag{C.3.4}$$

$$\langle:\bar{q}\gamma_\mu D_\nu q:\rangle = -\langle:\bar{q}q:\rangle i\frac{m_q}{4}g_{\mu\nu}$$
$$-\left[i\frac{m_q}{4}\langle:\bar{q}q:\rangle + \langle:\bar{q}\hat{v}\frac{(vD)}{v^2}q:\rangle\right]\frac{1}{3}\left(g_{\mu\nu} - 4\frac{v_\mu v_\nu}{v^2}\right), \tag{C.3.5}$$

$$\langle:G^A_{\mu\nu}G^B_{\kappa\lambda}:\rangle = \frac{\delta^{AB}}{96}\left(g_{\mu\kappa}g_{\nu\lambda} - g_{\mu\lambda}g_{\nu\kappa}\right)\langle:G^2:\rangle$$
$$-\frac{\delta^{AB}}{24}\langle:\left(\frac{(vG)^2}{v^2} - \frac{G^2}{4}\right):\rangle S_{\mu\nu\kappa\lambda}. \tag{C.3.6}$$

Again, terms of the form $\langle:\bar{q}\gamma_5\gamma_\mu D_\nu q:\rangle$, $\langle:\bar{q}\gamma_5 D_\mu q:\rangle$ do not have a projection due to the requirement of parity invariance. The same holds true for $\langle:\bar{q}\sigma_{\mu\nu}D_\kappa q:\rangle$, which can only be contracted with $\epsilon_{\mu\nu\kappa\lambda}v^\lambda$ giving an odd term with respect to parity.

(iii) Condensates of mass dimension 5:

$$\langle:\bar{q}D_\mu D_\nu q:\rangle = -\langle:\bar{q}q:\rangle\frac{m_q^2}{4}g_{\mu\nu} + \langle:\bar{q}g\sigma\mathcal{G}q:\rangle\frac{1}{8}g_{\mu\nu}$$
$$-\left[\frac{m_q^2}{4}\langle:\bar{q}q:\rangle - \frac{1}{8}\langle:\bar{q}g\sigma\mathcal{G}q:\rangle + \langle:\bar{q}\frac{(vD)^2}{v^2}q:\rangle\right]\frac{1}{3}\left(g_{\mu\nu} - 4\frac{v_\mu v_\nu}{v^2}\right), \tag{C.3.7}$$

$$\langle:\bar{q}\gamma_\mu D_\nu D_\alpha q:\rangle$$
$$= \frac{1}{v^4}\langle:\bar{q}\hat{v}(vD)^2 q:\rangle\left(\frac{2v_\mu v_\nu v_\alpha}{v^2} - \frac{1}{3}\left(v_\mu g_{\nu\alpha} + v_\nu g_{\mu\alpha} + v_\alpha g_{\mu\nu}\right)\right)$$
$$-\frac{1}{6v^2}\langle:\bar{q}\hat{v}g\sigma\mathcal{G}q:\rangle\left(\frac{v_\mu v_\nu v_\alpha}{v^2} - v_\mu g_{\nu\alpha}\right)$$
$$+\frac{m_q^2}{3v^2}\langle:\bar{q}\hat{v}q:\rangle\left(\frac{v_\mu v_\nu v_\alpha}{v^2} - v_\mu g_{\nu\alpha}\right)$$

C.3 Projection of color, Dirac and Lorentz indices

$$+ \frac{im_q}{3v^2} \langle :\bar{q}(vD)q: \rangle \left(\frac{2v_\mu v_\nu v_\alpha}{v^2} - v_\nu g_{\mu\alpha} - v_\alpha g_{\mu\nu} \right), \tag{C.3.8}$$

$$\langle :\bar{q}\gamma_5\gamma_\mu D_\nu D_\alpha q: \rangle$$
$$= -\frac{1}{6v^2} \epsilon_{\mu\nu\alpha\beta} v^\beta \left[\frac{i}{2} \langle :\bar{q}\hat{v} g\sigma G q: \rangle + im_q^2 \langle :\bar{q}\hat{v}q: \rangle + im_q^2 \langle :\bar{q}(vD)q: \rangle \right], \tag{C.3.9}$$

$$\langle :\bar{q}\sigma_{\mu\nu} D_\alpha D_\beta q: \rangle = -\frac{i}{24} \left(g_{\mu\alpha} g_{\nu\beta} - g_{\mu\beta} g_{\nu\alpha} \right) \langle :\bar{q}g\sigma\mathcal{G}q: \rangle$$
$$- \frac{1}{3} \left[\frac{i}{8} \langle :\bar{q}g\sigma\mathcal{G}q: \rangle - m_q \langle :\bar{q}\hat{v}\frac{(vD)}{v^2} q: \rangle - i \langle :\bar{q}\frac{(vD)^2}{v^2} q: \rangle \right] S_{\mu\nu\alpha\beta}, \tag{C.3.10}$$

$$\langle :\bar{q}\gamma_5\gamma_\alpha \mathcal{G}_{\mu\nu} q: \rangle = -\frac{1}{6v^2} \langle :\bar{q}\hat{v}g\sigma\mathcal{G}q: \rangle \epsilon_{\alpha\mu\nu\sigma} v^\sigma, \tag{C.3.11}$$

$$\langle :\bar{q}g\sigma_{\alpha\beta}\mathcal{G}_{\mu\nu} q: \rangle = \langle :\bar{q}g\sigma\mathcal{G}q: \rangle \frac{1}{12} \left(g_{\alpha\mu} g_{\beta\nu} - g_{\alpha\nu} g_{\beta\mu} \right)$$
$$+ \left[\frac{1}{12} \langle :\bar{q}g\sigma\mathcal{G}q: \rangle - i\frac{2m_q}{3} \langle :\bar{q}\hat{v}\frac{(vD)}{v^2} q: \rangle - \frac{2}{3} \langle :\bar{q}\frac{(vD)^2}{v^2} q: \rangle \right] S_{\alpha\beta\mu\nu}, \tag{C.3.12}$$

where we have defined

$$S_{\alpha\beta\mu\nu} = \left(g_{\alpha\mu} g_{\beta\nu} - g_{\alpha\nu} g_{\beta\mu} - 2 \left(g_{\alpha\mu} \frac{v_\beta v_\nu}{v^2} - g_{\alpha\nu} \frac{v_\beta v_\mu}{v^2} + g_{\beta\nu} \frac{v_\alpha v_\mu}{v^2} - g_{\beta\mu} \frac{v_\alpha v_\nu}{v^2} \right) \right). \tag{C.3.13}$$

We have explicitly separated the medium specific contributions from the vacuum projections. Medium specific contributions are either condensates that contain the medium four-velocity v or combinations of condensates that appear in vacuum and medium. The latter ones are always written with angled brackets. Applying vacuum projections to the medium specific terms makes them zero in the vacuum limit.

Note that Eq. (C.3.11) is actually not a projection and is valid for any four-vector. It is proven in the following. Due to Lorentz covariance we write

$$\langle :\bar{q}\gamma_5\gamma_\alpha \mathcal{G}_{\mu\nu} q: \rangle = A\epsilon_{\alpha\mu\nu\sigma} v^\sigma. \tag{C.3.14}$$

Our aim is to determine the Lorentz scalar A. Using Eq. (A.3.26), we can write for

C In-medium OPE for heavy-light mesons

any four-vector v_μ

$$\hat{v}\sigma^{\mu\nu}\mathcal{G}_{\mu\nu} = iv_\alpha \gamma^\alpha \gamma^\mu \gamma^\nu \mathcal{G}_{\mu\nu}. \tag{C.3.15}$$

Expanding the product of Dirac matrices in terms of the Clifford algebra, one obtains

$$\gamma^\alpha \gamma^\mu \gamma^\nu = g^{\mu\nu}\gamma^\alpha + g^{\nu\alpha}\gamma^\mu + g^{\alpha\mu}\gamma^\nu + i\epsilon^{\sigma\alpha\mu\nu}\gamma_5\gamma_\sigma. \tag{C.3.16}$$

Due to the EoM (A.3.23) and by the definition of the gluon field strength tensor, one can show that

$$\langle :\bar{d}v^\mu \gamma^\nu \mathcal{G}_{\mu\nu} d: \rangle = 0. \tag{C.3.17}$$

Altogether, this gives the relation

$$\langle :\bar{d}\hat{v}\sigma^{\mu\nu}\mathcal{G}_{\mu\nu} d: \rangle = -\langle :\bar{d}\gamma_5\gamma_\sigma \mathcal{G}_{\mu\nu} d: \rangle \epsilon^{\sigma\alpha\mu\nu} v_\alpha. \tag{C.3.18}$$

Contracting Eq. (C.3.14) with $\epsilon^{\alpha\mu\nu\sigma}$ and making use of $\epsilon^{\alpha\mu\nu\sigma}\epsilon_{\alpha\mu\nu}{}^\tau = 6g^{\sigma\tau}$, proves Eq. (C.3.11).

C.4 Operator product expansion for the D meson

In the following the OPE for heavy-light pseudo-scalar mesons is performed up to and including mass dimension 5 power corrections to the lowest-order term of the perturbative expansion. First, one applies Wick's theorem to the current-current correlation function $\Pi(q)$, with the current operators for the D^+ meson, $j_{D^+}(x) = i\bar{d}(x)\gamma_5 c(x) \equiv j(x)$, and the D^- meson, $j_{D^-}(x) = i\bar{c}(x)\gamma_5 d(x) = j_{D^+}^\dagger(x)$, in lowest order of the perturbative expansion

$$\Pi(q) = i\int d^4x\, e^{iqx} \langle T[j(x)j^\dagger(0)]\rangle \tag{C.4.1a}$$

$$= \Pi^{(0)}(q) + \Pi^{(2)}(q) + \Pi^{(4)}(q). \tag{C.4.1b}$$

Due to Eq. (A.2.4) the D^+ current-current correlator (C.4.1a) is related to the D^- correlator by $\Pi^{D^+}(q) = \Pi^{D^-}(-q)$. The OPE is performed for the D^+ meson. The label will be dropped in the following. The labels (0) and (2) denote the number of non-contracted quark fields, i.e., the number of quarks which participate in the

C.4 OPE for the D meson

formation of a condensate. The $\Pi^{(4)}$ term does not contribute because it contains only the soft contribution of all operators. There is no flow of hard momenta. Expanding the quark field operators according to Eq. (C.1.12) and performing n integrations by parts for the n-th order term gives the n-th derivative of Dirac's delta distribution. Therefore, at large $|q^2|$, i.e. in the OPE domain, there is no contribution from $\Pi^{(4)}$. Also, this term would lead to four-quark condensates which are of mass dimension 6. In general, for mesons four-quark condensates only contribute in first (or higher) order of α_s. Otherwise the diagrams would be disconnected and no momentum could flow through it as for the $\Pi^{(4)}$ term.

The first term in (C.4.1b), $\Pi^{(0)}(q)$, reads

$$\Pi^{(0)}(q) = -i \int d^4 x \, e^{iqx} \langle : \text{Tr}_{C,D} \left[\gamma_5 S_d(0,x) \gamma_5 S_c(x,0) \right] : \rangle \tag{C.4.2a}$$

$$= \Pi_{\text{per}}(q) + \Pi_{G^2}(q). \tag{C.4.2b}$$

It consists of the purely perturbative term $\Pi_{\text{per}}(q)$ and of $\Pi_{G^2}(q)$, which is the result of inserting the next-to-leading order propagator with the lowest order term of (C.1.23), and accounts for pure gluon condensates. Inserting higher orders in the quark propagator and/or the gluon field in (C.4.2a) leads to higher dimensional gluon condensates, such as $\langle G^3 \rangle$, or, by usage of the EoM, quark and mixed quark-gluon condensates. These are either of higher mass dimensions or higher orders in α_s.

The second term in (C.4.1b), $\Pi^{(2)}(q)$, is simplified by the assumption of vanishing charm-quark condensates $\langle : \bar{c} \ldots c : \rangle = 0$. This is motivated by the large charm-quark mass and the assumption that a heavy quark only interacts with the vacuum by emitting or absorbing gluons and not by annihilating with other heavy quarks. Hence, heavy quarks are considered as static or quenched quarks and only light-quark condensates remain. This becomes evident in the scope of the renormalization of condensates and absorption of infrared singularities. Indeed, introducing non-normal ordered condensates, heavy-quark condensates cancel out by virtue of the heavy quark mass expansion.

The perturbative contribution $\Pi_{\text{per}}(q)$, consists of the lowest order quark propagator insertion in (C.1.25), i.e. the one-loop diagram (see Fig. C.4.1). Additionally, the order α_s^1 contribution, i.e. two-loop diagrams, may be taken into account. It can be determined in terms of a two-fold subtracted dispersion relation [Ali83, Rei80, Bro81] (see App. B.5), where the imaginary part of $\Pi_{\text{per}}(q)$ is obtained by means of the Cutkosky cutting rules [Cut60, Das97]. In the $\overline{\text{MS}}$-scheme (cf. e.g. [Itz80]) it is

C In-medium OPE for heavy-light mesons

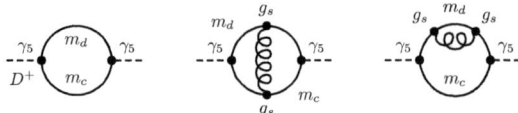

Figure C.4.1: Feynman diagrams for the perturbative contribution $\Pi^{\text{per}}(q)$. The fourth contribution can be obtained from the third by the replacement $d \leftrightarrow c$. Dashed: the D meson, solid: free quark propagator, curly: free gluon propagator.

infrared convergent and reads in Feynman gauge for the gluon propagator and $m_d \to 0$

$$\text{Im}\Pi_{\text{pert}}(s) = \frac{3}{8\pi}\frac{(s-m_c^2)^2}{s} + \frac{\alpha_s}{2\pi^2}\frac{(s-m_c^2)^2}{s}\left[\frac{9}{4} + 2\text{Li}_2\left(\frac{m_c^2}{s}\right) + \frac{3}{2}\ln\left(\frac{m_c^2}{s-m_c^2}\right)\right.$$
$$+ \ln\left(\frac{s}{m_c^2}\right)\ln\left(\frac{s}{s-m_c^2}\right) + \ln\left(\frac{s}{s-m_c^2}\right)$$
$$\left.+ \frac{m_c^2}{s}\ln\left(\frac{s-m_c^2}{m_c^2}\right) + \frac{m_c^2}{s-m_c^2}\ln\left(\frac{s}{m_c^2}\right)\right]. \quad \text{(C.4.3)}$$

Here, the Spence function is denoted as $\text{Li}_2(x) = -\int_0^x dt\, t^{-1}\ln(1-t)$.

For the gluonic contribution there are three terms, each stemming from a different number of background field insertions to the quark propagators (see Fig. C.4.2). We work in lowest order of the expansion (C.1.23). The three terms which have to be calculated at the one-loop level to obtain all contributions which are relevant for the vacuum gluon condensate $\langle:\frac{\alpha_s}{\pi}G^2:\rangle$ entering $\Pi_{G^2}(q)$ read:

$$\Pi_{G^2}(q) = \sum_{i+j=2}\left[-i\int\frac{d^4p}{(2\pi)^4}\langle:\text{Tr}_{C,D}\left[\gamma_5 S_c^{(i)}(p)\gamma_5 S_d^{(j)}(p-q)\right]:\rangle\right], \quad \text{(C.4.4)}$$

where the leading order gluon field is inserted. The following four integrals are needed to evaluate the loop integrals [Itz80]

$$\int\frac{d^4p}{(2\pi)^4}\frac{1}{\left[(p-q)^2+m_d^2\right]^n\left[p^2+m_c^2\right]^k}$$
$$= \frac{1}{(4\pi)^2}\frac{\Gamma(n+k-2)}{\Gamma(n)\Gamma(k)}I_{n+k-2}^{n-1,k-1}(q^2,m_d^2,m_c^2), \quad \text{(C.4.5a)}$$

$$\int \frac{\mathrm{d}^4 p}{(2\pi)^4} \frac{p_\mu}{\left[(p-q)^2 + m_d^2\right]^n \left[p^2 + m_c^2\right]^k}$$
$$= \frac{q_\mu}{(4\pi)^2} \frac{\Gamma(n+k-2)}{\Gamma(n)\Gamma(k)} I_{n+k-2}^{n,k-1}(q^2, m_d^2, m_c^2), \quad \text{(C.4.5b)}$$

$$\int \frac{\mathrm{d}^4 p}{(2\pi)^4} \frac{p_\mu p_\nu}{\left[(p-q)^2 + m_d^2\right]^n \left[p^2 + m_c^2\right]^k}$$
$$= \frac{g_{\mu\nu}}{(4\pi)^2} \frac{1}{2} \frac{\Gamma(n+k-3)}{\Gamma(n)\Gamma(k)} I_{n+k-3}^{n-1,k-1}(q^2, m_d^2, m_c^2)$$
$$+ \frac{q_\mu q_\nu}{(4\pi)^2} \frac{\Gamma(n+k-2)}{\Gamma(n)\Gamma(k)} I_{n+k-2}^{n+1,k-1}(q^2, m_d^2, m_c^2), \quad \text{(C.4.5c)}$$

$$\int \frac{\mathrm{d}^4 p}{(2\pi)^4} \frac{p_\mu p_\nu p_\kappa}{\left[(p-q)^2 + m_d^2\right]^n \left[p^2 + m_c^2\right]^k}$$
$$= \frac{g_{\mu\nu} q_\kappa + g_{\mu\kappa} q_\nu + g_{\nu\kappa} q_\mu}{(4\pi)^2} \frac{1}{2} \frac{\Gamma(n+k-3)}{\Gamma(n)\Gamma(k)} I_{n+k-3}^{n,k-1}(q^2, m_d^2, m_c^2)$$
$$+ \frac{q_\mu q_\nu q_\kappa}{(4\pi)^2} \frac{\Gamma(n+k-2)}{\Gamma(n)\Gamma(k)} I_{n+k-2}^{n+2,k-1}(q^2, m_d^2, m_c^2). \quad \text{(C.4.5d)}$$

Equation (C.4.5d) can be obtained from Eq. (C.4.5c) by applying a partial derivative w. r. t. q^κ. The master integral is given by

$$I_k^{i,j}(q^2, m_d^2, m_c^2) = \int_0^1 \mathrm{d}\alpha \frac{\alpha^i (1-\alpha)^j}{\left[\alpha(1-\alpha) q^2 + \alpha m_d^2 + (1-\alpha) m_c^2\right]^k}. \quad \text{(C.4.6)}$$

It is not possible to set $m_d = 0$ from the very beginning, because the results are infrared divergent. On the other hand, the expressions are rather cumbersome for both masses being nonzero. In order to obtain meaningful expressions, one may use

$$\arctan z = \frac{1}{2i} \ln \frac{1+iz}{1-iz} \quad \text{(C.4.7)}$$

to obtain

$$\arctan\left(i \frac{q^2 - m_d^2}{q^2 + m_d^2}\right) = \frac{1}{2i} \ln \frac{m_d^2}{q^2}, \quad \text{(C.4.8)}$$

C In-medium OPE for heavy-light mesons

which is a source of mass logarithms. Another source of terms $\propto \ln m_d$ arises from a slightly different expression, namely

$$\arctan\left(i\frac{q^2 + m_c^2 - m_d^2}{\sqrt{2q^2 m_d^2 + q^4 + 2q^2 m_c^2 + m_c^4 - 2m_c^2 m_d^2 + m_d^4}}\right) \stackrel{m_d=0}{\to} \arctan(i), \quad \text{(C.4.9)}$$

which is not well defined. Expanding the fraction in m_d and keeping only the lowest power, which is the dominant contribution for $m_d \to 0$, leads to

$$\arctan\left(i\frac{q^2 + m_c^2 - m_d^2}{\sqrt{2q^2 m_d^2 + q^4 + 2q^2 m_c^2 + m_c^4 - 2m_c^2 m_d^2 + m_d^4}}\right) \stackrel{m_d \approx 0}{=} \frac{1}{2i}\ln\frac{q^2 m_d^2}{(q^2 + m_c^2)^2}.$$

(C.4.10)

The last expression corresponds to the infrared term for the medium specific contribution, i. e. it is absent from the OPE in vacuum. Upon projecting color, Dirac and Lorentz indices, the following result can be verified

$$\Pi_{G^2}(q) = \langle : \frac{\alpha_s}{\pi} G^2 : \rangle \left(-\frac{1}{24}\frac{1}{q^2 - m_c^2} - \frac{1}{12}\frac{m_c}{m_d}\frac{1}{q^2 - m_c^2} - \frac{1}{24}\frac{m_c^2}{(q^2 - m_c^2)^2}\right)$$
$$+ \langle : \frac{\alpha_s}{\pi}\left(\frac{(vG)^2}{v^2} - \frac{G^2}{4}\right) : \rangle \left(q^2 - 4\frac{(vq)^2}{v^2}\right)\left[-\frac{1}{6q^2}\frac{1}{q^2 - m_c^2} - \frac{1}{6q^2}\frac{m_c^2}{(q^2 - m_c^2)^2}\right.$$
$$- \frac{1}{9q^2}\left(\frac{m_c^2}{(q^2 - m_c^2)^2} + \frac{1}{q^2 - m_c^2}\right)\ln\left(\frac{m_d^2}{m_c^2}\right)$$
$$\left. - \frac{2}{9q^2}\left(\frac{m_c^2}{(q^2 - m_c^2)^2} + \frac{1}{q^2 - m_c^2}\right)\ln\left(-\frac{m_c^2}{q^2 - m_c^2}\right)\right]. \quad \text{(C.4.11)}$$

The first line is the vacuum contribution, while the term $\propto \langle : \frac{\alpha_s}{\pi}\left(\frac{(vG)^2}{v^2} - \frac{G^2}{4}\right) : \rangle$ is a medium specific condensate. It vanishes in vacuum and is the symmetric and traceless part of the operator $\mathcal{G}_{\mu\nu}\mathcal{G}_{\alpha\beta}$.

One immediately observes terms $\propto m_d^{-1}$ and $\propto \ln m_d^2$. These are terms which diverge for $m_d \to 0$. They origin from the infrared part of the loop diagram with two gluon lines attached to the light-quark propagator, which results in three light-quark propagators with the same momentum because the background field is soft and does not transfer momenta (right panel of Fig. C.4.2), and are therefore called infrared divergences. Moreover, terms of the form $\ln(Q^2/\mu^2)$ and $\ln(m_d^2/\mu^2)$, which cannot be

228

C.4 OPE for the D meson

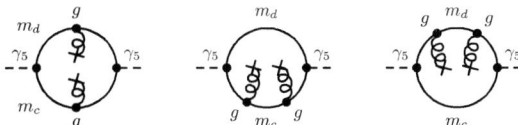

Figure C.4.2: Feynman diagrams for the gluon condensate contribution $\Pi^{G^2}(q)$. Line codes and meaning of crosses as in Fig. C.4.1.

made small at the same time for $Q^2 \gg m_d^2$, occur when calculating $\Pi^{(0)}$. As pointed out in [Tka83b], they are remnants of the large distance behavior, i.e. they origin from the small momentum contribution to the loop integrals. Their occurrence breaks the clean separation of scales, which is a necessary feature of every OPE, and must therefore be absorbed into the condensates.

In [Tka83b], the author argues that these logarithms do not occur when the Wilson coefficients are calculated within a minimal subtraction scheme. Moreover, they can be absorbed into the condensates if one re-expresses normal ordered condensates, which naturally emerge if one applies Wick's theorem when calculating Wilson coefficients perturbatively, by so-called non-normal ordered ones. This procedure has been known for a long time, cf. [Spi88, Che95, Jam93, Nar89] and references therein. An explicit formula could not be found in these references. Therefore, we elaborate the in-medium case.

After expanding the light quark fields and performing the Fourier transformation one obtains the general expression

$$\Pi^{(2)}(q) = \sum_{k=0}^{\infty} \frac{(-i)^k}{k!} \langle : \left(\bar{d}_i \overleftarrow{D}_{\alpha_1} \ldots \overleftarrow{D}_{\alpha_k} \right)^a \left(\gamma_5 \partial^{\alpha_1} \ldots \partial^{\alpha_k} S_c(q) \gamma_5 \right)^{ij}_{ab} d^b_j : \rangle . \quad (C.4.12)$$

As in Eq. (C.4.2) one can now insert higher orders in the quark fields, the heavy-quark propagator and/or higher orders of the gluon background field. In order to consider all mass dimension 5 contributions to the OPE in (C.4.12), $k + l(2 + m) \leq 2$ must be fulfilled, according to Eq. (C.1.35). The order of the quark field expansion is k, l is the number of gluon field insertions and m the order of (C.1.23). Hence, there are four terms in (C.4.12), namely $(k, l, m) \in \{(0, 0, 0), (1, 0, 0), (2, 0, 0), (0, 1, 0)\}$ and three terms in (C.4.2), which contribute to the OPE up to mass dimension 5. The diagrams for these terms are shown in Fig. C.4.3. Note that due to the additional medium structures there is no one-to-one correspondence between Eq. (C.4.12) and Feynman diagrams unless one introduces further building blocks.

Before going into details, we present the contribution of (C.4.12) to the OPE up to

C In-medium OPE for heavy-light mesons

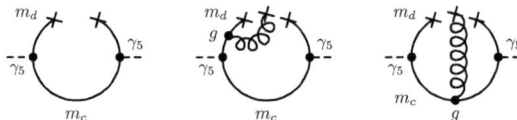

Figure C.4.3: Feynman diagrams for the quark and mixed quark-gluon condensate contribution $\Pi^{(2)}(q)$. Line codes as in Fig. C.4.1. Crosses symbolize creation or annihilation of quarks or gluons by virtual particles, i. e. condensates.

mass dimension 5. Separating the contribution of the free quark propagator from the sum in (C.4.12) one arrives at

$$\Pi^{(2)}(q) = \sum_{k=0}^{\infty} \frac{(-i)^k}{k!} \left[\langle :(\bar{d}\overleftarrow{D}_{a_1} \ldots \overleftarrow{D}_{a_k})d: \rangle \partial^{a_1} \ldots \partial^{a_k} \frac{m_c}{q^2 - m_c^2} \right.$$
$$\left. - \langle :(\bar{d}\gamma_\mu \overleftarrow{D}_{a_1} \ldots \overleftarrow{D}_{a_k})d: \rangle \partial^{a_1} \ldots \partial^{a_k} \frac{q^\mu}{q^2 - m_c^2} \right]$$
$$+ \sum_{k=0}^{\infty} \sum_{l=1}^{\infty} \frac{(-i)^k}{k!} \langle :(\bar{d}_i \overleftarrow{D}_{a_1} \ldots \overleftarrow{D}_{a_k})^a (\gamma_5 \partial^{a_1} \ldots \partial^{a_k} S_c^{(l)}(q) \gamma_5)_{ab}^{ij} d_j^b : \rangle, \quad \text{(C.4.13)}$$

where we use that the free propagator $S^{(0)}(q) = \frac{\slashed{q}+m}{q^2-m^2} \otimes \mathbb{1}_c$ is the unit operator in color space and does not contain Lorentz indices. In the first sum, we project color and Dirac indices, whereby the only non-vanishing traces over Dirac indices correspond to the projection of the free heavy-quark propagator onto $\{\mathbb{1}, \gamma_\mu\}$. Like in [Coh95] we recognize that the coefficients for terms which emerge from the insertion of the free propagator fulfill certain recursive relations. We will deepen this fact and give recursive relations for the Borel transformed Wilson coefficients in vacuum and medium. Obviously, this is also true for terms which emerge due to the insertion of higher order propagators.

For the above quoted tuples (k, l, m) which account for mass dimension 5 terms we obtain the following result at tree level

$$\Pi^{(2)}(q) = \langle :\bar{d}d: \rangle \left(\frac{m_c}{q^2 - m_c^2} - \frac{1}{2} \frac{m_d}{q^2 - m_c^2} + \frac{1}{2} \frac{m_d m_c^2}{(q^2 - m_c^2)^2} + \frac{m_d^2 m_c^3}{(q^2 - m_c^2)^3} \right)$$
$$- \langle :\bar{d} g \sigma \mathscr{G} d: \rangle \frac{1}{2} \left(\frac{m_c^3}{(q^2 - m_c^2)^3} + \frac{m_c}{(q^2 - m_c^2)^2} \right)$$
$$+ \left[\frac{m_d}{12} \langle :\bar{d}d: \rangle - \frac{1}{3} \langle :\bar{d}\hat{v} \frac{(ivD)}{v^2} d: \rangle \right] 2 \left(q^2 - 4 \frac{(vq)^2}{v^2} \right) \frac{1}{(q^2 - m_c^2)^2}$$

230

$$+\left[\frac{m_d^2}{12}\langle:\bar{d}d:\rangle+\frac{1}{3}\langle:\bar{d}\frac{(vD)^2}{v^2}d:\rangle-\frac{1}{24}\langle:\bar{d}g\sigma\mathcal{G}d:\rangle\right]$$
$$\times 4\left(q^2-4\frac{(vq)^2}{v^2}\right)\frac{m_c}{(q^2-m_c^2)^3}$$
$$-\langle:\bar{d}\hat{v}d:\rangle\frac{(vq)}{v^2}\left(\frac{1}{q^2-m_c^2}+2\frac{m_d m_c}{(q^2-m_c^2)^2}-\frac{m_d^2}{(q^2-m_c^2)^2}\right.$$
$$\left.+\frac{4}{3}\left(q^2-\frac{(vq)^2}{v^2}\right)\frac{m_d^2}{(q^2-m_c^2)^2}\right)$$
$$-\langle:\bar{d}\hat{v}(vD)^2d:\rangle 4\left(q^2-2\frac{(vq)^2}{v^2}\right)\frac{(vq)}{v^4}\frac{1}{(q^2-m_c^2)^3}$$
$$+\langle:\bar{d}\hat{v}g\sigma\mathcal{G}d:\rangle\frac{(vq)}{v^2}\left(\frac{2}{3}\left(q^2-\frac{(vq)^2}{v^2}\right)\frac{1}{(q^2-m_c^2)^3}-\frac{1}{(q^2-m_c^2)^2}\right)$$
$$+\langle:\bar{d}(ivD)d:\rangle\frac{(vq)}{v^2}\left(2\frac{m_d}{(q^2-m_c^2)^2}-\frac{8}{3}\left(q^2-\frac{(vq)^2}{v^2}\right)\frac{m_d}{(q^2-m_c^2)^3}\right).$$
(C.4.14)

Again, we explicitly separated the medium specific terms from terms which are also present in vacuum. The latter ones are merely the chiral condensate $\langle:\bar{d}d:\rangle$ and the mixed quark-gluon condensate $\langle:\bar{d}g\sigma\mathcal{G}d:\rangle$. All the other condensates or medium specific condensate combinations, which are written in squared brackets, vanish in vacuum. At this point of the evaluation we have to keep terms $\propto m_d$, which appear when applying the EoM, because they will be necessary to cancel the infrared divergences or give a finite contribution in the limit $m_d \to 0$ after absorption of the divergences. The limit $m_d \to 0$ will be taken in the next section after the absorption of these divergences.

C.5 Absorption of divergences

In order to ensure a consistent separation of scales, the infrared divergent terms have to be absorbed into the condensates, such that the Wilson coefficients are determined only by the perturbative dynamics, whereas the non-perturbative dynamics is parametrized by the condensates. In [Che82b, Tka83a, Gor83, LS88] it has been shown that the Wilson coefficients are polynomial in the mass only if they are evaluated within a minimal subtraction scheme. Normal ordering is not such a scheme. As

C In-medium OPE for heavy-light mesons

already noted above, all the mass logarithms can be merged into the condensates by re-expressing normal ordered condensates by non-normal ordered ones. We use the following relation between normal ordered and non-normal ordered condensates

$$\langle \bar{q} \mathcal{O}[D_\mu] q \rangle = \langle : \bar{q} \mathcal{O}[D_\mu] q : \rangle - i \int \frac{d^4 p}{(2\pi)^4} \langle \text{Tr}_{C,D} [\mathcal{O}[-ip_\mu - i\tilde{A}_\mu] S(p)] \rangle, \quad (C.5.1)$$

which we will evaluate up to one-loop level. The operator $\mathcal{O}[D_\mu]$ is a function of the covariant derivative, the gluon fields and possible Dirac structures. The index μ stands as label for all possible Lorentz indices. The proofs partly follows [Gro95]. One applies Wick's theorem to the time-ordered product

$$T[\bar{q}(0) \mathcal{O}[D_\mu] q(y)] =: \bar{q}(0) \mathcal{O}[D_\mu] q(y) : -i : \text{Tr}_{C,D}(\mathcal{O}[D_\mu] S(y,0)) :, \quad (C.5.2)$$

inserts the Fourier transform of the operator function $\mathcal{O}$ and sets $y = 0$. As $\mathcal{O}$ in Eq. (C.5.1) is now a derivative operator acting to the right, the order of Lorentz indices is important. This equation reproduces the vacuum expressions but also gives the correct ones for the in-medium case, i. e. all the mass singularities are absorbed into the condensates. Equation (C.5.1) may be defined as the relation between normal ordered and physical condensates. A renormalization scale dependence naturally occurs upon introducing non-normal ordered condensates and delineates the region between perturbative and non-perturbative physics.

Up to the considered order in α_s the following set of equations is obtained in the $\overline{\text{MS}}$ scheme

$$\langle \bar{q} q \rangle = \langle : \bar{q} q : \rangle + \frac{3}{4\pi^2} m_q^3 \left(\ln \frac{\mu^2}{m_q^2} + 1 \right) - \frac{1}{12 m_q} \langle \frac{\alpha_s}{\pi} G^2 \rangle, \quad (C.5.3a)$$

$$\langle \bar{q} g \sigma \mathcal{G} q \rangle = \langle : \bar{q} g \sigma \mathcal{G} q : \rangle - \frac{1}{2} m_q \ln \frac{\mu^2}{m_q^2} \langle \frac{\alpha_s}{\pi} G^2 \rangle, \quad (C.5.3b)$$

$$\langle \bar{q} \gamma_\mu i D_\nu q \rangle = \langle : \bar{q} \gamma_\mu i D_\nu q : \rangle + \frac{3}{16\pi^2} m_q^4 g_{\mu\nu} \left(\ln \frac{\mu^2}{m_q^2} + 1 \right) - \frac{g_{\mu\nu}}{48} \langle \frac{\alpha_s}{\pi} G^2 \rangle$$
$$+ \frac{1}{18} \left(g_{\mu\nu} - 4 \frac{v_\mu v_\nu}{v^2} \right) \left(\ln \frac{\mu^2}{m_q^2} - \frac{1}{3} \right) \langle \frac{\alpha_s}{\pi} \left(\frac{(vG)^2}{v^2} - \frac{G^2}{4} \right) \rangle, \quad (C.5.3c)$$

$$\langle \bar{q} i D_\mu i D_\nu q \rangle = \langle : \bar{q} i D_\mu i D_\nu q : \rangle + \frac{3 m_q^5}{16 \pi^2} g_{\mu\nu} \left(\ln \frac{\mu^2}{m_q^2} + 1 \right)$$

C.5 Absorption of divergences

$$+ \frac{m_q}{16} g_{\mu\nu} \left(\ln \frac{\mu^2}{m_q^2} - \frac{1}{3} \right) \langle \frac{\alpha_s}{\pi} G^2 \rangle$$

$$- \frac{m_q}{36} \left(g_{\mu\nu} - 4 \frac{v_\mu v_\nu}{v^2} \right) \left(\ln \frac{\mu^2}{m_q^2} + \frac{2}{3} \right) \langle \frac{\alpha_s}{\pi} \left(\frac{(vG)^2}{v^2} - \frac{G^2}{4} \right) \rangle.$$

(C.5.3d)

Note that $S(p)$ is merely the product of $S^{(0)}(p)$ and $(\gamma \tilde{A})$, where each $(\gamma \tilde{A})$ accounts for two additional Dirac matrices, cf. (C.1.30).

When working in lowest order of the expansion (C.1.23), $\langle :\bar{q}\gamma_\mu q: \rangle$, $\langle :\bar{q}i\vec{D}_\mu q: \rangle$, and $\langle :\bar{q}\gamma_5\gamma_\alpha G_{\mu\nu}q: \rangle$ are not renormalized due to the following reason. In order to get a non-vanishing integral, the integrand must be even in p, requiring an even number of Dirac matrices. Therefore, it is clear that $\int d^4p \, p_\mu \text{Tr}_{C,D}[S(p)] = 0$ and similar terms are also zero. Each $\tilde{A}_\mu$ accounts for one additional Dirac matrix, and by this $\int d^4p \, \text{Tr}_{C,D}[\tilde{A}_\mu S(p)]$ is zero. Hence, only if the number of Lorentz indices (i.e. the sum of Dirac matrices and covariant derivatives in the condensates) is even, physical and normal ordered condensates differ.

The last two equations differ in vacuum and medium and account for the proper cancellation of mass logarithms in the medium. One immediately recognizes the operator mixing, which affects the pure gluonic vacuum and medium condensates. Moreover, for vanishing normal ordered condensates, Eq. (C.5.1) leads to the well-known heavy-quark mass expansion, first suggested in [Shi79b] (further terms are calculated in [Gen84]). This is why there are no heavy quark condensates. Introducing non-normal ordered heavy quark condensates to perform the renormalization of the normal ordered ones and inserting the heavy-quark mass expansion, cancels all heavy-quark condensates and effectively means setting the normal ordered condensates to zero.

Inserting the non-normal ordered condensates into (C.4.14) one obtains the desired infrared stable Wilson coefficients and the limit $m_d \to 0$ can safely be taken. From a physical point of view, the OPE is only meaningful if the $m_d \to 0$ limit is well defined. Any infrared divergence represents a sensitivity of the OPE to small momenta. These, however, are cut off by confinement at the typical QCD scale $\Lambda_{\text{QCD}} \ll m_d$. Therefore, any OPE must be finite and infrared stable. The final OPE up to mass dimension 5, in the rest frame of nuclear matter $v = (1, \vec{0})$ and for the meson at rest $q = (q_0, \vec{0})$ reads

$$\Pi^e(q_0) = c_0(q_0^2) + \langle \bar{d}d \rangle \frac{m_c}{q_0^2 - m_c^2} - \langle \bar{d}g\sigma\mathscr{G}d \rangle \frac{1}{2} \left(\frac{m_c^3}{(q_0^2 - m_c^2)^3} + \frac{m_c}{(q_0^2 - m_c^2)^2} \right)$$

C In-medium OPE for heavy-light mesons

Table C.5.1: Comparison of vacuum Wilson coefficients entering Eq. (3.1.1a) with the literature. In [Hay00] only operators up to mass dimension 4 were considered.

	$\langle \bar{d}d \rangle$	$\langle \frac{\alpha_s}{\pi} G^2 \rangle$	$\langle \bar{d} g_s \sigma \mathcal{G} d \rangle$
Eq. (3.1.1a)	$-m_c$	$\frac{1}{12}$	$-\frac{1}{2}\frac{1}{M^2}\left(1 - \frac{1}{2}\frac{m_c^2}{M^2}\right)m_c$
Ref. [Hay00]	$-m_c$	$\frac{1}{12} - \frac{1}{24}\frac{m_c^2}{M^2}$	
Ref. [Mor01]	$-m_c$	$\frac{1}{12}$	$-\frac{1}{2}\frac{1}{M^2}\left(1 - \frac{1}{2}\frac{m_c^2}{M^2}\right)m_c$
Ref. [Hay04]	$-m_c$	$\frac{1}{12}\left(\frac{3}{2} - \frac{m_c^2}{M^2}\right)$	$-\frac{1}{2}\frac{1}{M^2}\left(1 - \frac{1}{2}\frac{m_c^2}{M^2}\right)m_c$

$$
\begin{aligned}
&-\langle \frac{\alpha_s}{\pi} G^2 \rangle \frac{1}{12} \frac{1}{q_0^2 - m_c^2} + \langle d^\dagger i D_0 d \rangle 2 \left(\frac{m_c^2}{(q_0^2 - m_c^2)^2} + \frac{1}{q_0^2 - m_c^2} \right) \\
&+ \langle \frac{\alpha_s}{\pi} \left(\frac{(vG)^2}{v^2} - \frac{G^2}{4} \right) \rangle \left(\frac{7}{18} + \frac{1}{3} \ln \frac{\mu^2}{m_c^2} + \frac{2}{3} \ln\left(-\frac{m_c^2}{q_0^2 - m_c^2}\right) \right) \\
&\qquad\qquad\qquad \times \left(\frac{m_c^2}{(q_0^2 - m_c^2)^2} + \frac{1}{q_0^2 - m_c^2} \right) \\
&- \left[\frac{1}{3} \langle \bar{d} D_0^2 d \rangle - \frac{1}{24} \langle \bar{d} g \sigma \mathcal{G} d \rangle \right] 12 \left(\frac{m_c^3}{(q_0^2 - m_c^2)^3} + \frac{m_c}{(q_0^2 - m_c^2)^2} \right)
\end{aligned}
$$
(C.5.4a)

and

$$
\Pi^o(q_0^2) = -\langle d^\dagger d \rangle \frac{1}{q_0^2 - m_c^2} + \langle d^\dagger D_0^2 d \rangle 4 \left(\frac{m_c^2}{(q_0^2 - m_c^2)^3} + \frac{1}{(q_0^2 - m_c^2)^2} \right)
$$
$$
- \langle d^\dagger g \sigma \mathcal{G} d \rangle \frac{1}{(q_0^2 - m_c^2)^2}.
$$
(C.5.4b)

We separate the even and odd parts of the OPE according to $\Pi(q_0, \vec{q}) = \Pi^e(q_0^2, \vec{q}) + q_0 \Pi^o(q_0^2, \vec{q})$, and $c_0(q_0^2)$ denotes the perturbative contribution to the current-current correlator with subtracted mass-singularities. In Tabs. C.5.1 and C.5.2 the Borel transform of Eq. (C.5.4a), given in Eq. (3.1.1a), is compared to results reported in the literature. There is no odd OPE given in the literature.

Table C.5.2: Comparison of the medium specific Wilson coefficients entering Eq. (3.1.1b) with the literature.

	$\langle d^\dagger i \vec{D}_0 d \rangle$	$\langle \frac{\alpha_s}{\pi} \left(\frac{(vG)^2}{v^2} - \frac{G^2}{4} \right) \rangle$
Eq. (3.1.1a)	$2\left(\frac{m_c^2}{M^2} - 1 \right)$	$\left(\frac{7}{18} + \frac{1}{3} \ln \frac{\mu^2 m_c^2}{M^4} - \frac{2}{3} \gamma_E \right) \left(\frac{m_c^2}{M^2} - 1 \right) - \frac{2}{3} \frac{m_c^2}{M^2}$
Ref. [Hay00]	$2\left(\frac{m_c^2}{M^2} - 1 \right)$	$\frac{1}{3} \left[\frac{4}{3} - \frac{1}{6} \frac{m_c^2}{M^2} + \frac{1}{2} \frac{m_c^6}{M^6} + \left(1 - \frac{m_c^2}{M^2} \right) \ln \left(\frac{m_c^2}{4\pi\mu^2} \right) + \right.$ $\left. e^{m_c^2/M^2} \left(-2\gamma_E - \ln \frac{m_c^2}{M^2} + \int_0^{m_c^2/M^2} dt \frac{1-e^{-t}}{t} \right) \right]$

C.6 Recurrence relations

As stated in section C.4 the Wilson coefficients of condensates which differ only by the degree of the non-local quark field expansion, i. e. they emerge from the same order propagator and gluon field insertion, fulfill certain recursive relations. We illustrate this for the example of the first sum in (C.4.13), which corresponds to all diagrams with a free heavy-quark propagator and the light-quark line being cut in order to form a light quark condensate, i. e. to all orders of the expansion (C.1.12).

Let us start with the vacuum case, because there is no odd part. The first sum in (C.4.13) reduces to

$$\Pi^{(2)}(q) = \sum_{l=0}^{\infty} \left[\frac{(-i)^{2l}}{(2l)!} \langle : \left(\bar{d} \overleftarrow{D}_{\alpha_1} \ldots \overleftarrow{D}_{\alpha_{2l}} \right) d : \rangle \partial^{\alpha_1} \ldots \partial^{\alpha_{2l}} \frac{m_c}{q^2 - m_c^2} \right.$$
$$\left. - \frac{(-i)^{2l+1}}{(2l+1)!} \langle : \left(\bar{d} \gamma_\mu \overleftarrow{D}_{\alpha_1} \ldots \overleftarrow{D}_{\alpha_{2l+1}} \right) d : \rangle \partial^{\alpha_1} \ldots \partial^{\alpha_{2l+1}} \frac{q^\mu}{q^2 - m_c^2} \right]. \quad (C.6.1)$$

The projection of the Lorentz indices of the non-local condensates has the following well known form

$$\langle : \left(\bar{d} \overleftarrow{D}_{\alpha_1} \ldots \overleftarrow{D}_{\alpha_{2l}} \right) d : \rangle = a_1 g_{\alpha_1 \alpha_2} \ldots g_{\alpha_{2l-1} \alpha_{2l}} + \text{permutations of } \{\alpha_1, \ldots, \alpha_{2l}\}, \quad (C.6.2)$$

where the coefficients a_i can be found by solving the system of linear equations which emerges from contracting (C.6.2) with each product of metric tensors that appears on the r. h. s.. The key observation now is that inserting (C.6.2) into (C.6.1) gives the same result for all permutations in (C.6.2). One ends up with

C In-medium OPE for heavy-light mesons

$$\Pi^{(2)}(q) = \sum_{l=0}^{\infty} \left[\frac{(-i)^{2l}}{(2l)!} A_{2l} \left(\partial_\mu \partial^\mu \right)^l \frac{m_c}{q^2 - m_c^2} \right.$$
$$\left. - \frac{(-i)^{2l+1}}{(2l+1)!} B_{2l+1} \left(\partial_\mu \partial^\mu \right)^l \left(\frac{2}{q^2 - m_c^2} - \frac{2m_c^2}{(q^2 - m_c^2)^2} \right) \right], \quad (\text{C.6.3})$$

where we explicitly performed one partial derivative in the second term and A (B) denote the sum of the coefficients a_i (b_i). Furthermore, we employ the following identity

$$\partial_\mu \partial^\mu f(q^2) = \frac{\partial}{\partial q_\mu} \frac{\partial}{\partial q^\mu} f(q^2) = \left[8 \frac{\partial}{\partial q^2} + 4q^2 \left(\frac{\partial}{\partial q^2} \right)^2 \right] f(q^2) \equiv \bar{\partial} f(q^2) \quad (\text{C.6.4})$$

which can be proven for functions $f = f(q^2)$ by the chain rule. If we define the following sequence of functions

$$C_l(q^2) \equiv \bar{\partial}^l \frac{m_c}{q^2 - m_c^2} \equiv \bar{\partial}^l C_0(q^2) = \bar{\partial} C_{l-1}(q^2) \quad (\text{C.6.5})$$

and use

$$\bar{\partial} C_0^n(q^2) = \frac{4n(n-1)}{m_c} C_0^{n+1}(q^2) + 4n(n+1) C_0^{n+2}(q^2) \quad (\text{C.6.6})$$

we recognize that each function $C_l(q^2)$ must be a polynomial in $C_0(q^2)$:

$$C_l(q^2) = \sum_{n=l+1}^{2l+1} \alpha_n^{(l)} C_0^n(q^2). \quad (\text{C.6.7})$$

Application of $\bar{\partial}$ then gives the following recurrence relations for the coefficients $\alpha_n^{(l)}$

$$\alpha_n^{(l)} = 4(n-1)(n-2) \left(\frac{1}{m_c} \alpha_{n-1}^{(l-1)} + \alpha_{n-1}^{(l-2)} \right) \quad (\text{C.6.8})$$

with the boundary conditions $\alpha_n^{(n)} = 0$, $\alpha_n^{(2n+2)} = 0$ and the initial values $\alpha_1^{(0)} = 1$, $\alpha_l^{(0)} = 0 \, \forall l \neq 1$. Actually, if there is only one nonzero initial value satisfying the boundary conditions, the boundary conditions are automatically fulfilled for all $\alpha_n^{(l)}$.

C.6 Recurrence relations

In the same way we define

$$\tilde{C}_l(q^2) \equiv \bar{\partial}^l C_0^2(q^2) = \sum_{n=l+2}^{2l+2} \tilde{a}_n^{(l)} C_0^n(q^2) \tag{C.6.9}$$

and employ the initial values $\tilde{a}_2^{(0)} = 1$, $\tilde{a}_l^{(0)} = 0 \, \forall l \neq 2$, whereas the recurrence relations remain the same (of course, the boundary conditions change with changing initial values but remain similar to the first case and will be satisfied automatically).

Inserting these expressions into (C.6.3) we end up with

$$\Pi^{(2)}(q) = \sum_{l=0}^{\infty} \left[\frac{(-i)^{2l}}{(2l)!} A_{2l} \sum_{n=l+1}^{2l+1} a_n^{(l)} C_0^n(q^2) \right.$$
$$\left. - \frac{(-i)^{2l+1}}{(2l+1)!} B_{2l+1} \left(\frac{2}{m_c} \sum_{n=l+1}^{2l+1} a_n^{(l)} C_0^n(q^2) - 2 \sum_{n=l+2}^{2l+2} \tilde{a}_n^{(l)} C_0^n(q^2) \right) \right]. \tag{C.6.10}$$

This expression can easily be transformed into Euclidean space by virtue of $C_0^n(q^2) \to (-1)^n m_c^n/(q_E^2 + m_c^2)^n$. Finally, applying

$$\mathcal{B}\left[C_0^n(q^2)\right](M^2) = e^{-m_c^2/M^2} m_c^n/[(n-1)!(M^2)^{n-1}], \tag{C.6.11}$$

where the translation property of the Laplace transformation is employed, the Borel transformed result reads

$$\Pi^{(2)}(q) = e^{-m_c^2/M^2} \sum_{l=0}^{\infty} \left[\frac{(-i)^{2l}}{(2l)!} A_{2l} \sum_{n=l+1}^{2l+1} (-1)^n a_n^{(l)} \frac{m_c^n}{(n-1)!(M^2)^{n-1}} \right.$$
$$- \frac{(-i)^{2l+1}}{(2l+1)!} B_{2l+1} \left(\frac{2}{m_c} \sum_{n=l+1}^{2l+1} (-1)^n a_n^{(l)} \frac{m_c^n}{(n-1)!(M^2)^{n-1}} \right.$$
$$\left. \left. - 2 \sum_{n=l+2}^{2l+2} (-1)^n \tilde{a}_n^{(l)} \frac{m_c^n}{(n-1)!(M^2)^{n-1}} \right) \right]. \tag{C.6.12}$$

C In-medium OPE for heavy-light mesons

This can be written as

$$\Pi^{(2)}(q) \equiv e^{-m_c^2/M^2} \sum_{l=0}^{\infty} \left[\frac{(-i)^{2l}}{(2l)!} A_{2l} c_l(M^2) \right.$$
$$\left. - \frac{(-i)^{2l+1}}{(2l+1)!} B_{2l+1} \left(\frac{2}{m_c} c_l(M^2) - 2\tilde{c}_l(M^2) \right) \right]$$
$$\equiv e^{-m_c^2/M^2} \sum_{l=0}^{\infty} \left[\frac{(-i)^{2l}}{(2l)!} A_{2l} c_l(M^2) + \frac{(-i)^{2l+1}}{(2l+1)!} B_{2l+1} d_l(M^2) \right]. \quad \text{(C.6.13)}$$

Here, we defined $d_l(M^2) \equiv -2\left(\frac{c_l(M^2)}{m_c} - \tilde{c}_l(M^2) \right)$. Using the recurrence relations for the coefficients $\alpha_n^{(l)}$ ($\tilde{\alpha}_n^{(l)}$) and the respective boundary conditions, the following recurrence relations can be given for the coefficient functions defined above

$$c_{l+1}(M^2) = 4 \left[\frac{\partial}{\partial M^2} + \frac{m_c^2}{M^4} \right] c_l(M^2) = \frac{4}{M^4} \left[m_c^2 - \frac{\partial}{\partial 1/M^2} \right] c_l(M^2)$$
$$= \left(\frac{4}{M^4} \left[m_c^2 - \frac{\partial}{\partial 1/M^2} \right] \right)^{l+1} c_0(M^2) \quad \text{(C.6.14)}$$

for $c_l(M^2)$ as well as $\tilde{c}_l(M^2)$ and $d_l(M^2)$. The boundary conditions now read $c_0(M^2) = -m_c$, $\tilde{c}_l(M^2) = \frac{m_c^2}{M^2}$ and $d_0(M^2) = 2\left(1 + \frac{m_c^2}{M^2}\right)$.

In a similar way we proceed with the in-medium case. The first step is to separate the even (e) and odd (o) parts of the OPE:

$$\Pi^{(2)e}(q) = \sum_{l=0}^{\infty} \left[\frac{(-i)^{2l}}{(2l)!} \langle :(\bar{d}\overleftarrow{D}_{\alpha_1} \ldots \overleftarrow{D}_{\alpha_{2l}}) d: \rangle \partial^{\alpha_1} \ldots \partial^{\alpha_{2l}} \frac{m_c}{q^2 - m_c^2} \right.$$
$$\left. - \frac{(-i)^{2l+1}}{(2l+1)!} \langle :(\bar{d}\gamma_\mu \overleftarrow{D}_{\alpha_1} \ldots \overleftarrow{D}_{\alpha_{2l+1}}) d: \rangle \partial^{\alpha_1} \ldots \partial^{\alpha_{2l+1}} \frac{q^\mu}{q^2 - m_c^2} \right], \quad \text{(C.6.15a)}$$

$$\Pi^{(2)o}(q) = \sum_{l=0}^{\infty} \left[\frac{(-i)^{2l+1}}{(2l+1)!} \langle :(\bar{d}\overleftarrow{D}_{\alpha_1} \ldots \overleftarrow{D}_{\alpha_{2l+1}}) d: \rangle \partial^{\alpha_1} \ldots \partial^{\alpha_{2l+1}} \frac{m_c}{q^2 - m_c^2} \right.$$
$$\left. - \frac{(-i)^{2l}}{(2l)!} \langle :(\bar{d}\gamma_\mu \overleftarrow{D}_{\alpha_1} \ldots \overleftarrow{D}_{\alpha_{2l}}) d: \rangle \partial^{\alpha_1} \ldots \partial^{\alpha_{2l}} \frac{q^\mu}{q^2 - m_c^2} \right]. \quad \text{(C.6.15b)}$$

Concerning the projection of Lorentz indices of the condensates the main difference to the vacuum case is the appearance of another Lorentz structure which has to be included referring to the medium velocity v_μ. Hence, the general form of the

C.6 Recurrence relations

in-medium Lorentz projection reads

$$\langle:\left(\bar{d}\overleftarrow{D}_{\alpha_1}\ldots\overleftarrow{D}_{\alpha_{2l+1}}\right)d:\rangle = \sum_{m=1}^{2l+1}\sum_{\kappa\in K_m(\alpha)} a_\kappa v_{\kappa_1}\ldots v_{\kappa_m} g_{\sigma_1\sigma_2}\ldots g_{\sigma_{2l-m-1}\sigma_{2l-m}}, \quad (C.6.16)$$

where the first sum runs over the numbers of occurring factors v_μ starting from at least one factor in the case above and from $m = 0$ in case of an even number of indices on the l.h.s., the second sum runs over all permutations of the indices $\alpha \equiv \{\alpha_1,\ldots,\alpha_{2l+1}\}$, where m indices have been assigned to v, $\sigma \equiv \alpha/\{\kappa_1,\ldots,\kappa_m\}$ and a_κ is the corresponding coefficient. Of course, only coefficients which belong to an odd number of factors v can be nonzero in the above case.

Analog to the vacuum considerations, contraction of (C.6.16) with $\partial^{\alpha_1}\ldots\partial^{\alpha_{2l+1}}$ gives identical expressions for terms with the same number of factors v_μ:

$$\langle:\left(\bar{d}\overleftarrow{D}_{\alpha_1}\ldots\overleftarrow{D}_{\alpha_{2l+1}}\right)d:\rangle\partial^{\alpha_1}\ldots\partial^{\alpha_{2l+1}} = \sum_{n=0}^{l} A_n^o(v\partial)(v\partial)^{2n}\left(\partial_\mu\partial^\mu\right)^{l-n}, \quad (C.6.17a)$$

$$\langle:\left(\bar{d}\overleftarrow{D}_{\alpha_1}\ldots\overleftarrow{D}_{\alpha_{2l}}\right)d:\rangle\partial^{\alpha_1}\ldots\partial^{\alpha_{2l}} = \sum_{n=0}^{l} A_n^o(v\partial)^{2n}\left(\partial_\mu\partial^\mu\right)^{l-n}, \quad (C.6.17b)$$

where we defined the sum of coefficients which belong to the same number $(2n+1)$ of factors v_μ as $\sum_{\kappa\in K_m(\alpha)} a_\kappa \equiv A_n^o$. For the second expression it is clear, that it must contain an even number of factors $(v\partial)$.

Considering condensates which are projected onto γ_μ, i.e. $\langle:\left(\bar{d}\gamma_\mu\overleftarrow{D}_{\alpha_1}\ldots\overleftarrow{D}_{\alpha_k}\right)d:\rangle$, and contracted with $\partial^{\alpha_1}\ldots\partial^{\alpha_k}$, two different structures appear within each sum over κ. Depending on whether k is even or odd these are $(v\partial)^m\left(\partial^2\right)^{\frac{2l-m}{2}}\partial_\mu\frac{q^\mu}{q^2-m_c^2}$, for $k = 2l+1$, and $(v\partial)^{m-1}\left(\partial^2\right)^{\frac{2l-m+2}{2}}\frac{vq}{q^2-m_c^2}$, for $k = 2l$. Fortunately, we have $(v\partial)\partial_\mu\frac{q^\mu}{q^2-m_c^2} = \partial^2\frac{vq}{q^2-m_c^2}$ and the contracted partial derivatives applied to $\frac{q^\mu}{q^2-m_c^2}$ can again be factored out from the sum over all permutations κ with $m = 2n$ factors[33] of the medium velocity:

$$\langle:\left(\bar{d}\gamma_\mu\overleftarrow{D}_{\alpha_1}\ldots\overleftarrow{D}_{\alpha_{2l+1}}\right)d:\rangle\partial^{\alpha_1}\ldots\partial^{\alpha_{2l+1}}\frac{q^\mu}{q^2-m_c^2}$$

$$= \sum_{n=0}^{l+1} B_n^e (v\partial)^{2n}\left(\partial^2\right)^{l-n}\partial_\mu\frac{q^\mu}{q^2-m_c^2}, \quad (C.6.18a)$$

---

[33] $m = 2n + 1$ for an odd number of Lorentz indices, respectively

239

C In-medium OPE for heavy-light mesons

$$\langle :(\bar{d}\gamma_\mu \overleftarrow{D}_{\alpha_1} \ldots \overleftarrow{D}_{\alpha_{2l}})d:\rangle \partial^{\alpha_1} \ldots \partial^{\alpha_{2l}} \frac{q^\mu}{q^2 - m_c^2}$$
$$= \sum_{n=0}^{l} B_n^o (v\partial)^{2n+1} (\partial^2)^{l-n-2} \partial_\mu \frac{q^\mu}{q^2 - m_c^2}. \quad \text{(C.6.18b)}$$

Again, we defined $\sum_{\kappa \in K_m(\alpha)} b_\kappa^{(e,o)} \equiv B_m^{(e,o)}$. Unfortunately, we had to use the special form of the Dirac trace of the heavy quark propagator multiplied with γ_μ which occurs in this propagator order, whereas the considerations which led to (C.6.3) do not rely on this.

Putting all together, we arrive at

$$\Pi^{(2)e}(q) = \sum_{l=0}^{\infty} \left[\frac{(-i)^{2l}}{(2l)!} \sum_{n=0}^{l} A_n^e (v\partial)^{2n} (\partial_\mu \partial^\mu)^{l-n} C_0(q^2) \right.$$
$$\left. - \frac{(-i)^{2l+1}}{(2l+1)!} \sum_{n=0}^{l+1} B_n^e (v\partial)^{2n} (\partial_\mu \partial^\mu)^{l-n-2} 2\left(\frac{C_0(q^2)}{m_c} - C_0^2(q^2) \right) \right], \quad \text{(C.6.19a)}$$

$$q_0 \Pi^{(2)o}(q) = \sum_{l=0}^{\infty} \left[\frac{(-i)^{2l+1}}{(2l+1)!} \sum_{n=0}^{l} A_n^o (v\partial)^{2n+1} (\partial_\mu \partial^\mu)^{l-n} C_0(q^2) \right.$$
$$\left. - \frac{(-i)^{2l}}{(2l)!} \sum_{n=0}^{l} B_n^o (v\partial)^{2n+1} (\partial_\mu \partial^\mu)^{l-n-2} 2\left(\frac{C_0(q^2)}{m_c} - C_0^2(q^2) \right) \right]. \quad \text{(C.6.19b)}$$

One now easily recovers the functions $C_l(q^2)$ and $\tilde{C}_l(q^2)$, which are already known in the vacuum case. In order to proceed we make explicitly use of the rest frame of nuclear matter, $v = (1, \vec{0})$ and the considered meson at rest, $q = (q_0, \vec{0})$, to obtain

$$\partial_0^2 C_0^j(q_0^2) = \frac{2j(2j+1)}{m_c} C_0^{j+1}(q_0^2) + 4j(j+1) C_0^{j+2}(q_0^2), \quad \text{(C.6.20)}$$

leading to

$$\partial_0^{2i} C_0^j(q_0^2) \equiv K^{i,j}(q_0^2) = \sum_{k=j+1}^{j+2i} f_k^{ij} C_0^k(q_0^2) \quad \text{(C.6.21)}$$

with

$$f_k^{ij} = \frac{(2k-1)(2k-2)}{m_c} f_{k-1}^{i-1,j} + 4(k-1)(k-2) f_{k-2}^{i-1,j} \tag{C.6.22}$$

and initial and boundary conditions given by

$$f_j^{0,j} = 1, \quad f_k^{i,j} = 0 : \begin{cases} k \leq i+j-1, \\ 2i+j+1 \leq k. \end{cases} \tag{C.6.23}$$

Inserting (C.6.5) and (C.6.21) into (C.6.19) we obtain

$$\Pi^{(2)e}(q_0^2) = \sum_{l=0}^{\infty} \left[\frac{(-i)^{2l}}{(2l)!} \sum_{n=0}^{l} A_n^e \sum_{i=l-n+1}^{2(l-n)+1} \alpha_i^{(l-n)} \sum_{k=i+n}^{i+2n} f_k^{n,i} C_0^k(q_0^2) \right.$$
$$- \frac{(-i)^{2l+1}}{(2l+1)!} \sum_{n=0}^{l+1} 2B_n^e \left(\sum_{i=l-n-1}^{2(l-n)-3} \frac{\alpha_i^{(l-n-2)}}{m_c} \sum_{k=i+n}^{i+2n} f_k^{n,i} C_0^k(q_0^2) \right.$$
$$\left. \left. - \sum_{i=l-n}^{2(l-n)-2} \frac{\tilde{\alpha}_i^{(l-n-2)}}{m_c} \sum_{k=i+n}^{i+2n} f_k^{n,i} C_0^k(q_0^2) \right) \right], \tag{C.6.24a}$$

$$\Pi^{(2)o}(q) = -\sum_{l=0}^{\infty} \left[\frac{(-i)^{2l+1}}{(2l+1)!} \sum_{n=0}^{l} A_n^o \sum_{i=l-n+1}^{2(l-n)+1} \alpha_i^{(l-n)} \sum_{k=i+n}^{i+2n} f_k^{n,i} \frac{2k}{m_c} C_0^{k+1}(q_0^2) \right.$$
$$- \frac{(-i)^{2l}}{(2l)!} \sum_{n=0}^{l} 2B_n^o \left(\sum_{i=l-n-1}^{2(l-n)-3} \frac{\alpha_i^{(l-n-2)}}{m_c} \sum_{k=i+n}^{i+2n} f_k^{n,i} \frac{2k}{m_c} C_0^{k+1}(q_0^2) \right.$$
$$\left. \left. - \sum_{i=l-n}^{2(l-n)-2} \frac{\tilde{\alpha}_i^{(l-n-2)}}{m_c} \sum_{k=i+n}^{i+2n} f_k^{n,i} \frac{2k}{m_c} C_0^{k+1}(q_0^2) \right) \right], \tag{C.6.24b}$$

where we eliminated a factor q_0 in (C.6.24b). After a rather elaborate calculation, which proceeds analogously to the vacuum case, one can give the following expressions

$$\Pi^{(2)e}(q_0^2) = \sum_{l=0}^{\infty} \left[\frac{(-i)^{2l}}{(2l)!} \sum_{n=0}^{l} A_n^e c_{l,n}^e(M^2) \right.$$

C In-medium OPE for heavy-light mesons

$$\Pi^{(2)o}(q) = -\sum_{l=0}^{\infty}\left[\frac{(-i)^{2l+1}}{(2l+1)!}\sum_{n=0}^{l+1}2B_n^e\left(c_{l-2,n}^e(M^2) - \tilde{c}_{l-2,n}^e(M^2)\right)\right], \quad \text{(C.6.25a)}$$

$$\frac{(-i)^{2l+1}}{(2l+1)!}\sum_{n=0}^{l}A_n^o c_{l,n}^o(M^2)$$

$$-\frac{(-i)^{2l}}{(2l)!}\sum_{n=0}^{l}2B_n^o\left(c_{l-2,n}^o(M^2) - \tilde{c}_{l-2,n}^o(M^2)\right)\right], \quad \text{(C.6.25b)}$$

where the coefficient functions satisfy the recurrence relations

$$c_{l,n}^e(M^2) = \frac{4}{M^2}\left(\frac{3}{2} + \frac{m_c^2}{M^2} + \frac{1}{M^2}\frac{\partial}{\partial 1/M^2}\right)c_{l-1,n-1}^e(M^2), \quad \text{(C.6.26a)}$$

$$c_{l,n}^o(M^2) = \frac{4}{M^2}\left(\frac{1}{2} + \frac{m_c^2}{M^2} + \frac{1}{M^2}\frac{\partial}{\partial 1/M^2}\right)c_{l-1,n-1}^o(M^2) \quad \text{(C.6.26b)}$$

with initial conditions $c_{l,0}^e(M^2) = c_l^e(M^2)$, as well as $c_{l,0}^o(M^2) = \sum_{i=l+1}^{2l+1}\alpha_i^{(l)}\frac{m_c^i}{(i-1)!(M^2)^i}$ and $\tilde{c}_{l,0}^o(M^2) = \sum_{i=l+1}^{2l+1}\tilde{\alpha}_i^{(l)}\frac{m_c^i}{(i-1)!(M^2)^i}$, for both types of functions $c_{l,n}^{(e,o)}(M^2)$ and $\tilde{c}_{l,n}^{(e,o)}(M^2)$. The expressions for the initial conditions of the odd coefficient functions are the same as for the even coefficient functions. But because there are no odd condensates in vacuum the corresponding coefficient can be discarded.

C.7 Sum rule analysis for heavy-light mesons

Equation (3.2.1) is nonlinear in the masses $m_\pm$. In order to express $m_\pm$ in terms of the functions e and o one must assume that $m_\pm$ and $F_\pm$ are Borel mass independent, which allows to eliminate $F_\pm$. In vacuum, where $m = m_+ = m_-$, explicit equations for m and F can be derived immediately. In the medium, however, the nonlinear coupling of the equations necessitates more advanced solutions of Eq. (3.2.1). As discussed in Sec. 3.2 right below Eq. (3.2.1), approximate solutions for $m_\pm$ and $F_\pm$ may be obtained by assuming that mass-splitting and splitting of $F_\pm$ are small, as was done in [Hay00, Mor01, Mor99, Zsc06]. It turns out that such an approximation is not necessary and in the following a nonlinear coupled system of equations for $m_\pm$ is derived, which can in principle be solved numerically. Furthermore, we find an analytic solution of the nonlinear coupled system of equations giving explicit equations for $m_\pm$ and $F_\pm$. These equations will in turn be used to show the consistency of the

C.7 Sum rule analysis for heavy-light mesons

strategy which is applied to analyze the sum rule.

To begin with, Eq. (3.2.1) can be written as

$$(m_+ + m_-)F_+ e^{-m_+^2/M^2} = e(M) + m_- o(M), \tag{C.7.1a}$$
$$(m_+ + m_-)F_- e^{-m_-^2/M^2} = e(M) - m_+ o(M). \tag{C.7.1b}$$

Assuming $dm_\pm/dM = dF_\pm/dM = 0$ and taking a derivative of (C.7.1) w.r.t. $1/M^2$ and dividing each resulting equation by the corresponding equation of (C.7.1) eliminates $F_\pm$ and yields the following system of coupled, nonlinear equations for the masses $m_\pm$ in medium

$$-m_+^2 = \frac{\frac{d}{d1/M^2}e(M) + m_- \frac{d}{d1/M^2}o(M)}{e(M) + m_- o(M)}, \tag{C.7.2a}$$

$$-m_-^2 = \frac{\frac{d}{d1/M^2}e(M) - m_+ \frac{d}{d1/M^2}o(M)}{e(M) - m_+ o(M)}. \tag{C.7.2b}$$

As already stated, Eq. (C.7.2) uniquely determines $m_\pm$ and can in principle be solved numerically. However, the relation will be used to show the consistency of the employed evaluation strategy.

In the limit of vanishing density, the vacuum sum rules are deduced from Eq. (C.7.2). In this case, one has $o = 0$ and e reduces to its vacuum expression. From Eq. (3.2.1b) and the Borel mass independence of $m_\pm$ and $F_\pm$, $m_+ = m_-$ and $F_+ = F_-$ follows. The r.h.s. of Eq. (C.7.2) reduces to de/dM^{-2}, which is the known vacuum sum rule [Raf84].

Rewriting these equations as polynomials in the masses m_+ and m_-

$$0 = -m_+^2 e(M) - m_+^2 m_- o(M) - \frac{d}{d1/M^2}e(M) - m_- \frac{d}{d1/M^2}o(M), \tag{C.7.3a}$$

$$0 = -m_-^2 e(M) + m_-^2 m_+ o(M) - \frac{d}{d1/M^2}e(M) + m_+ \frac{d}{d1/M^2}o(M), \tag{C.7.3b}$$

subtracting (C.7.3b) from (C.7.3a) on the one hand and summing the products of (C.7.3a) with m_- and (C.7.3b) with m_+ on the other hand, one derives the following system of linear equations for Δm and $m_+ m_-$

$$0 = -2\Delta m\, e(M) - m_+ m_- o(M) - \frac{d}{d1/M^2}o(M), \tag{C.7.4a}$$

C In-medium OPE for heavy-light mesons

$$0 = -m_+ m_- e(M) + 2\Delta m \frac{\mathrm{d}}{\mathrm{d}1/M^2} o(M) - \frac{\mathrm{d}}{\mathrm{d}1/M^2} e(M). \tag{C.7.4b}$$

Here, the following quantities have been introduced

$$\Delta m = \frac{1}{2}(m_+ - m_-), \quad m = \frac{1}{2}(m_+ + m_-). \tag{C.7.5}$$

Therefore, one has

$$m_\pm = m \pm \Delta m, \quad m^2 = \Delta m^2 + m_+ m_- \tag{C.7.6}$$

and Eq. (C.7.4) can be solved w. r. t. Δm and $m_+ m_-$:

$$\Delta m = \frac{1}{2} \frac{o(M) \frac{\mathrm{d}}{\mathrm{d}1/M^2} e(M) - e(M) \frac{\mathrm{d}}{\mathrm{d}1/M^2} o(M)}{e^2(M) + o(M) \frac{\mathrm{d}}{\mathrm{d}1/M^2} o(M)}, \tag{C.7.7}$$

$$m_+ m_- = -\frac{e(M) \frac{\mathrm{d}}{\mathrm{d}1/M^2} e(M) + \left(\frac{\mathrm{d}}{\mathrm{d}1/M^2} o(M)\right)^2}{e^2(M) + o(M) \frac{\mathrm{d}}{\mathrm{d}1/M^2} o(M)}. \tag{C.7.8}$$

Given $m_\pm$, the strengths $F_\pm$ may be evaluated by virtue of Eq. (C.7.1)

$$F_\pm = e^{m_\pm^2/M^2} \frac{e \pm m_\mp o}{m_+ + m_-} \tag{C.7.9a}$$

$$= e^{(m \pm \Delta m)^2/M^2} \frac{e \pm (m \mp \Delta m)o}{2m}, \tag{C.7.9b}$$

which can be transformed into Eq. (3.2.3).

In deriving (3.2.2) we made use of $\mathrm{d}m_\pm/\mathrm{d}M = 0$. Therefore, it is convenient to adjust the continuum thresholds such that $m_\pm$ and $F_\pm$ are Borel mass independent. Unfortunately, varying the threshold parameters in order to require maximal flatness of the Borel curves leads to too small vacuum masses of $\approx 1.6\,\mathrm{GeV}$. Thresholds $s_0^2 = (s_0^\pm)^2 \approx 6\,\mathrm{GeV}^2$ are commonly chosen in vacuum [Nar01] and reproduce the vacuum D meson mass of $\approx 1.9\,\mathrm{GeV}$. Furthermore, a careful consideration of current and pole mass of the charm quark is necessary, cf. e. g. [Nar05, Nar04]. As the focus of this work is put on medium modifications, the center of the thresholds s_0^2, defined in Eq. (3.3.1), and the charm quark mass is chosen such that the vacuum mass is reproduced. To satisfy the requirement of Borel mass independence, the common evaluation strategy is modified. At finite densities, $(s_0^\pm)^2$ stays fixed to the vacuum value, while we demand that the minima of the respective Borel curves $m_+(M)$ and

C.8 Condensates near the deconfinement transition

$m_-(M)$ shall be at a common Borel mass M_0 by varying the splitting of the thresholds Δs_0^2. These minima are taken as the physical parameters $m_\pm$. As a byproduct, Δs_0^2 becomes density dependent.

It must now be clarified whether the strengths $F_\pm$ have their minima at the same Borel mass M_0, as demanded by the initial assumption. Indeed

$$\frac{d}{d1/M^2} F_\pm(M)\bigg|_{M=M_0} = e^{m_\pm^2/M^2} \frac{e' \pm o'}{m_+ + m_-} \left[m_\pm^2(M_0)\left(e \pm m_\mp o\right) + e'(M_0) \pm m_\mp o' \right]\bigg|_{M_0}, \qquad \text{(C.7.10)}$$

where a "prime" denotes a derivative w.r.t. M^{-2} and $m'_\pm|_{M_0} = 0$ has been used. Inserting Eq. (C.7.2) it follows that $F_\pm$ is extremal if $m_\pm$ is extremal.

C.8 Condensates near the deconfinement transition

As discussed in Secs. 1 and 2 the method of QSRs may, in principle, be applied to high densities and temperatures provided the medium dependence of the condensates is known. At finite temperatures and zero densities the in-medium condensates may be obtained from, e.g., lattice QCD. Also the Dyson-Schwinger approach is in principle capable of determining condensates at nonzero temperature and densities [Lan03, Zha05, Zon03a, Cha07]. However, there are many more possibilities, cf. e.g. [Jin93], and one of the most interesting one is given in [Coh92], which is used in the following to determine gluon condensates at large chemical potentials and temperatures next to the deconfinement transition.

According to the consideration in [Mor08] one may relate the gluon condensate at finite temperature to the QCD trace-anomaly (see Sec. A.2.13) for $N_f = 3$ flavors and the equation of state via

$$\langle \frac{\alpha_s}{\pi} G^2 \rangle_T = \langle \frac{\alpha_s}{\pi} G^2 \rangle_0 - \frac{8}{9}(e - 3p), \qquad \text{(C.8.1a)}$$

$$\langle \frac{\alpha_s}{\pi} \left(\frac{(vG)^2}{v^2} - \frac{G^2}{4} \right) \rangle_T = -\frac{3}{4} \frac{\alpha_s}{\pi}(e + p), \qquad \text{(C.8.1b)}$$

whereas contributions from light quarks to (C.8.1a) have been omitted in a first step, as we focus on the continuation to finite densities. The energy density e and the pressure p are accessible by QCD lattice evaluations [Baz09] at large temperatures but zero chemical potential. In [Mor08] these relations where used to analyze the charmonium

C In-medium OPE for heavy-light mesons

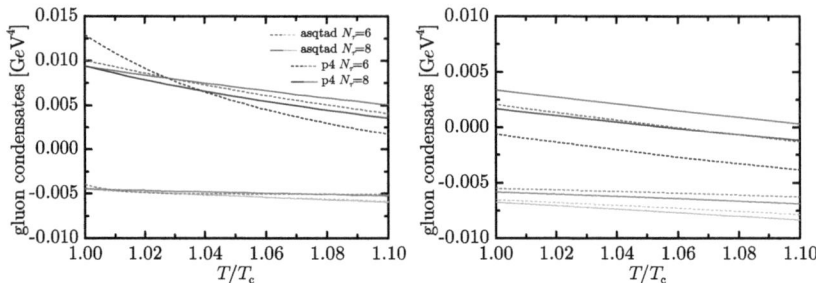

Figure C.8.1: Temperature dependence of the gluon condensates (C.8.1a) (upper curves) and (C.8.1b) (lower curves) at $\mu_q = 0$ (left panel) and $\mu_q = 270\,\text{MeV}$ (right panel, same line and color code as in left panel).

QSR above the critical temperature. This is possible because hadrons consisting of heavy quarks only, such as charmonium, only depend on gluon condensates, as heavy-quark condensates are supposed to be zero or at least small enough to be negligible or can be evaluated by virtue of the heavy-quark-mass expansion in terms of gluon condensates. Such quarks are called static quarks. However, energy density and pressure may be extrapolated to finite baryon densities by employing the Rossendorf quasiparticle model [Sch08, Sch09b, Sch09a] which is here adjusted to [Baz09]. Both condensates (C.8.1) are depicted in Fig. C.8.1 in the region near T_c at quark chemical potential $\mu_q = 0$ (left panel) and $\mu_q = 270\,\text{MeV}$ (right panel); energy density and pressure have been employed from the quasiparticle model which allows for a thermodynamically consistent extrapolation to finite densities. The curves for the condensate (C.8.1a) are flattened with increasing lattice temporal extend N_τ; [Mor08] is reproduced by the p4 action [Baz09] for $N_\tau = 6$. On the other hand, the condensate (C.8.1b) seems not to be affected by such a choice. At nonzero densities, the gluon condensate for $\langle \frac{\alpha_s}{\pi} G^2 \rangle_T$ drops significantly due to the non-vanishing chemical potential. The symmetric and traceless gluon condensate $\langle \frac{\alpha_s}{\pi} \left(\frac{(vG)^2}{v^2} - \frac{G^2}{4} \right) \rangle_T$ is much less influenced by density effects. These effects have to be studied in detail for the J/ψ, which depends essentially on the considered gluon condensates [Mor08].

As mentioned above (see also Eq. (A.2.13)), finite current quark masses modify the trace of the energy momentum tensor to [Coh92] ($N_f = 3$)

$$T_\mu^{\ \mu} = -\frac{9\alpha_s}{8\pi} G^A_{\mu\nu} G^{\mu\nu}_A + m_u \bar{u}u + m_d \bar{d}d + m_s \bar{s}s + \dots, \qquad (C.8.2)$$

where higher order α_s terms have been neglected. (The contributions of heavy

C.8 Condensates near the deconfinement transition

quarks can be absorbed into the first term by using the heavy-quark mass expansion.) A precise analysis would hence also have to include the density and temperature dependence of the quark condensates. However, as the chiral condensate is supposed to vanish above T_c due to the restoration of chiral symmetry, it is justified to neglect their influence on the gluon condensates in Eq. (C.8.1). Furthermore, if light quark contributions can be neglected in specific QSR evaluations due to their numerical suppression by light-quark mass terms, it is consistent with neglecting such terms in Eq. (C.8.1), too. This offers the avenue towards the evaluation of QSRs near/above the deconfinement transition at nonzero baryon densities.

D Chiral partner QCD sum rules addendum

D.1 Dirac projection

Consider the trace over Dirac indices $\text{Tr}_D[S_2\gamma_5\{S_1,\gamma_5\}]$, which naturally arises in $\Pi^{\text{P-S},(0)}$. To write this as a product of two simpler traces, one can project the commutator $\{S,\gamma_5\} = \sum_{\hat{O}} A_{\hat{O}} \hat{O}$ with coefficients $A_{\hat{O}} = \frac{1}{4}\text{Tr}_D[\{S,\gamma_5\}\hat{O}]$ and $\hat{O}$ being an element of the Clifford algebra. A similar expression applies for $\text{Tr}_D[S_2\gamma_5\{\gamma_\mu S_1 \gamma^\mu, \gamma_5\}]$, which is the corresponding expression in $\Pi^{\text{V-A},(0)}$, with coefficients listed in Tab. D.1.1. Alternatively, the coefficients for projecting $\Gamma^C S \Gamma_C = \sum_{\hat{O}} A_{\hat{O}} \hat{O}$, where $\Gamma^C \in \{\mathbb{1}, \gamma_5, \gamma_\mu, \gamma_5\gamma_\mu\}$, are given in Tab. D.1.2. Note the occurrence of γ_5 in the $A^{\mu<\nu}$ coefficient. Equation (4.1.5) can be derived by using

$$\gamma_5 \sigma^{\mu\nu} = \frac{i}{2}\epsilon_{\mu\nu\alpha\beta}\sigma^{\alpha\beta}, \tag{D.1.1}$$

which leads to

$$\text{Tr}_D[S_1\gamma_5\sigma_{\mu\nu}]\text{Tr}_D[S_2\gamma_5\sigma^{\mu\nu}]$$
$$= \left(\frac{i}{2}\right)^2 \epsilon_{\mu\nu}{}^{\alpha\beta}\epsilon^{\mu\nu\kappa\lambda}\text{Tr}_D[S_1\sigma_{\alpha\beta}]\text{Tr}_D[S_2\sigma_{\kappa\lambda}]$$

Table D.1.1: Coefficients of the projection of $\{S,\gamma_5\}$.

$A_{\hat{O}}$	$\{S,\gamma_5\}$	$\{\gamma_\mu S\gamma^\mu,\gamma_5\}$
$A_{\mathbb{1}}$	$\frac{1}{2}\text{Tr}_D[S\gamma_5]$	$-2\text{Tr}_D[S\gamma_5]$
A^ν_-	0	0
$A^{\mu<\nu}$	$\frac{1}{2}\text{Tr}_D[S\gamma_5\sigma^{\mu<\nu}]$	0
A^ν_+	0	0
A_5	$\frac{1}{2}\text{Tr}_D[S]$	$2\text{Tr}_D[S]$

D Chiral partner QCD sum rules addendum

Table D.1.2: Coefficients of the projection of $\Gamma^C S \Gamma_C$.

Γ^C	$\mathbb{1}$	γ_5	γ_μ	$\gamma_5\gamma_\mu$
A_1	$\frac{1}{4}\text{Tr}_D[S]$	$\frac{1}{4}\text{Tr}_D[S]$	$\text{Tr}_D[S]$	$-\text{Tr}_D[S]$
A_-^ν	$\frac{1}{4}\text{Tr}_D[S\gamma_\nu]$	$-\frac{1}{4}\text{Tr}_D[S\gamma_\nu]$	$-\frac{1}{2}\text{Tr}_D[S\gamma_\nu]$	$-\frac{1}{2}\text{Tr}_D[S\gamma_\nu]$
$A^{\mu<\nu}$	$\frac{1}{4}\text{Tr}_D[S\sigma^{\mu<\nu}]$	$\frac{1}{4}\text{Tr}_D[S\sigma^{\mu<\nu}]$	0	0
A_+^ν	$\frac{1}{4}\text{Tr}_D[S\gamma_5\gamma_\mu]$	$-\frac{1}{4}\text{Tr}_D[S\gamma_5\gamma_\mu]$	$\frac{1}{2}\text{Tr}_D[S\gamma_5\gamma_\mu]$	$\frac{1}{2}\text{Tr}_D[S\gamma_5\gamma_\mu]$
A_5	$\frac{1}{4}\text{Tr}_D[S\gamma_5]$	$\frac{1}{4}\text{Tr}_D[S\gamma_5]$	$-\text{Tr}_D[S\gamma_5]$	$\text{Tr}_D[S\gamma_5]$

$$= \left(\frac{i}{2}\right)^2 \left(-2\left(g^{\alpha\kappa}g^{\beta\lambda} - g^{\alpha\lambda}g^{\beta\kappa}\right)\right)\text{Tr}_D\left[S_1\sigma^{\alpha\beta}\right]\text{Tr}_D\left[S_2\sigma^{\kappa\lambda}\right]$$
$$= \text{Tr}_D\left[S_1\sigma_{\mu\nu}\right]\text{Tr}_D\left[S_2\sigma^{\mu\nu}\right]. \tag{D.1.2}$$

Equation (D.1.1) is obtained by expanding $\gamma_5\sigma^{\mu\nu}$ over the Clifford basis. Obviously, the only non-vanishing trace is that for $\sigma_{\alpha\beta}$:

$$\frac{1}{4}\text{Tr}_D\left[\gamma_5\sigma^{\mu\nu}\sigma^{\alpha\beta}\right] = \frac{1}{4}\left(\frac{i}{2}\right)^2 \text{Tr}_D\left[\gamma_5[\gamma^\mu,\gamma^\nu][\gamma^\alpha,\gamma^\beta]\right]$$
$$= \frac{1}{4}\left(\frac{i}{2}\right)^2 (-4i)\epsilon^{[\mu,\nu][\alpha,\beta]}$$
$$= \frac{i}{4}2^2\epsilon^{\mu\nu\alpha\beta}$$
$$= i\epsilon^{\mu\nu\alpha\beta}, \tag{D.1.3}$$

where the antisymmetry of the Levi-Cevita symbol has been used.

D.2 The perturbative quark propagator

According to Eq. (C.1.26), the perturbative quark propagator in momentum space in a weak (classical) gluonic background field can be written as $S(p) = \sum_{n=0}^{\infty} S^{(n)}(p)$ with $S^{(n)}(p) = -S^{(0)}(p) \times (\gamma\tilde{A}) S^{(n-1)}(p)$ and conventions given in Sec. C.1. By expanding the operator $-\Gamma S^{(0)}\gamma^\mu$ over the Clifford basis the following recursion relation is obtained for the traces of each propagator term

$$\text{Tr}_D\left[\Gamma S^{(n)}(p)\right] = -\frac{1}{4}\sum_{\Gamma'}\text{Tr}_D\left[\Gamma'\Gamma S^{(0)}(p)\gamma^\mu\right]\sum_k \left(\tilde{D}_{\bar{\alpha}_k}\mathcal{G}_{\mu\nu}(0)\right)\partial^\nu\partial^{\bar{\alpha}_k}\text{Tr}_D\left[\Gamma' S^{(n-1)}(p)\right]. \tag{D.2.1}$$

D.2 The perturbative quark propagator

For the sake of a concise notation we have defined $\tilde{D}_{\vec{\alpha}_k} \mathscr{G}_{\mu\nu} \equiv -g \frac{(-i)^{k+1}}{k!(k+2)} D_{\alpha_1} \times \ldots \times D_{\alpha_k} \mathscr{G}_{\mu\nu}\big|_{x=0}$ and $\partial^{\vec{\alpha}_k} \equiv \partial^{\alpha_1} \ldots \partial^{\alpha_k}$. Successive application of (D.2.1) reveals that the trace of each term $S^{(n)}(p)$ can be written as

$$\begin{aligned}
&\mathrm{Tr}_{\mathrm{D}}[\Gamma S^{(n)}(p)] \\
&= (-1)^n \left(\frac{1}{4}\right)^n \sum_{k_1,\ldots,k_n} \left(\tilde{D}_{\vec{\alpha}_{k_n}} \mathscr{G}_{\mu_n \nu_n}\right) \ldots \left(\tilde{D}_{\vec{\alpha}_{k_1}} \mathscr{G}_{\mu_1 \nu_1}\right) \sum_{\Gamma_1,\ldots,\Gamma_n} \mathrm{Tr}_{\mathrm{D}}[\Gamma_n \Gamma S^{(0)}(p) \gamma^{\mu_n}] \\
&\quad \times \left(\left(\partial^{\nu_n} \partial^{\vec{\alpha}_{k_n}} \mathrm{Tr}_{\mathrm{D}}[\Gamma_{n-1} \Gamma_n S^{(0)}(p) \gamma^{\mu_{n-1}}]\right) \ldots \ldots \left(\partial^{\nu_1} \partial^{\vec{\alpha}_{k_1}} \mathrm{Tr}_{\mathrm{D}}[\Gamma_1 S^{(0)}(p)]\right)\right).
\end{aligned}$$
(D.2.2)

The sum runs over elements of the Clifford basis. With these reduction formulas only traces of the free propagator $S^{(0)} = (\hat{p} + m)/(p^2 - m^2)$ need to be considered. Moreover, the single traces form a chain making the sum over Γ_i dependent on traces surviving from the sum over Γ_{i-1}. In the limit $m \to 0$ and $p^2 \neq 0$ we have $\mathrm{Tr}_{\mathrm{D}}[\Gamma_1 S^{(0)}(p, m=0)] = 0\ \forall \Gamma \neq \gamma_\mu$. Hence, from the last trace in Eq. (D.2.2) only γ_μ is passed to the next trace. For the latter one we therefore have $\mathrm{Tr}_{\mathrm{D}}[\Gamma_1 \Gamma_2 S^{(0)}(p, m=0)\gamma_\mu] = 0\ \forall \Gamma_2 \notin \{\gamma_\alpha, \gamma_5\gamma_\alpha\}$. As $\mathrm{Tr}_{\mathrm{D}}[\gamma_5\gamma_\alpha \Gamma_n S^{(0)}(p, m=0)\gamma_\mu] = 0\ \forall \Gamma_n \notin \{\gamma_\alpha, \gamma_5\gamma_\alpha\}$ also holds, each sum merely contains γ_μ and $\gamma_5\gamma_\mu$. Finally, we obtain for the first trace in the second line of Eq. (D.2.2) $\mathrm{Tr}_{\mathrm{D}}[\Gamma_n \Gamma S^{(0)}(p, m=0)\gamma_\mu] = 0\ \forall \Gamma \notin \{\gamma_\alpha, \gamma_5\gamma_\alpha\}$, which means that $\mathrm{Tr}_{\mathrm{D}}[\Gamma S^{(n)}(p, m=0)] = 0\ \forall \Gamma \notin \{\gamma_\mu, \gamma_5\gamma_\mu\}$. As this is true for all orders of the perturbative sum, we can conclude that $\mathrm{Tr}_{\mathrm{D}}[\Gamma S(p, m=0)] = 0\ \forall \Gamma \notin \{\gamma_\mu, \gamma_5\gamma_\mu\}$.

The same property follows from a different representation of the propagator traces (D.2.2). Upon projection of each $S(0)$ in $S^{(n)}$ onto elements of the Clifford base the trace reads

$$\begin{aligned}
&\mathrm{Tr}_{\mathrm{D}}\left[\Gamma S^{(n)}(p)\right] \\
&= \sum_{\Gamma_1,\ldots,\Gamma_n}^{1,\gamma} \mathrm{Tr}_{\mathrm{D}}\left[\Gamma S^{(0)} \gamma^{\mu_1} \Gamma_1 \ldots \Gamma_{n-1} \gamma^{\mu_n} \Gamma_n\right] \sum_{k_1,\ldots,k_n} \left[\tilde{D}_{\vec{\alpha}_{k_1}} \mathscr{G}_{\mu_1 \nu_1}\right] \ldots \left[\tilde{D}_{\vec{\alpha}_{k_n}} \mathscr{G}_{\mu_n \nu_n}\right] \\
&\quad \times \left(-\frac{1}{4}\right)^n \left(\partial^{\nu_1} \partial^{\vec{\alpha}_{k_1}} \frac{\mathrm{Tr}_{\mathrm{D}}\left[\Gamma^1 (\hat{p} + m)\right]}{p^2 - m^2}\right) \ldots \left(\partial^{\nu_n} \partial^{\vec{\alpha}_{k_n}} \frac{\mathrm{Tr}_{\mathrm{D}}\left[\Gamma^n (\hat{p} + m)\right]}{p^2 - m^2}\right).
\end{aligned}$$
(D.2.3)

For $m = 0$ and $p^2 \neq 0$, only γ_μ remains in the sum over the Clifford base. Thus, only for $\Gamma \in \{\gamma_\mu, \gamma_5\gamma_\mu\}$ there is an even number of Dirac matrices in the trace of the first

D Chiral partner QCD sum rules addendum

line of (D.2.3). For $\Gamma \notin \{\gamma_\mu, \gamma_5\gamma_\mu\}$ the trace is zero.

D.3 Cancellation of IR divergences

This section documents the cancellation of infrared divergences for chiral partner OPEs in Eq. (4.1.6). For the contributions to (4.1.6a) up to and including mass dimension 5, which restricts the quark propagator to next-to-next-to-leading order and the gluon field to lowest order (cf. App. C.1 Eq. (C.1.35)), the following traces are evaluated

$$\mathrm{Tr}_D\left[S^{(0)}(p)\right] = \frac{4m}{p^2 - m^2}, \tag{D.3.1a}$$

$$\mathrm{Tr}_D\left[S^{(0)}(p)\gamma_5\sigma_{\mu\nu}\right] = 0, \tag{D.3.1b}$$

$$\mathrm{Tr}_D\left[S^{(0)}(p)\gamma_5\right] = 0, \tag{D.3.1c}$$

$$\mathrm{Tr}_D\left[S^{(1)}(p)\right] = 0, \tag{D.3.1d}$$

$$\mathrm{Tr}_D\left[S^{(1)}(p)\gamma_5\sigma_{\mu\nu}\right] = -ig\frac{2m}{(p^2-m^2)^2}\epsilon_{\mu\nu\kappa\lambda}\mathscr{G}^{\kappa\lambda}, \tag{D.3.1e}$$

$$\mathrm{Tr}_D\left[S^{(1)}(p)\gamma_5\right] = 0, \tag{D.3.1f}$$

$$\mathrm{Tr}_D\left[S^{(2)}(p)\right] = 8g^2\frac{mp_\mu p_\alpha}{(p^2-m^2)^4}\mathscr{G}^{\mu\nu}\mathscr{G}^\alpha_\nu. \tag{D.3.1g}$$

It is not necessary to consider traces of the second order quark propagator with γ_5 and $\sigma_{\mu\nu}$: Up to mass dimension 5 they can only be multiplied with their lowest order counterparts (D.3.1c) and (D.3.1b), which are zero. A combination of second and first order propagator leads to mass dimension 6 terms. Using (D.1.2) the following contributions have to be considered

$$\Pi^{P-S(0)}(q) = -i\int\frac{d^4p}{(2\pi)^4}\langle:\frac{1}{2}\mathrm{Tr}_C\Big\{\mathrm{Tr}_D[S_c^{(0)}(p+q)]\mathrm{Tr}_D[S_d^{(0)}(p)]$$
$$+ \mathrm{Tr}_D[S_c^{(2)}(p+q)]\mathrm{Tr}_D[S_d^{(0)}(p)] + \mathrm{Tr}_D[S_c^{(0)}(p+q)]\mathrm{Tr}_D[S_d^{(2)}(p)]$$
$$+ \frac{1}{2}\mathrm{Tr}_D[S_c^{(1)}(p+q)\sigma_{\mu\nu}]\mathrm{Tr}_D[S_d^{(1)}(p)\sigma^{\mu\nu}]\Big\}:\rangle. \tag{D.3.2}$$

Due to $\mathrm{Tr}_D[S^{(0)}\gamma_5] = \mathrm{Tr}_D[S^{(1)}\gamma_5] = 0$ for arbitrary quark masses there is no γ_5 contribution up to this mass dimension. Likewise $\mathrm{Tr}_D[S^{(0)}\gamma_5\sigma_{\mu\nu}] = \mathrm{Tr}_D[S^{(1)}] = 0$.

D.3 Cancellation of IR divergences

Lorentz invariance requires that terms which contain only one gluon field are zero. Therefore, a first order quark propagator must be combined with a propagator of at least the same order. Keeping both masses finite these four terms give rise to the following integral

$$\Pi^{P-S(0)}(q) = -i \int \frac{d^4p}{(2\pi)^4} \langle : \frac{1}{2} \frac{4m_c}{(p+q)^2 - m_c^2} \frac{4m_d}{p^2 - m_d^2} \text{Tr}_C[\mathbb{1}_C]$$
$$+ \frac{g^2}{2} \frac{8m_c}{[(p+q)^2 - m_c^2]^4} \frac{4m_d}{p^2 - m_d^2} p^\mu p^\alpha G^A_{\mu\nu} G^B{}_\alpha{}^\nu \text{Tr}_C[t^A t^B]$$
$$+ \frac{g^2}{2} \frac{4m_c}{(p+q)^2 - m_c^2} \frac{8m_d}{[p^2 - m_d^2]^4} p^\mu p^\alpha G^A_{\mu\nu} G^B{}_\alpha{}^\nu \text{Tr}_C[t^A t^B]$$
$$+ \frac{g^2}{4} \frac{4m_c}{[(p+q)^2 - m_c^2]^2} \frac{4m_d}{[p^2 - m_d^2]^2} G^A_{\mu\nu} G^{B\mu\nu} \text{Tr}_C[t^A t^B] : \rangle. \quad \text{(D.3.3)}$$

Analyzing the integral in Euclidean space reveals the following results in terms of the integral (C.4.6). The first term is $\propto m_d m_c I_0^{0,0}(q^2, m_d^2, m_c^2)$. The second gives rise to two terms which are $\propto m_d m_c \times I_2^{0,3}(q^2, m_d^2, m_c^2)$ and $\propto m_d m_c I_3^{2,3}(q^2, m_d^2, m_c^2)$, respectively. The last term is $\propto m_d m_c I_2^{1,1}(q^2, m_d^2, m_c^2)$. These terms are all zero in the limit $m_d \to 0$. On the other hand, the third term does not vanish for $m_d \to 0$. It gives rise to a term $\propto m_d m_c I_2^{3,0}(q^2, m_d^2, m_c^2)$, which diverges with m_d^{-1}. Using (C.3.6) the required projection of the gluon fields reads

$$\langle : \delta^{AB} G^A_{\mu\nu} G^B{}_\alpha{}^\nu : \rangle = \frac{g^{\mu\alpha}}{4} \langle : G^2 : \rangle - \frac{1}{3}\left(g^{\mu\alpha} - 4\frac{v^\mu v^\alpha}{v^2}\right) \langle : \left(\frac{(vG)^2}{v^2} - \frac{G^2}{4}\right) : \rangle. \quad \text{(D.3.4)}$$

Note that the diagonal elements of (D.3.4) are vacuum specific, whereas the medium specific contribution is traceless. The integral of the third term of Eq. (D.3.2) can be evaluated in Euclidean space:

$$\int \frac{d^4p}{(2\pi)^4} \frac{m_d m_c}{[(p-q)^2 - m_d^2]^4} \frac{(p-q)_\mu (p-q)_\nu}{p^2 - m_c^2}$$
$$\xrightarrow{\text{W.R.}} \frac{2}{3} \frac{m_c m_d}{(4\pi)^2} \left(\frac{g_{\mu\alpha}}{2} I_{3,0,2}(q^2, m_d^2, m_c^2) + 2q_\mu q_\alpha I_{3,2,3}(q^2, m_d^2, m_c^2)\right). \quad \text{(D.3.5)}$$

Here, the relation $I_{5,0,3} - 2I_{4,0,3} + I_{3,0,3} = I_{3,2,3}$ has been used and "W.R." indicates that a Wick rotation has been performed. By virtue of $I_{ijk}(q^2, m_d^2, m_c^2) \to I_{jik}(q^2, m_d^2, m_c^2)$ the corresponding integral of the second term in Eq. (D.3.3) can be derived. The limit

D Chiral partner QCD sum rules addendum

$m_d \to 0$ for both terms in Eq. (D.3.5) is

$$\lim_{m_d \to 0} m_d I_{3,0,2}(q^2, m_d^2, m_c^2) = \frac{1}{m_d} \frac{1}{q^2 + m_c^2}, \quad \text{(D.3.6a)}$$

$$\lim_{m_d \to 0} m_d I_{3,2,3}(q^2, m_d^2, m_c^2) = 0, \quad \text{(D.3.6b)}$$

where Eq. (D.3.6a) reveals the famous infrared singularity from vacuum D meson sum rules (cf. Eq. (C.4.11)). As the medium specific contribution to (D.3.4) is traceless with respect to Lorentz indices and, due to the vanishing of the second term in Eq. (D.3.5), there is no medium specific infrared divergent term. Hence, the only terms that have to be absorbed into the condensates by virtue of the introduction of non-normal ordered condensates are vacuum specific. This is in line with the cancellation of $\langle : \bar{d}\gamma_\mu D_\nu d : \rangle$ in Eq. (4.1.6) which would have to absorb the medium specific divergences. Moreover, infrared divergent terms which enter through the medium specific gluon condensate, therefore, must be part of the γ_μ or $\gamma_5 \gamma_\mu$ parts of the quark propagator.

Owing to Eqs. (D.3.4), (D.3.5) and (D.3.6) the limit $m_d \to 0$ of (D.3.2) in Euclidean space is

$$\Pi^{\text{P-S}(0)}(q) = -\frac{i}{2} \int \frac{d^4p}{(2\pi)^4} \langle : \text{Tr}_D [S_c^{(0)}(p+q)] \text{Tr}_{C,D}[S_d^{(2)}(p)] : \rangle \quad \text{(D.3.7)}$$

$$\xrightarrow{\text{W.R.}} -\frac{i}{6} i \langle : \frac{\alpha_s}{\pi} G^2 : \rangle I_{3,0,2}(q^2, m_d^2, m_c^2) m_c m_d \quad \text{(D.3.8)}$$

$$= -\frac{i}{6} i \langle : \frac{\alpha_s}{\pi} G^2 : \rangle \frac{m_c}{m_d} \frac{1}{q^2 + m_c^2}. \quad \text{(D.3.9)}$$

To arrive at the first equality we use the fact that the lowest order quark propagator is a unit in color space and the additional imaginary unit stems from the Wick rotation.

The normal ordered chiral condensate in lowest order of the light quark mass enters via the following expression

$$\Pi^{(2)}_{\langle \bar{d}d \rangle}(q) = \frac{1}{2} \langle : \bar{d}d : \rangle \text{Tr}_D [S_c(q)]. \quad \text{(D.3.10)}$$

Here, $\text{Tr}_D [S_c(q)]$ is the Wilson coefficient of the chiral condensate. Expressing the normal ordered condensate by the non-normal ordered condensate (cf. Eq. (C.5.1))

$$\langle : \bar{d}d : \rangle = \langle \bar{d}d \rangle + i \int \frac{d^4p}{(2\pi)^4} \langle : \text{Tr}_{C,D} [S_d(p)] : \rangle \quad \text{(D.3.11)}$$

D.3 Cancellation of IR divergences

leads to

$$\Pi^{(2)}_{\langle \bar{d}d \rangle}(q) = \frac{1}{2}\langle \bar{d}d \rangle \text{Tr}_\text{D}\left[S_c(q)\right] + \frac{1}{2}\text{Tr}_\text{D}\left[S_c(q)\right] i \int \frac{d^4p}{(2\pi)^4} \langle :\text{Tr}_{\text{C,D}}\left[S_d(p)\right]:\rangle. \quad \text{(D.3.12)}$$

Despite the striking similarity, revealed by the projection onto elements of the Clifford basis in Eqs. (4.1.5) and (4.1.6), of the third term in Eq. (D.3.2) which has to be canceled and the additional term in Eq. (D.3.12), a general proof cannot be given and the necessity for a precise evaluation is obvious. Upon insertion of (C.5.1) and in the limit $m_d \to 0$ the anticipated cancellation of infrared divergences in terms of light quark masses is revealed

$$\Pi^{(2)}_{\langle \bar{d}d \rangle}(q) \xrightarrow{\text{W.R.}} \frac{1}{2}\langle \bar{d}d \rangle \text{Tr}_\text{D}\left[S_c(q)\right] - \frac{1}{6}\frac{m_c}{m_d}\frac{1}{q^2 + m_c^2}\langle :\frac{\alpha_s}{\pi}G^2:\rangle . \quad \text{(D.3.13)}$$

Adding (D.3.6) and (D.3.13) the infrared divergent term cancels out. Furthermore, Eq. (4.1.6) and the explicit evaluation [Hil08] of the renormalization of normal ordered condensates shows that, in chiral partner sum rules up to and including mass dimension 5 in the limit $m_d \to 0$, only the chiral condensate mixes with other condensates by virtue of introducing non-normal ordered condensates. Thus, apart from the term which cancels the infrared divergence no additional chirally even terms enter and the final OPE is chirally odd.

So far, the investigation was carried out for the spin-0 case. Fortunately, as the only formal difference between the spin-0 and spin-1 cases is the cancellation of the projections of quark propagators onto the tensor $\sigma_{\mu\nu}$ in the spin-1 case, see Eqs. (4.1.5) and (4.1.6), from the previous evaluation it is clear that the terms of interest are the same in both cases up to mass dimension 5.

Acronyms

BJL	Bjorken-Johnson-Low
BSA	Bethe-Salpeter amplitude
BSE	Bethe-Salpeter equation
BSM	Bethe-Salpeter matrix
BSV	Bethe-Salpeter vertex function
DCSB	dynamical/spontaneous chiral symmetry breaking
DSE	Dyson-Schwinger equation
EoM	equations of motion
ETC	equal-time commutator
ETCCR	equal-time current commutation relation
OPE	operator product expansion
QCD	quantum chromodynamics
QSR	QCD sum rule
VOC	vanishing of chirally odd condensates

List of Figures

1.0.1	Illustrative picture depicting hadrons in the constituent quark model	2
1.2.1	Illustrative picture of a D^- meson in vacuum and placed in nuclear matter	7
1.3.1	Representation of some QCD condensates by Feynman diagrams	9
1.5.1	Expected spectral changes of D mesons in a cold nuclear medium	16
2.0.1	Schematic view of a spectral density	20
2.1.1	Integration contours in the complex plane for the derivation of dispersion relations	25
2.2.1	Graphical visualization of an OPE for a current-current correlation function	32
3.3.1	$D-\bar{D}$ mass pattern for $\langle q^\dagger g\sigma\mathcal{G}q\rangle = +0.33\,\text{GeV}^2\,n$	44
3.3.2	$D-\bar{D}$ strength pattern	44
3.3.3	$D-\bar{D}$ mass pattern for $\langle q^\dagger g\sigma\mathcal{G}q\rangle = -0.33\,\text{GeV}^2\,n$	45
3.3.4	$D-\bar{D}$ Borel curves $m_\pm(M)$ for different densities and chiral condensates	46
3.3.5	$D_s-\bar{D}_s$ mass pattern	47
3.3.6	Mass-width correlation for $D-\bar{D}$ mesons at finite densities	48
3.3.7	Low density approximation of the mass splitting as function of the Borel mass for different splittings of the thresholds	51
3.3.8	Low density approximation of the mass splitting as function of the threshold splitting at $M=1.37\,\text{GeV}$	52
3.3.9	$D^*-\bar{D}^*$ mass pattern	52
3.4.1	$B-\bar{B}$ mass pattern	54
4.1.1	$\Pi^{(0)}$ and $\Pi^{(2)}$ OPE contributions	58
4.2.1	OPE for chiral partner sum rules as a function of the Borel mass in vacuum and at nuclear saturation density	67

List of Figures

5.2.1	ρ meson Borel curves for various thresholds in vacuum and for the VOC scenario	75
5.2.2	ρ meson Borel curves with maximal flatness within the Borel window in vacuum and for the VOC scenario	75
5.3.1	Vacuum: $\langle \mathcal{O}_4^{\text{even}} \rangle$ from moment analysis; VOC: parametrization (i) . .	82
5.3.2	Vacuum: $\langle \mathcal{O}_4^{\text{even}} \rangle$ fitted to $\Gamma_0 = 149.4\,\text{MeV}$ and $\overline{m_0} = 775.5\,\text{MeV}$ for parametrization (i); VOC: parametrization (i)	83
5.3.3	Vacuum: $\langle \mathcal{O}_4^{\text{even}} \rangle$ fitted to $\Gamma_0 = 149.4\,\text{MeV}$ and $\overline{m_0} = 775.5\,\text{MeV}$ for parametrization (ii); VOC: parametrization (i)	85
5.3.4	Vacuum: $\langle \mathcal{O}_4^{\text{even}} \rangle$ fitted to $\Gamma_0 = 149.4\,\text{MeV}$, $m_{\text{peak}} = 775.5\,\text{MeV}$ for parametrization (ii); VOC: parametrization (i)	86
5.3.5	Comparison of experimental ρ spectral function to various fits and QSR results	87
5.3.6	Comparison of experimental ρ spectral function to the best fit and the corresponding QSR result for fitted $\langle \mathcal{O}_4^{\text{even}} \rangle$	89
5.3.7	VOC scenario for $\langle \mathcal{O}_4^{\text{even}} \rangle$ obtained from experimental data and (iii): $m_{\text{peak}}(\Gamma_{\text{FWHM}})$ and $\overline{F_0}(\Gamma_0)$	90
5.3.8	VOC scenario for $\langle \mathcal{O}_4^{\text{even}} \rangle$ obtained from experimental data and (iii): $s_+(\Gamma_0)$ and $\text{Im}\Pi(s)$	90
6.1.1	The quark Dyson-Schwinger equation	97
6.1.2	The quark Dyson-Schwinger equation in rainbow approximation . .	99
6.1.3	The first iteration of the quark Dyson-Schwinger equation in rainbow approximation	100
6.1.4	A selection of diagrams for radiative corrections of the quark propagator which contribute to the solution of the Dyson-Schwinger equation in rainbow approximation	100
6.2.1	The exact four-point function	100
6.2.2	Perturbative expansion of the two-particle propagator	102
6.2.3	Two-particle irreducibility	102
6.2.4	DSE for the exact four-point function	103
6.2.5	Iterated solution of the DSE for the four-point function	103
6.2.6	Typical contributions to the homogeneous BSE in ladder approximation	104
6.2.7	A diagram contributing to the BSE beyond the ladder approximation	105
6.2.8	Diagrammatic representation for the BSE in ladder approximation .	106
7.1.1	Solutions of the DSE	116

List of Figures

7.1.2	Comparison of dynamically generated masses	117
7.1.3	Solutions of the DSE in the Nambu-Goldstone and Wigner-Weyl phase for zero bare quark mass	117
7.1.4	Real and imaginary part of the quark wave function A in the complex plane	118
7.1.5	Real and imaginary part of B in the complex plane	119
7.1.6	Real and imaginary part of the quark mass function M along different rays in the complex plane	120
7.1.7	Real and imaginary part of the quark wave function A along different rays in the complex plane	120
7.1.8	Parabolic integration contour in the complex plane	122
7.1.9	The roots of $\sigma_{v,s}^{-1}$ in the complex plane	127
7.1.10	Propagator functions $\sigma_{v,s}$ in the complex plane	127
7.2.1	Parabolic integration domain and poles of DSE and BSE for asymmetric momentum partition	132
7.2.2	Maximal values of χ as a function of the bare quark mass	132
7.2.3	Pseudo-scalar and scalar meson masses	135
7.3.1	Comparison of scalar and pseudo-scalar g_{1j} and g_{2j} partial amplitudes	137
7.3.2	Comparison of scalar and pseudo-scalar g_{3j} and g_{4j} partial amplitudes	138
7.3.3	Propagator functions in the Wigner-Weyl phase up to the critical mass	139
7.3.4	Bound state mass in the Wigner-Weyl phase up to the critical mass .	140
C.1.1	Perturbative quark propagator in a weak glounic background field .	210
C.4.1	Feynman diagrams for the perturbative contribution $\Pi^{per}(q)$ to the D meson OPE	226
C.4.2	Feynman diagrams for the gluon condensate contribution $\Pi^{G^2}(q)$ to the D meson OPE	229
C.4.3	Feynman diagrams for the quark and mixed quark-gluon condensate contribution $\Pi^{(2)}(q)$ to the D meson OPE	230
C.8.1	Temperature dependence of the gluon condensates at densities and temperatures near the phase transition	246

List of Tables

3.3.1 List of employed in-medium condensate parameters 43

5.3.1 Results of the finite width VOC scenario 81

7.1.1 Position of the propagator poles and maximum values of χ 126
7.2.1 Commutators and anticommutators of the elements of the base defined in (7.2.4) with $\hat{q}$ 129
7.2.2 Maximal bound state masses and corresponding momentum partitioning parameter 133

A.1.1 Transformation of meson currents under infinitesimal axial and vector transformations 159

B.2.1 Symmetry properties of the correlation functions 191
B.3.1 Contraction table of the projectors 192

C.5.1 Comparison of vacuum Wilson coefficients with the literature 234
C.5.2 Comparison of the medium specific Wilson coefficients with the literature 235

D.1.1 Coefficients of the projection of $\{S, \gamma_5\}$ 249
D.1.2 Coefficients of the projection of $\Gamma^C S \Gamma_C$ 250

Bibliography

[Abr72] M. ABRAMOWITZ AND I. A. STEGUN: *Handbook of Mathematical functions*, US Department of Commerce, 1972.

[Ack99] K. ACKERSTAFF ET AL.: *Measurement of the strong coupling constant α_s and the vector and axial-vector spectral functions in hadronic tau decays*, Eur. Phys. J. C **7**, 571 (1999).

[Ada08] D. ADAMOVA ET AL.: *Modification of the rho meson detected by low-mass electron-positron pairs in central Pb-Au collisions at 158 A GeV/c*, Phys. Lett. B **666**, 425 (2008).

[Adl68] S. ADLER AND R. F. DASHEN: *Current Algebras and Applications to Particle Physics*, W. A. Benjamin, Inc., 1968.

[ALI] ALICE Collaboration, *A Large Ion Collider Experiment*, http://aliceinfo.cern.ch. http://cms.web.cern.ch/news/cms-search-standard-model-higgs-boson-lhc-data-2010-and-2011 .

[Ali83] T. M. ALIEV AND V. L. ELETSKY: *On Leptonic Decay Constants of Pseudoscalar D and B Mesons*, Sov. J. Nucl. Phys. **38**, 936 (1983), [Yad. Fiz. **38**, 1537-1541 (1983)].

[Alk02] R. ALKOFER, P. WATSON AND H. WEIGEL: *Mesons in a Poincare covariant Bethe-Salpeter approach*, Phys. Rev. D **65**, 094026 (2002).

[Ams08] C. AMSLER ET AL.: *Review of particle physics*, Phys. Lett. B **667**, 1 (2008).

[Arn06] R. ARNALDI ET AL.: *First measurement of the rho spectral function in high-energy nuclear collisions*, Phys. Rev. Lett. **96**, 162302 (2006).

[Arn09] R. ARNALDI ET AL.: *NA60 results on thermal dimuons*, Eur. Phys. J. C **61**, 711 (2009).

[Asa01]   M. Asakawa, T. Hatsuda and Y. Nakahara: *Maximum entropy analysis of the spectral functions in lattice QCD*, Prog. Part. Nucl. Phys. **46**, 459 (2001).

[ATL]   ATLAS Collaboration, *A Toroidal LHC ApparatuS,* http://www.atlas.ch/. http://www.atlas.ch/news/2011/status-report-dec-2011.html.

[Bar69]   W. A. Bardeen: *Anomalous Ward identities in spinor field theories*, Phys. Rev. **184**, 1848 (1969).

[Baz09]   A. Bazavov, T. Bhattacharya, M. Cheng, N. H. Christ, C. DeTar et al.: *Equation of state and QCD transition at finite temperature*, Phys. Rev. D **80**, 014504 (2009).

[BD71]   Q. Bui-Duy: *Canonical formalism and schwinger terms*, Lett. Nuovo Cim. **2**, 515 (1971).

[Bec74]   C. Becchi, A. Rouet and R. Stora: *The Abelian Higgs-Kibble Model. Unitarity of the S Operator*, Phys. Lett. B **52**, 344 (1974).

[Bec75]   C. Becchi, A. Rouet and R. Stora: *Renormalization of the Abelian Higgs-Kibble Model*, Commun. Math. Phys. **42**, 127 (1975).

[Ber87]   V. Bernard, U. G. Meissner and I. Zahed: *Properties of the scalar sigma meson at finite density*, Phys. Rev. Lett. **59**, 966 (1987).

[Ber88]   R. A. Bertlmann, C. A. Dominguez, M. Loewe, M. Perrottet and E. de Rafael: *Determination of the Gluon Condensate and the Four Quark Condensate via FESR*, Z. Phys. C **39**, 231 (1988).

[Bic06]   P. Bicudo: *Mesons and tachyons with confinement and chiral restoration*, Phys. Rev. D **74**, 065001 (2006).

[Bir96]   M. C. Birse and B. Krippa: *Chiral symmetry and QCD sum rules for the nucleon mass in matter*, Phys. Lett. B **381**, 397 (1996).

[Bjo65]   J. D. Bjorken and S. D. Drell: *Relativistic quantum fields*, McGraw-Hill, 1965.

[Bjo66]   J. D. Bjorken: *Applications of the Chiral U(6) × U(6) Algebra of Current Densities*, Phys. Rev. **148**, 1467 (1966).

[Boc84] A. I. BOCHKAREV AND M. E. SHAPOSHNIKOV: *The first order quark - hadron phase transition in QCD*, Phys. Lett. B **145**, 276 (1984).

[Boc86] A. I. BOCHKAREV AND M. E. SHAPOSHNIKOV: *Spectrum of the Hot Hadronic Matter and Finite Temperature QCD Sum Rules*, Nucl. Phys. B **268**, 220 (1986).

[Bor05] B. BORASOY, R. NISSLER AND W. WEISE: *Chiral dynamics of kaon nucleon interactions, revisited*, Eur. Phys. J. A **25**, 79 (2005).

[Bor06] J. BORDES, C. A. DOMINGUEZ, J. PENARROCHA AND K. SCHILCHER: *Chiral condensates from tau decay: A critical reappraisal*, JHEP **02**, 037 (2006).

[Bou70] D. G. BOULWARE AND S. DESER: *Classical Schwinger terms*, Commun. Math. Phys. **19**, 327 (1970).

[Boy00] J. P. BOYD: *Chebyshev and Fourier Spectral Methods*, Dover Publications, 2000.

[Bra68] R. A. BRANDT: *Approach to Equal-Time Commutators in Quantum Field Theory*, Phys. Rev. **166**, 1795 (1968).

[Bra93] V. M. BRAUN, P. GORNICKI, L. MANKIEWICZ AND A. SCHÄFER: *Gluon formfactor of the proton from QCD sum rules*, Phys. Lett. B **302**, 291 (1993).

[Bro81] D. J. BROADHURST: *Chiral Symmetry Breaking and Perturbative QCD*, Phys. Lett. B **101**, 423 (1981).

[Bro91] G. E. BROWN AND M. RHO: *Scaling effective Lagrangians in a dense medium*, Phys. Rev. Lett. **66**, 2720 (1991).

[Bro92] B. BROWN: *Quantum field theory*, Cambridge University Press, 1992.

[Bro01] I. N. BRONSTEIN AND K. A. SEMENDJAJEW: *Taschenbuch der Mathematik*, Harri Deutsch, 2001.

[Bur00] C. P. BURGESS: *Goldstone and pseudo-Goldstone bosons in nuclear, particle and condensed matter physics*, Phys. Rept. **330**, 193 (2000).

[Cao10] T. Y. CAO: *From Current Algebra to Quantum Chromodynamics*, Cambridge University Press, 2010.

[CBM]   CBM Collaboration, *Compressed Baryonic Matter Experiment*, http://www.fair-center.eu/public/experiment-program/nuclear-matter-physics/cbm.html.

[Cha70]   M. S. CHANOWITZ: *Schwinger terms in fermion electrodynamics*, Phys. Rev. D **2**, 3016 (1970).

[Cha93]   G. CHANFRAY AND P. SCHUCK: *The Rho meson in dense matter and its influence on dilepton production rates*, Nucl. Phys. A **555**, 329 (1993).

[Cha07]   L. CHANG, H. CHEN, B. WANG, F.-Y. HOU, Y.-X. LIU ET AL.: *Chemical potential dependence of chiral quark condensate in Dyson-Schwinger equation approach*, Phys. Lett. B **644**, 315 (2007).

[Che82a]   T.-P. CHENG AND L.-F. LI: *Gauge Theory of Elementary Particle Physics*, Clarendon Press, Oxford, 1982.

[Che82b]   K. G. CHETYRKIN, F. V. TKACHOV AND S. G. GORISHNII: *Operator product expansion in the minimal subtraction scheme*, Phys. Lett. B **119**, 407 (1982).

[Che95]   K. G. CHETYRKIN, C. A. DOMINGUEZ, D. PIRJOL AND K. SCHILCHER: *Mass singularities in light quark correlators: The Strange quark case*, Phys. Rev. D **51**, 5090 (1995).

[Chi77]   S. A. CHIN: *A Relativistic Many Body Theory of High Density Matter*, Annals Phys. **108**, 301 (1977).

[Coh92]   T. D. COHEN, R. J. FURNSTAHL AND D. K. GRIEGEL: *Quark and gluon condensates in nuclear matter*, Phys. Rev. C **45**, 1881 (1992).

[Coh95]   T. D. COHEN, R. J. FURNSTAHL, D. K. GRIEGEL AND X.-M. JIN: *QCD sum rules and applications to nuclear physics*, Prog. Part. Nucl. Phys. **35**, 221 (1995).

[Col84]   J. C. COLLINS: *Renormalization*, Cambridge University Press, 1984.

[Col00]   P. COLANGELO AND A. KHODJAMIRIAN: *QCD sum rules, a modern perspective*, in M. SHIFMAN (Editor), *Boris Ioffe Festschrift 'At the Frontier of Particle Physics / Handbook of QCD'*, World Scientific, 2000.

[Col01]   G. COLANGELO, J. GASSER AND H. LEUTWYLER: *The quark condensate from K_{e_4} decays*, Phys. Rev. Lett. **86**, 5008 (2001).

[Cro00] P. CROCHET ET AL.: *Sideward flow of K^+ mesons in Ru + Ru and Ni + Ni reactions near threshold*, Phys. Lett. B **486**, 6 (2000).

[Cut54] R. E. CUTKOSKY: *Solutions of a Bethe-Salpeter equations*, Phys. Rev. **96**, 1135 (1954).

[Cut60] R. E. CUTKOSKY: *Singularities and discontinuities of Feynman amplitudes*, J. Math. Phys. **1**, 429 (1960).

[Das67] T. DAS, V. S. MATHUR AND S. OKUBO: $A_1 \to \rho\pi$ *and* $\rho \to 2\pi$ *decays*, Phys. Rev. Lett. **19**, 1067 (1967).

[Das97] A. K. DAS: *Finite temperature field theory*, World Scientific, 1997.

[Dja08] C. DJALALI, M. WOOD, R. NASSERIPOUR AND D. WEYGAND: *Vector meson modification in nuclear matter at CLAS*, Mod. Phys. Lett. A **23**, 2417 (2008).

[Doi04] T. DOI, N. ISHII, M. OKA AND H. SUGANUMA: *Thermal effects on quark gluon mixed condensate* $g\langle\bar{q}\sigma_{\mu\nu}G_{\mu\nu}q\rangle$ *from lattice QCD*, Phys. Rev. D **70**, 034510 (2004).

[Dom88] C. A. DOMINGUEZ AND J. SOLA: *Determination of quark and gluon vacuum condensates from tau lepton decay data*, Z. Phys. C **40**, 63 (1988).

[Dom09] C. A. DOMINGUEZ, N. F. NASRALLAH AND K. SCHILCHER: *Confronting QCD with the experimental hadronic spectral functions from tau-decay*, Phys. Rev. D **80**, 054014 (2009).

[Don96] S. J. DONG, J. F. LAGAE AND K. F. LIU: $\pi N \sigma$ *Term , $\bar{s}s$ in Nucleon, and Scalar Form Factor — a Lattice Study*, Phys. Rev. D **54**, 5496 (1996).

[Dor08] S. M. DORKIN, M. BEYER, S. S. SEMIKH AND L. KAPTARI: *Two-Fermion Bound States within the Bethe-Salpeter Approach*, Few Body Syst. **42**, 1 (2008).

[Dor10] S. M. DORKIN, T. HILGER, B. KÄMPFER AND L. P. KAPTARI: *A Combined Solution of the Schwinger-Dyson and Bethe-Salpeter Equations for Mesons as $q\bar{q}$ Bound States*, arxiv:nucl-th/1012.5372 (2010).

[Dor11] S. M. DORKIN, T. HILGER, L. P. KAPTARI AND B. KÄMPFER: *Heavy pseudoscalar mesons in a Schwinger-Dyson–Bethe-Salpeter approach*, Few Body Syst. **49**, 247 (2011).

[Dos89] H. G. Dosch, M. Jamin and S. Narison: *Baryon masses and flavor symmetry breaking of chiral condensates*, Phys. Lett. B **220**, 251 (1989).

[Dru91] E. G. Drukarev and E. M. Levin: *Structure of nuclear matter and QCD sum rules*, Prog. Part. Nucl. Phys. **27**, 77 (1991).

[Dys49] F. J. Dyson: *The S matrix in quantum electrodynamics*, Phys. Rev. **75**, 1736 (1949).

[Ele93] V. L. Eletsky: *Four quark condensates at $T \neq 0$*, Phys. Lett. B **299**, 111 (1993).

[Erd55a] A. Erdelyi: *Higher Transcendental functions, Volume II*, McGraw-Hill, 1955.

[Erd55b] A. Erdelyi: *Higher Transcendental functions, Volume III*, McGraw-Hill, 1955.

[FAI] *Facility for Antiproton and Ion Research,* http://www.fair-center.eu.

[Fet71] A. L. Fetter and J. D. Walecka: *Quantum Theory of Many-Particle Systems*, McGraw-Hill, 1971.

[Fis05] C. S. Fischer, P. Watson and W. Cassing: *Probing unquenching effects in the gluon polarisation in light mesons*, Phys. Rev. D **72**, 094025 (2005).

[Fis08] C. S. Fischer and R. Williams: *Beyond the rainbow: Effects from pion back-coupling*, Phys. Rev. D **78**, 074006 (2008).

[Fis09a] C. S. Fischer, D. Nickel and R. Williams: *On Gribov's supercriticality picture of quark confinement*, Eur. Phys. J. C **60**, 47 (2009).

[Fis09b] C. S. Fischer and R. Williams: *Probing the gluon self-interaction in light mesons*, Phys. Rev. Lett. **103**, 122001 (2009).

[För07] A. Förster, F. Uhlig, I. Bottcher, D. Brill, M. Debowski et al.: *Production of K^+ and of K^- Mesons in Heavy-Ion Collisions from 0.6 to 2.0-A-GeV Incident Energy*, Phys. Rev. C **75**, 024906 (2007).

[Fra96] M. R. Frank and C. D. Roberts: *Model gluon propagator and pion and rho meson observables*, Phys. Rev. C **53**, 390 (1996).

[Fri97] B. Friman and H. J. Pirner: *P-wave polarization of the ρ meson and the dilepton spectrum in dense matter*, Nucl. Phys. A **617**, 496 (1997).

[Fri11] B. Friman, (ed.) et al.: *The CBM physics book: Compressed baryonic matter in laboratory experiments*, Lect. Notes Phys. **814**, 1 (2011).

[Fuj87] K. Fujikawa: *Schwinger terms and Slavnov-Taylor identities*, Phys. Lett. B **188**, 115 (1987).

[Fur90] R. J. Furnstahl, T. Hatsuda and S. H. Lee: *Applications of QCD sum rules at finite temperature*, Phys. Rev. D **42**, 1744 (1990).

[Fur92] R. J. Furnstahl, D. K. Griegel and T. D. Cohen: *QCD sum rules for nucleons in nuclear matter*, Phys. Rev. C **46**, 1507 (1992).

[Fur96] R. J. Furnstahl, X.-M. Jin and D. B. Leinweber: *New QCD sum rules for nucleons in nuclear matter*, Phys. Lett. B **387**, 253 (1996).

[Gal91] C. Gale and J. I. Kapusta: *Vector dominance model at finite temperature*, Nucl. Phys. B **357**, 65 (1991).

[Gen84] S. C. Generalis and D. J. Broadhurst: *The heavy quark mass expansion and QCD sum rules for light quarks*, Phys. Lett. B **139**, 85 (1984).

[Ger89] P. Gerber and H. Leutwyler: *Hadrons Below the Chiral Phase Transition*, Nucl. Phys. B **321**, 387 (1989).

[Gim91] V. Gimenez, J. Bordes and J. Penarrocha: *A method to calculate ratios among QCD condensates*, Nucl. Phys. B **357**, 3 (1991).

[GM51] M. Gell-Mann and F. Low: *Bound states in quantum field theory*, Phys. Rev. **84**, 350 (1951).

[GM68] M. Gell-Mann, R. J. Oakes and B. Renner: *Behavior of current divergences under SU(3) × SU(3)*, Phys. Rev. **175**, 2195 (1968).

[Gol50] H. Goldstein: *Classical Mechanics*, Addison-Wesley, 1950.

[Gol61] J. Goldstone: *Field Theories with Superconductor Solutions*, Nuovo Cim. **19**, 154 (1961).

[Gol62] J. Goldstone, A. Salam and S. Weinberg: *Broken Symmetries*, Phys. Rev. **127**, 965 (1962).

[Gor83] S. G. Gorishnii, S. A. Larin and F. V. Tkachov: *The algorithm for OPE coefficient functions in the MS scheme*, Phys. Lett. B **124**, 217 (1983).

[Got55] T. Goto and T. Imamura: *Note on the Non-Perturbation-Approach to Quantum Field Theory*, Progr. Theor. Phys. **3**, 296 (1955).

[Gre92] W. Greiner and J. Reinhardt: *Quantum Electrodynamics*, Springer, 1992.

[Gre96] W. Greiner and J. Reinhardt: *Field quantization*, Springer, 1996.

[Gro73] D. Gross and F. Wilczek: *Ultraviolet Behavior of Nonabelian Gauge Theories*, Phys. Rev. Lett. **30**, 1343 (1973).

[Gro95] A. G. Grozin: *Methods of calculation of higher power corrections in QCD*, Int. J. Mod. Phys. A **10**, 3497 (1995).

[Gub10] P. Gubler and M. Oka: *A Bayesian approach to QCD sum rules*, Prog. Theor. Phys. **124**, 995 (2010).

[Gub11] P. Gubler, K. Morita and M. Oka: *Charmonium spectra at finite temperature from QCD sum rules with the maximum entropy method*, Phys. Rev. Lett. **107**, 092003 (2011).

[Ham30] A. Hammerstein: *Nichtlineare Integralgleichungen nebst Anwendungen*, Acta Math. **54**, 117 (1930).

[Har03] M. Harada and K. Yamawaki: *Hidden local symmetry at loop: A New perspective of composite gauge boson and chiral phase transition*, Phys. Rept. **381**, 1 (2003).

[Hat92] T. Hatsuda and S. H. Lee: *QCD sum rules for vector mesons in nuclear medium*, Phys. Rev. C **46**, 34 (1992).

[Hat93] T. Hatsuda, Y. Koike and S.-H. Lee: *Finite temperature QCD sum rules reexamined: ρ, ω and A_1 mesons*, Nucl. Phys. B **394**, 221 (1993).

[Hat95] T. Hatsuda, S. H. Lee and H. Shiomi: *QCD sum rules, scattering length and the vector mesons in nuclear medium*, Phys. Rev. C **52**, 3364 (1995).

[Hat98] B. Hatfield: *Quantum Field Theory of Point Particles and Strings*, Westview Press, 1998.

[Hay00] A. Hayashigaki: *Mass modification of D meson at finite density in QCD sum rule*, Phys. Lett. B **487**, 96 (2000).

[Hay04]   A. HAYASHIGAKI AND K. TERASAKI: *Charmed-meson spectroscopy in QCD sum rule*, arxiv:hep-ph/0411285 (2004).

[Hay10]   R. S. HAYANO AND T. HATSUDA: *Hadron properties in the nuclear medium*, Rev. Mod. Phys. **82**, 2949 (2010).

[Hay12]   T. HAYATA: *New QCD sum rules based on canonical commutation relations*, Prog. Part. Nucl. Phys. **67**, 136 (2012).

[Hee06]   H. VAN HEES AND R. RAPP: *Comprehensive interpretation of thermal dileptons at the SPS*, Phys. Rev. Lett. **97**, 102301 (2006).

[Hee08]   H. VAN HEES AND R. RAPP: *Dilepton Radiation at the CERN Super Proton Synchrotron*, Nucl. Phys. A **806**, 339 (2008).

[Hel67]   W. S. HELLMAN AND P. ROMAN: *Schwinger terms from local currents*, Il Nuovo Cim. A Series 10 **52**, 1341 (1967).

[Hel95]   J. HELGESSON AND J. RANDRUP: *Dilepton production from pion annihilation in a realistic delta hole model*, Phys. Rev. C **52**, 427 (1995).

[Her93]   M. HERRMANN, B. L. FRIMAN AND W. NÖRENBERG: *Properties of ρ mesons in nuclear matter*, Nucl. Phys. A **560**, 411 (1993).

[Hil08]   T. HILGER: *QCD sum rules for D mesons in nuclear matter*, Diploma Thesis, Technische Universität Dresden (2008).

[Hil09]   T. HILGER, R. THOMAS AND B. KÄMPFER: *QCD sum rules for D and B mesons in nuclear matter*, Phys. Rev. C **79**, 025202 (2009).

[Hil10a] T. HILGER AND B. KÄMPFER: *Chiral symmetry and open-charm mesons*, in *XLVII International Winter Meeting on Nuclear Physics*, volume 99 of *Conference Proceedings*, Italien Physical Society, 2010.

[Hil10b] T. HILGER AND B. KÄMPFER: *In-Medium Modifications of Scalar Charm Mesons in Nuclear Matter*, Nucl. Phys. Proc. Suppl. **207-208**, 277 (2010).

[Hil10c] T. HILGER, R. SCHULZE AND B. KÄMPFER: *QCD sum rules for D mesons in dense and hot nuclear matter*, J. Phys. G **37**, 094054 (2010).

[Hil11]   T. HILGER, B. KÄMPFER AND S. LEUPOLD: *Chiral QCD sum rules for open charm mesons*, Phys. Rev. C **84**, 045202 (2011).

[Hil12a]   T. HILGER, T. BUCHHEIM, B. KÄMPFER AND S. LEUPOLD: *Four-quark condensates in open-charm chiral QCD sum rules*, Prog. Part. Nucl. Phys. **67**, 188 (2012).

[Hil12b]   T. HILGER, R. THOMAS, B. KÄMPFER AND S. LEUPOLD: *The impact of chirally odd condensates on the rho meson*, Phys. Lett. B **709**, 200 (2012).

[Hin95]   E. J. HINCH: *Perturbation methods*, Cambridge University Press, 1995.

[Hsi92]   A. HSIEH AND E. YEHUDAI: *HIP: Symbolic high-energy physics calculations*, Comput. Phys. **6**, 253 (1992).

[Ioa91]   N. I. IOAKIMIDIS, K. E. PAPADAKIS AND E. A. PERDIOS: *Numerical evaluation of analytic functions by Cauchy's theorem*, BIT Numer. Math. **31**, 276 (1991).

[Iof81]   B. L. IOFFE: *Calculation of Baryon Masses in Quantum Chromodynamics*, Nucl. Phys. B **188**, 317 (1981).

[Iof06]   B. L. IOFFE: *Axial anomaly: the modern status*, Int. J. Mod. Phys. A **21**, 6249 (2006).

[Isg89]   N. ISGUR AND M. B. WISE: *Weak Decays of Heavy Mesons in the Static Quark Approximation*, Phys. Lett. B **232**, 113 (1989).

[Itz80]   C. ITZYKSON AND J. B. ZUBER: *Quantum Field Theory*, McGraw-Hill, 1980.

[Jai07]   B. JAIN, I. MITRA AND H. SHARATCHANDRA: *Criterion for dynamical chiral symmetry breaking*, AIP Conf. Proc. **939**, 355 (2007).

[Jam93]   M. JAMIN AND M. MÜNZ: *Current correlators to all orders in the quark masses*, Z. Phys. C **60**, 569 (1993).

[Jin93]   X.-M. JIN, T. D. COHEN, R. J. FURNSTAHL AND D. K. GRIEGEL: *QCD sum rules for nucleons in nuclear matter. 2*, Phys. Rev. C **47**, 2882 (1993).

[Jin94]   X.-M. JIN, M. NIELSEN, T. D. COHEN, R. J. FURNSTAHL AND D. K. GRIEGEL: *QCD sum rules for nucleons in nuclear matter. 3.*, Phys. Rev. C **49**, 464 (1994).

[Joh66]   K. JOHNSON AND F. E. LOW: *Current algebras in a simple model*, Prog. Theor. Phys. Suppl. **37**, 74 (1966).

[Kai95]   N. KAISER, P. B. SIEGEL AND W. WEISE: *Chiral dynamics and the $S_11(1535)$ nucleon resonance*, Phys. Lett. B **362**, 23 (1995).

[Kai01]   N. KAISER AND W. WEISE: *Systematic calculation of s-wave pion and kaon self- energies in asymmetric nuclear matter*, Phys. Lett. B **512**, 283 (2001).

[Kai08]   N. KAISER, P. DE HOMONT AND W. WEISE: *In-medium chiral condensate beyond linear density approximation*, Phys. Rev. C **77**, 025204 (2008).

[Kak93]   M. KAKU: *Quantum Field Theory*, Oxford University Press, 1993.

[Kap94]   J. I. KAPUSTA AND E. V. SHURYAK: *Weinberg type sum rules at zero and finite temperature*, Phys. Rev. D **49**, 4694 (1994).

[Kap06]   J. I. KAPUSTA AND C. GALE: *Finite-temperature field theory: Principles and applications*, Cambridge University Press, 2006.

[Kit86]   C. KITTEL: *Introduction to Solid State Physics*, John Wiley & Sons, 1986.

[Käl52]   G. KÄLLÉN: *On the definition of the Renormalization Constants in Quantum Electrodynamics*, Helv. Phys. Acta **25**, 417 (1952).

[Kli96]   F. KLINGL AND W. WEISE: *Spectral distributions of current correlation functions in baryonic matter*, Nucl. Phys. A **606**, 329 (1996).

[Kli97]   F. KLINGL, N. KAISER AND W. WEISE: *Current correlation functions, QCD sum rules and vector mesons in baryonic matter*, Nucl. Phys. A **624**, 527 (1997).

[Käm10]   B. KÄMPFER, T. HILGER, H. SCHADE, R. SCHULZE AND G. WOLF: *Chiral symmetry, di-electrons and charm*, PoS **BORMIO 2010**, 045 (2010).

[Koc95]   V. KOCH: *Introduction to Chiral Symmetry*, arXiv:nucl-th/9512029 (1995).

[Koc97]   V. KOCH: *Aspects of chiral symmetry*, Int. J. Mod. Phys. E **6**, 203 (1997).

[Kol04]   E. E. KOLOMEITSEV AND M. F. M. LUTZ: *On baryon resonances and chiral symmetry*, Phys. Lett. B **585**, 243 (2004).

[Kon06]   Y. KONDO, O. MORIMATSU AND T. NISHIKAWA: *Coupled QCD sum rules for positive and negative-parity nucleons*, Nucl. Phys. A **764**, 303 (2006).

[Kra08]   A. KRASSNIGG: *Excited mesons in a Bethe-Salpeter approach*, PoS **Confinement 8**, 075 (2008).

[Kub72]   J. J. KUBIS: *Partial-wave analysis of spinor bethe-salpeter equations for single-particle exchange*, Phys. Rev. D **6**, 547 (1972).

[Kug91]   T. KUGO: *Basic concepts in dynamical symmetry breaking and bound state problems*, in Nagoya 1991, Proceedings, Dynamical symmetry breaking, 1991.

[Kug97]   T. KUGO: *Eichtheorie*, Springer, 1997.

[Kwo08]   Y. KWON, M. PROCURA AND W. WEISE: *QCD sum rules for rho mesons in vacuum and in-medium, re- examined*, Phys. Rev. C **78**, 055203 (2008).

[Lan86]   L. D. LANDAU AND E. M. LIFSCHITZ: *Lehrbuch der Theoretischen Physik 4: Quantenelektrodynamik*, Akademie-Verlag, 1986.

[Lan03]   K. LANGFELD, H. MARKUM, R. PULLIRSCH, C. D. ROBERTS AND S. M. SCHMIDT: *Concerning the quark condensate*, Phys. Rev. C **67**, 065206 (2003).

[Lau84]   G. LAUNER, S. NARISON AND R. TARRACH: *Nonperturbative QCD vacuum from $e^+ e^- \to I = 1$ hadrons data*, Z. Phys. C **26**, 433 (1984).

[LE07]   F. J. LLANES-ESTRADA, T. VAN CAUTEREN AND A. P. MARTIN: *Fermion family recurrences in the Dyson-Schwinger formalism*, Eur. Phys. J. C **51**, 945 (2007).

[Leh54]   H. LEHMANN: *On the Properties of propagation functions and renormalization constants of quantized fields*, Nuovo Cim. **11**, 342 (1954).

[Lei97]   D. B. LEINWEBER: *QCD sum rules for skeptics*, Annals Phys. **254**, 328 (1997).

[Leu98a]   S. LEUPOLD AND U. MOSEL: *On QCD sum rules for vector mesons in nuclear medium*, Phys. Rev. C **58**, 2939 (1998).

[Leu98b]   S. LEUPOLD, W. PETERS AND U. MOSEL: *What QCD sum rules tell about the rho meson*, Nucl. Phys. A **628**, 311 (1998).

[Leu01]   S. LEUPOLD: *QCD sum rule analysis for light vector and axial-vector mesons in vacuum and nuclear matter*, Phys. Rev. C **64**, 015202 (2001).

[Leu04]   S. LEUPOLD: *Rho meson properties from combining QCD-based models*, Nucl. Phys. A **743**, 283 (2004).

[Leu05a] S. LEUPOLD: *Factorization and non-factorization of in-medium four-quark condensates*, Phys. Lett. B **616**, 203 (2005).

[Leu05b] S. LEUPOLD AND M. POST: *QCD sum rules at finite density in the large N_c limit: The coupling of the rho nucleon system to the D_{13}(1520)*, Nucl. Phys. A **747**, 425 (2005).

[Leu06a] S. LEUPOLD: *Fate of QCD sum rules or fate of vector meson dominance in a nuclear medium*, arXiv:hep-ph/0604055 (2006).

[Leu06b] S. LEUPOLD: *Four-quark condensates and chiral symmetry restoration in a resonance gas model*, J. Phys. G **32**, 2199 (2006).

[Leu07]   S. LEUPOLD: *Selfconsistent approximations, symmetries and choice of representation*, Phys. Lett. B **646**, 155 (2007).

[Leu09a] S. LEUPOLD: *Information on the structure of the rho meson from the pion form-factor*, Phys. Rev. D **80**, 114012 (2009).

[Leu09b] S. LEUPOLD, M. F. M. LUTZ AND M. WAGNER: *Chiral Partners and their Electromagnetic Radiation – Ingredients for a systematic in-medium calculation*, Porg. Part. Nucl. Phys. **62**, 305 (2009).

[Leu10]   S. LEUPOLD, V. METAG AND U. MOSEL: *Hadrons in strongly interacting matter*, Int. J. Mod. Phys. E **19**, 147 (2010).

[LS88]   C. H. LLEWELLYN SMITH AND J. P. DE VRIES: *The operator product expansion for minimally subtracted operators*, Nucl. Phys. B **296**, 991 (1988).

[Lur68]   D. LURIÉ: *Particles and Fields*, Interscience Publishers, 1968.

[Lut00]   M. F. M. LUTZ, B. FRIMAN AND C. APPEL: *Saturation from nuclear pion dynamics*, Phys. Lett. B **474**, 7 (2000).

[Lut02]   M. F. M. LUTZ, G. WOLF AND B. FRIMAN: *Scattering of vector mesons off nucleons*, Nucl. Phys. A **706**, 431 (2002).

[Lut04]   M. F. M. LUTZ AND E. E. KOLOMEITSEV: *On meson resonances and chiral symmetry*, Nucl. Phys. A **730**, 392 (2004).

[Lut06]   M. F. M. LUTZ AND C. L. KORPA: *Open-charm systems in cold nuclear matter*, Phys. Lett. B **633**, 43 (2006).

[Man55]   S. MANDELSTAM: *Dynamical variables in the Bethe-Salpeter formalism*, Proc. Roy. Soc. Lond. A **233**, 248 (1955).

[Mar92]   P. MARIS AND H. A. HOLTIES: *Determination of the singularities of the Dyson-Schwinger equation for the quark propagator*, Int. J. Mod. Phys. A **7**, 5369 (1992).

[Mar97]   P. MARIS AND C. D. ROBERTS: *π- and K-meson Bethe-Salpeter amplitudes*, Phys. Rev. C **56**, 3369 (1997).

[Mar99]   P. MARIS AND P. C. TANDY: *Bethe-Salpeter study of vector meson masses and decay constants*, Phys. Rev. C **60**, 055214 (1999).

[Mar03]   P. MARIS AND C. D. ROBERTS: *Dyson-Schwinger equations: A Tool for hadron physics*, Int. J. Mod. Phys. E **12**, 297 (2003).

[Mar06]   P. MARIS AND P. C. TANDY: *QCD modeling of hadron physics*, Nucl. Phys. Proc. Suppl. **161**, 136 (2006).

[Mei02]   U. G. MEISSNER, J. A. OLLER AND A. WIRZBA: *In-medium chiral perturbation theory beyond the mean-field approximation*, Annals Phys. **297**, 27 (2002).

[Met08]   V. METAG: *Medium Modifcations of Mesons in Elementary Reactions and Heavy-Ion Collisions*, Prog. Part. Nucl. Phys. **61**, 245 (2008).

[Met11]   V. METAG ET AL.: *Experimental approaches for determining in-medium properties of hadrons from photo-nuclear reactions*, arXiv:nucl-ex/1111.6004 (2011).

[Miz06]   T. MIZUTANI AND A. RAMOS: *D mesons in nuclear matter: A DN coupled-channel equations approach*, Phys. Rev. C **74**, 065201 (2006).

[Mor99]   P. MORATH, W. WEISE AND S. H. LEE: *QCD sum rules and heavy quark systems in the nuclear medium*, in *17th Autumn School on QCD : Perturbative or Nonperturbative?*, 1999.

[Mor01]   P. MORATH: *Schwere Quarks in dichter Materie*, Ph.D. Thesis, Technische Universität München (2001).

[Mor08] K. MORITA AND S. H. LEE: *Mass shift and width broadening of J/ψ in QGP from QCD sum rule*, Phys. Rev. Lett. **100**, 022301 (2008).

[Mos99] U. MOSEL: *Fields, Symmetries and Quarks*, Springer, 1999.

[Mue06] P. MUEHLICH, V. SHKLYAR, S. LEUPOLD, U. MOSEL AND M. POST: *The spectral function of the ω meson in nuclear matter from a coupled-channel resonance model*, Nucl. Phys. A **780**, 187 (2006).

[Mun92] H. J. MUNCZEK AND P. JAIN: *Relativistic pseudoscalar $q\bar{q}$ bound states: Results on Bethe-Salpeter wave functions and decay constants*, Phys. Rev. D **46**, 438 (1992).

[Mut87] T. MUTA: *Foundations of Quantum Chromodynamics*, World Scientific, 1987.

[Nak69] N. NAKANISHI: *A general survey of the theory of the Bethe-Salpeter equation*, Prog. Theor. Phys. Suppl. **43**, 1 (1969).

[Nak10] K. NAKAMURA ET AL.: *Review of particle physics*, J. Phys. G **37**, 075021 (2010).

[Nam60] Y. NAMBU: *Quasiparticles and Gauge Invariance in the Theory of Superconductivity*, Phys. Rev. **117**, 648 (1960).

[Nar83] S. NARISON AND R. TARRACH: *Higher dimensional renormalization group invariant vacuum condensates in quantum chromodynamics*, Phys. Lett. B **125**, 217 (1983).

[Nar89] S. NARISON: *QCD spectral sum rules*, volume 26 of *Lect. Notes Phys.*, World Scientific, 1989.

[Nar99] S. NARISON: *On the strange quark mass from combined e^+e^- and τ decay data: test of the isospin symmetry and implications on ϵ'/ϵ and $m_{u,d}$*, Phys. Lett. B **466**, 345 (1999).

[Nar01] S. NARISON: *c,b quark masses and $f_{D_{(s)}}$, $f_{B_{(s)}}$ decay constants from pseudoscalar sum rules in full QCD to order α_s^2*, Phys. Lett. B **520**, 115 (2001).

[Nar02] S. NARISON: *QCD as a theory of hadrons: from partons to confinement*, volume 17 of *Camb. Monogr. Part. Phys. Nucl. Phys. Cosmol.*, Cambridge University Press, 2002.

[Nar04]   S. NARISON: *Heavy-light $\bar{Q}q$ mesons in QCD*, Nucl. Phys. B Proc. Suppl. **152**, 217 (2004).

[Nar05]   S. NARISON: *Open charm and beauty chiral multiplets in QCD*, Phys. Lett. B **605**, 319 (2005).

[Nav05]   F. S. NAVARRA, M. NIELSEN AND K. TSUSHIMA: *A QCD sum rule study of Θ^+ in nuclear matter*, Phys. Lett. B **606**, 335 (2005).

[Neg88]   J. W. NEGELE AND H. ORLAND: *Quantum many particle systems*, volume 68 of *Frontiers in physics*, Addison-Wesley, 1988.

[Neu92]   M. NEUBERT: *Heavy meson form-factors from QCD sum rules*, Phys. Rev. D **45**, 2451 (1992).

[New55]   W. A. NEWCOMB AND E. E. SALPETER: *Mass Corrections to the Hyperfine Structure in Hydrogen*, Phys. Rev. **97**, 1146 (1955).

[Noe18]   E. NOETHER: *Invariant Variation Problems*, Gött. Nachr. **1918**, 235 (1918).

[Nol05]   W. NOLTING: *Grundkurs Theoretische Physik: Viel-Teilchen-Theorie*, Springer, 2005.

[Nov84a] V. A. NOVIKOV, M. A. SHIFMAN, A. I. VAINSHTEIN, M. B. VOLOSHIN AND V. I. ZAKHAROV: *Use and Misuse of QCD Sum Rules, Factorization and Related Topics*, Nucl. Phys. B **237**, 525 (1984).

[Nov84b] V. A. NOVIKOV, M. A. SHIFMAN, A. I. VAINSHTEIN AND V. I. ZAKHAROV: *Calculations in External Fields in Quantum Chromodynamics. Technical Review*, Fortschr. Phys. **32**, 585 (1984).

[Oht11a] K. OHTANI, P. GUBLER AND M. OKA: *A Bayesian analysis of the nucleon QCD sum rules*, Eur. Phys. J. A **47**, 114 (2011).

[Oht11b] K. OHTANI, P. GUBLER AND M. OKA: *A Bayesian analysis of the nucleon QCD sum rules*, AIP Conf. Proc. **1343**, 343 (2011).

[PAN]   PANDA Collaboration, *Antiproton Annihilation at Darmstadt,* http://www-panda.gsi.de/auto/phy/_home.htm,
M.F.M. LUTZ ET AL., arXiv:hep-ex/0903.3905 [hep-ex].

[Par05]   H.-J. PARK, C.-H. LEE AND G. E. BROWN: *The problem of mass: Mesonic bound states above T_c*, Nucl. Phys. A **763**, 197 (2005).

[Pas84]   P. PASCUAL AND R. TARRACH: *QCD: Renormalization for the practitioner*, Lect. Notes Phys. **194**, 1 (1984).

[Pes95]   M. E. PESKIN AND D. V. SCHRÖDER: *An Introduction to Quantum Field Theory*, Addison-Wesley, 1995.

[Pet98a] W. PETERS, H. LENSKE AND U. MOSEL: *Coherent photoproduction of eta mesons on spin-zero nuclei in a relativistic, non-local model*, Nucl. Phys. A **642**, 506 (1998).

[Pet98b] W. PETERS, H. LENSKE AND U. MOSEL: *Coherent photoproduction of pions on nuclei in a relativistic, non-local model*, Nucl. Phys. A **640**, 89 (1998).

[Pet98c] W. PETERS, M. POST, H. LENSKE, S. LEUPOLD AND U. MOSEL: *The spectral function of the rho meson in nuclear matter*, Nucl. Phys. A **632**, 109 (1998).

[Pfa06]   J. P. PFANNMÖLLER: *Properties of D-mesons from QCD sum rules with an improved spectral function*, Diploma Thesis, Technische Universität Darmstadt (2006).

[Pog66]   W. POGORZELSKI: *Integral equations and their applications*, Pergamon Press, 1966.

[Pok00]   S. POKORSKI: *Gauge Field Theories*, Cambridge University Press, 2000.

[Pos01]   M. POST, S. LEUPOLD AND U. MOSEL: *The rho spectral function in a relativistic resonance model*, Nucl. Phys. A **689**, 753 (2001).

[Pos04]   M. POST, S. LEUPOLD AND U. MOSEL: *Hadronic spectral functions in nuclear matter*, Nucl. Phys. A **741**, 81 (2004).

[Raf81]   E. DE RAFAEL: *Current algebra quark masses in QCD*, in *Marseille 1981, Proceedings, Theoretical Aspects Of Quantum Chromodynamics*, 1981.

[Raf84]   E. DE RAFAEL: *Some comments on QCD sum rules*, in *Dubrovnik 1983, Proceedings, Phenomenology Of Unified Theories*, 1984.

[Raf89]   E. DE RAFAEL: *Introduction to weak decays and QCD: Short distance corrections*, Nucl. Phys. B Proc. Suppl. 7, 1 (1989).

[Raf97]   E. DE RAFAEL: *An Introduction to sum rules in QCD: Course*, arXiv:hep-ph/9802448 (1997).

[Ran09]   M. RANGAMANI: *Gravity & Hydrodynamics: Lectures on the fluid-gravity correspondence*, Class. Quant. Grav. **26**, 224003 (2009).

[Rap97]   R. RAPP, G. CHANFRAY AND J. WAMBACH: *Rho meson propagation and dilepton enhancement in hot hadronic matter*, Nucl. Phys. A **617**, 472 (1997).

[Rap99]   R. RAPP AND J. WAMBACH: *Low mass dileptons at the CERN-SPS: Evidence for chiral restoration?*, Eur. Phys. J. A **6**, 415 (1999).

[Rap00]   R. RAPP AND J. WAMBACH: *Chiral symmetry restoration and dileptons in relativistic heavy-ion collisions*, Adv. Nucl. Phys. **25**, 1 (2000).

[Rap09]   R. RAPP, J. WAMBACH AND H. VAN HEES: *The Chiral Restoration Transition of QCD and Low Mass Dileptons*, arXiv:hep-ph/0901.3289 (2009).

[Rap11]   R. RAPP, B. KÄMPFER, A. ANDRONIC, D. BLASCHKE, C. FUCHS, M. HARADA, T. HILGER ET AL.: *In-medium excitations*, Lect. Notes Phys. **814**, 335 (2011).

[Rat04]   C. RATTI AND W. WEISE: *Thermodynamics of two-colour QCD and the Nambu Jona-Lasinio model*, Phys. Rev. D **70**, 054013 (2004).

[Rei80]   L. J. REINDERS, H. R. RUBINSTEIN AND S. YAZAKI: *QCD contribution to vacuum polarization 2. The pseudoscalar unequal mass case*, Phys. Lett. B **97**, 257 (1980).

[Rei84]   L. J. REINDERS: *QCD sum rules: An introduction and some applications*, Acta Phys. Polon. B **15**, 329 (1984).

[Rei85]   L. J. REINDERS, H. RUBINSTEIN AND S. YAZAKI: *Hadron Properties from QCD Sum Rules*, Phys. Rept. **127**, 1 (1985).

[Rob94]   C. D. ROBERTS AND A. G. WILLIAMS: *Dyson-Schwinger equations and their application to hadronic physics*, Prog. Part. Nucl. Phys. **33**, 477 (1994).

[Rob00]   C. D. ROBERTS AND S. M. SCHMIDT: *Dyson-Schwinger equations: Density, temperature and continuum strong QCD*, Prog. Part. Nucl. Phys. **45**, S1 (2000).

[Rob07]   C. D. ROBERTS, M. S. BHAGWAT, A. HOLL AND S. V. WRIGHT: *Aspects of hadron physics*, Eur. Phys. J. ST **140**, 53 (2007).

[Roc05]   L. ROCA, E. OSET AND J. SINGH: *Low lying axial-vector mesons as dynamically generated resonances*, Phys. Rev. D **72**, 014002 (2005).

[Rom65]   P. ROMAN: *Advanced Quantum Theory*, Addison Wesley, 1965.

[Rom69]   P. ROMAN: *Introduction to Quantum Field Theory*, Wiley, 1969.

[Rup06]   J. RUPPERT, T. RENK AND B. MÜLLER: *Mass and width of the ρ meson in a nuclear medium from Brown-Rho scaling and QCD sum rules*, Phys. Rev. C **73**, 034907 (2006).

[Ryd85]   L. H. RYDER: *Quantum field theory*, Cambridge University Press, 1985.

[Sai07]   K. SAITO, K. TSUSHIMA AND A. W. THOMAS: *Nucleon and hadron structure changes in the nuclear medium and impact on observables*, Prog. Part. Nucl. Phys. **58**, 1 (2007).

[Sak73]   J. J. SAKURAI: *Currents and Mesons*, The University of Chicago Press Ltd., 1973.

[Sal51]   E. E. SALPETER AND H. A. BETHE: *A Relativistic equation for bound state problems*, Phys. Rev. **84**, 1232 (1951).

[Sar05]   S. SARKAR, E. OSET AND M. J. VICENTE VACAS: *Baryonic resonances from baryon decuplet-meson octet interaction*, Nucl. Phys. A **750**, 294 (2005).

[Sch51a]   J. S. SCHWINGER: *On the Green's functions of quantized fields. 1.*, Proc. Nat. Acad. Sci. **37**, 452 (1951).

[Sch51b]   J. S. SCHWINGER: *On the Green's functions of quantized fields. 2.*, Proc. Nat. Acad. Sci. **37**, 455 (1951).

[Sch59]   J. S. SCHWINGER: *Field theory commutators*, Phys. Rev. Lett. **3**, 296 (1959).

[Sch61]   S. S. SCHWEBER: *An introduction to relativistic quantum field theory*, Row, Peterson and Company, 1961.

[Sch05]   S. SCHAEL ET AL.: *Branching ratios and spectral functions of tau decays: Final ALEPH measurements and physics implications*, Phys. Rept. **421**, 191 (2005).

[Sch06]   W. SCHEINAST ET AL.: *First observation of in-medium effects on phase space distributions of antikaons measured in proton nucleus collisions*, Phys. Rev. Lett. **96**, 072301 (2006).

[Sch08]   R. SCHULZE, M. BLUHM AND B. KÄMPFER: *Plasmons, plasminos and Landau damping in a quasiparticle model of the quark-gluon plasma*, Eur. Phys. J. ST **155**, 177 (2008).

[Sch09a] R. SCHULZE AND B. KÄMPFER: *Cold quark stars from hot lattice QCD*, arXiv:nucl-th/0912.2827 (2009).

[Sch09b] R. SCHULZE AND B. KÄMPFER: *Equation of state for QCD matter in a quasiparticle model*, Prog. Part. Nucl. Phys. **962**, 386 (2009).

[Shi79a] M. A. SHIFMAN, A. I. VAINSHTEIN AND V. I. ZAKHAROV: *QCD and Resonance Physics: Applications*, Nucl. Phys. B **147**, 448 (1979).

[Shi79b] M. A. SHIFMAN, A. I. VAINSHTEIN AND V. I. ZAKHAROV: *QCD and Resonance Physics. Sum Rules*, Nucl. Phys. B **147**, 385 (1979).

[Shi79c] M. A. SHIFMAN, A. I. VAINSHTEIN AND V. I. ZAKHAROV: *QCD and Resonance Physics. The $\rho - \omega$ Mixing*, Nucl. Phys. B **147**, 519 (1979).

[Shi79d] M. A. SHIFMAN, A. I. VAINSHTEIN AND V. I. ZAKHAROV: *Resonance properties in quantum chromodynamics*, Phys. Rev. Lett. **42**, 297 (1979).

[Shi99]   M. A. SHIFMAN: *ITEP lectures on particle physics and field theory.*, World Sci. Lect. Notes Phys. **62**, 1 (1999).

[Sou05]   N. A. SOUCHLAS: *Quark dynamics and constituent masses in heavy quark systems*, Ph.D. Thesis, Kent State University (2005).

[Sou10a] N. SOUCHLAS: *A dressed quark propagator representation in the Bethe-Salpeter description of mesons*, J. Phys. G **37**, 115001 (2010).

[Sou10b] N. SOUCHLAS AND D. STRATAKIS: *Bethe-Salpeter dynamics and the constituent mass concept for heavy quark mesons*, Phys. Rev. D **81**, 114019 (2010).

[Spi88]   V. P. SPIRIDONOV AND K. G. CHETYRKIN: *Nonleading mass corrections and renormalization of the operators $m\psi\bar{\psi}$ and $G^2_{\mu\nu}$*, Sov. J. Nucl. Phys. **47**, 522 (1988).

[Sta14]   J. STARK: *Beobachtungen über den Effekt des elektrischen Feldes auf Spektrallinien I. Quereffekt*, Ann. Phys. **43**, 965 (1914).

[Sta92]   S. J. STAINSBY AND R. T. CAHILL: *The Analytic structure of quark propagators*, Int. J. Mod. Phys. A **7**, 7541 (1992).

[Ste06]   B. STEINMÜLLER AND S. LEUPOLD: *Weighted finite energy sum rules for the omega meson in nuclear matter*, Nucl. Phys. A **778**, 195 (2006).

[Sug61]   M. SUGAWARA AND A. KANAZAWA: *Subtractions in Dispersion Relations*, Phys. Rev. **123**, 1895 (1961).

[Tho05]   R. THOMAS, S. ZSCHOCKE AND B. KÄMPFER: *Evidence for in-medium changes of four-quark condensates*, Phys. Rev. Lett. **95**, 232301 (2005).

[Tho07]   R. THOMAS, T. HILGER AND B. KÄMPFER: *Four-Quark Condensates in Nucleon QCD Sum Rules*, Nucl. Phys. A **795**, 19 (2007).

[Tho08a] R. THOMAS: *In-Medium QCD Sum Rules for ω Meson, Nucleon and D Meson*, Ph.D. Thesis, Technische Universität Dresden (2008).

[Tho08b] R. THOMAS, T. HILGER AND B. KÄMPFER: *Role of Four-Quark Condensates in QCD Sum Rules*, Prog. Part. Nucl. Phys. **61**, 297 (2008).

[Tka83a] F. V. TKACHOV: *On the operator product expansion in the MS scheme*, Phys. Lett. B **124**, 212 (1983).

[Tka83b] F. V. TKACHOV: *The Limit $m_q \to 0$ of perturbative QCD*, Phys. Lett. B **125**, 85 (1983).

[Tol04]   L. TOLOS, J. SCHAFFNER-BIELICH AND A. MISHRA: *Properties of D-mesons in nuclear matter within a self- consistent coupled-channel approach*, Phys. Rev. C **70**, 025203 (2004).

[Tol05]   L. TOLOS, J. SCHAFFNER-BIELICH AND A. MISHRA: *D-mesons in dense nuclear matter*, Eur. Phys. J. C **43**, 127 (2005).

[Tol06a] L. TOLOS, A. RAMOS AND E. OSET: *Chiral approach to antikaon s- and p-wave interactions in dense nuclear matter*, Phys. Rev. C **74**, 015203 (2006).

[Tol06b] L. TOLOS, J. SCHAFFNER-BIELICH AND H. STÖCKER: *D-mesons: In-medium effects at FAIR*, Phys. Lett. B **635**, 85 (2006).

[Tol08]   L. Tolos, A. Ramos and T. Mizutani: *Open charm in nuclear matter at finite temperature*, Phys. Rev. C **77**, 015207 (2008).

[Tre85]   S. B. Treiman, R. Jackiw, B. Zumino and E. Witten: *Current Algebra and Anomalies*, World Scientific, 1985.

[Tri85]   F. G. Tricomi: *Integral equations*, Dover Publications, 1985.

[Tse09]   I. Tserruya: *Electromagnetic Probes*, arxiv:nucl-ex/0903.0415 (2009).

[Waa96]   T. Waas, N. Kaiser and W. Weise: *Effective kaon masses in dense nuclear and neutron matter*, Phys. Lett. B **379**, 34 (1996).

[Waa97]   T. Waas, M. Rho and W. Weise: *Effective kaon mass in dense baryonic matter: Role of correlations*, Nucl. Phys. A **617**, 449 (1997).

[Wag08a] M. Wagner and S. Leupold: *Information on the structure of the a_1 from τ decay*, Phys. Rev. D **78**, 053001 (2008).

[Wag08b] M. Wagner and S. Leupold: *τ decay and the structure of the a_1*, Phys. Lett. B **670**, 22 (2008).

[Wei67]   S. Weinberg: *Precise relations between the spectra of vector and axial vector mesons*, Phys. Rev. Lett. **18**, 507 (1967).

[Wei95]   S. Weinberg: *The Quantum theory of fields. Vol. 1: Foundations*, Cambridge University Press, 1995.

[Wei96]   S. Weinberg: *The Quantum theory of fields. Vol. 2: Modern applications*, Cambridge University Press, 1996.

[Wey85]   J. Weyers: *An introduction to QCD sum rules*, in *Cargese 1985, Proceedings, Particle physics*, pages 171–915, 1985.

[Wic50]   G. C. Wick: *The evaluation of the collision matrix*, Phys. Rev. **80**, 268 (1950).

[Wid46]   D. V. Widder: *The Laplace Transform*, Princeton Mathematical Series, 1946.

[Wil69]   K. G. Wilson: *Nonlagrangian models of current algebra*, Phys. Rev. **179**, 1499 (1969).

[Wil07a] R. WILLIAMS: *Schwinger-Dyson equations in QED and QCD: The Calculation of fermion-antifermion condensates*, Ph.d. thesis, University of Durham (2007).

[Wil07b] R. WILLIAMS, C. S. FISCHER AND M. R. PENNINGTON: *$\bar{q}q$ condensate for light quarks beyond the chiral limit*, Phys. Lett. B **645**, 167 (2007).

[Wil07c] R. WILLIAMS, C. S. FISCHER AND M. R. PENNINGTON: *Extracting the $\bar{q}q$ condensate for light quarks beyond the chiral limit in models of QCD*, arxiv:hep-ph/0704.2296 (2007).

[Wil07d] R. WILLIAMS, C. S. FISCHER AND M. R. PENNINGTON: *Quark condensates: Flavour dependence*, Acta Phys. Polon. B **38**, 2803 (2007).

[Win06] S. WINITZKI: *Introduction to asymptotic series*, https://sites.google.com/site/winitzki (2006).

[Wol98] G. WOLF, B. FRIMAN AND M. SOYEUR: *In-medium ω meson broadening and s-wave pion annihilation into e^+e^- pairs*, Nucl. Phys. A **640**, 129 (1998).

[Wu57] C. S. WU, E. AMBLER, R. W. HAYWARD, D. D. HOPPES AND R. P. HUDSON: *Experimental test of parity conservation in beta decay*, Phys. Rev. **105**, 1413 (1957).

[Yeh92] E. YEHUDAI: *HIP: Symbolic high-energy physics calculations using maple*, in *La Londe-les-Maures 1992, New computing techniques in physics research*, pages 721–726, 1992.

[Ynd06] F. J. YNDURAIN: *The Theory of Quark and Gluon Interactions*, Springer, 2006.

[Zee97a] P. ZEEMAN: *Doubles and triplets in the spectrum produced by external magnetic forces*, Phil. Mag. **44**, 55 (1897).

[Zee97b] P. ZEEMAN: *On the influence of Magnetism on the Nature of the Light emitted by a Substance*, Phil. Mag. **43**, 226 (1897).

[Zee97c] P. ZEEMAN: *The Effect of Magnetisation on the Nature of Light Emitted by a Substance*, Nature **55**, 347 (1897).

[Zha05] Z. ZHANG AND W.-Q. ZHAO: *Landau gauge condensates from global color model*, Phys. Lett. B **617**, 157 (2005).

[Zho08] L.-J. ZHOU, Q. WU AND W.-X. MA: *Quark gluon condensate, virtuality and susceptibility of QCD vacuum*, Commun. Theor. Phys. **50**, 161 (2008).

[Zia09] R. K. P. ZIA, E. F. REDISH AND S. R. MCKAY: *Making sense of the Legendre transform*, Am. J. Phys. **77**, 614 (2009).

[Zim73] W. ZIMMERMANN: *Normal products and the short distance expansion in the perturbation theory of renormalizable interactions*, Annals Phys. **77**, 570 (1973).

[Zon03a] H.-S. ZONG, J.-L. PING, H.-T. YANG, X.-F. LU AND F. WANG: *The Calculation of vacuum properties from the global color symmetry model*, Phys. Rev. D **67**, 074004 (2003).

[Zon03b] H.-S. ZONG, S. QI, W. CHEN, W.-M. SUN AND E.-G. ZHAO: *Pion susceptibility of the QCD vacuum from an effective quark-quark interaction*, Phys. Lett. B **576**, 289 (2003).

[Zsc02] S. ZSCHOCKE, O. P. PAVLENKO AND B. KÄMPFER: *Evaluation of QCD sum rules for light vector mesons at finite density and temperature*, Eur. Phys. J. A **15**, 529 (2002).

[Zsc03] S. ZSCHOCKE, O. P. PAVLENKO AND B. KÄMPFER: *In-medium spectral change of ω mesons as a probe of QCD four-quark condensate*, Phys. Lett. B **562**, 57 (2003).

[Zsc04] S. ZSCHOCKE AND B. KÄMPFER: *$\rho-\omega$ splitting and mixing in nuclear matter*, Phys. Rev. C **70**, 035207 (2004).

[Zsc06] S. ZSCHOCKE: *Open charm mesons in nuclear matter within QCD sum rule approach* (2006), unpublished.

[Zsc11] S. ZSCHOCKE, T. HILGER AND B. KÄMPFER: *In-medium operator product expansion for heavy-light-quark pseudoscalar mesons*, Eur. Phys. J. A **47**, 151 (2011).

i want morebooks!

Buy your books fast and straightforward online - at one of world's fastest growing online book stores! Environmentally sound due to Print-on-Demand technologies.

Buy your books online at

www.get-morebooks.com

Kaufen Sie Ihre Bücher schnell und unkompliziert online – auf einer der am schnellsten wachsenden Buchhandelsplattformen weltweit! Dank Print-On-Demand umwelt- und ressourcenschonend produziert.

Bücher schneller online kaufen

www.morebooks.de

VDM Verlagsservicegesellschaft mbH
Heinrich-Böcking-Str. 6-8   Telefon: +49 681 3720 174   info@vdm-vsg.de
D - 66121 Saarbrücken   Telefax: +49 681 3720 1749   www.vdm-vsg.de

Printed by Books on Demand GmbH, Norderstedt / Germany